DENTAL FUNCTIONAL MORPHOLOGY
How Teeth Work

Dental Functional Morphology offers an innovative alternative to the received wisdom that teeth merely crush, cut, shear or grind food, and shows how teeth adapt to diet. Providing an analysis of tooth action based on an understanding of how food particles break, it shows how tooth form from the earliest mammals to modern-day humans can be understood using very basic considerations about fracture. It outlines the theoretical basis step by step, explaining the factors governing tooth shape and size, and provides an allometric analysis that will revolutionize attitudes to the evolution of the human face and the impact of cooked foods on our dentition. In addition, the basis of the mechanics behind the fracture of different types of food, and methods of measurement are given in an easy-to-use appendix. It will be an important sourcebook for physical anthropologists, dental and food scientists, palaeontologists, and those interested in feeding ecology.

PETER W. LUCAS is a professor of anatomy at the University of Hong Kong. He is particularly interested in the function and evolution of mammalian teeth, but he also studies the factors involved in food choice, particularly in primates.

DENTAL FUNCTIONAL MORPHOLOGY
How Teeth Work

PETER W. LUCAS

University of Hong Kong

CAMBRIDGE
UNIVERSITY PRESS

PUBLISHED BY THE PRESS SYNDICATE OF THE UNIVERSITY OF CAMBRIDGE
The Pitt Building, Trumpington Street, Cambridge, United Kingdom

CAMBRIDGE UNIVERSITY PRESS
The Edinburgh Building, Cambridge CB2 2RU, UK
40 West 20th Street, New York, NY 10011-4211, USA
477 Williamstown Road, Port Melbourne, VIC 3207, Australia
Ruiz de Alarcón 13, 28014 Madrid, Spain
Dock House, The Waterfront, Cape Town 8001, South Africa

http://www.cambridge.org

First published 2004

Printed in the United Kingdom at the University Press, Cambridge

Typeface Adobe Garamond 11/12.5 pt   *System* LATEX 2ε   [TB]

A catalogue record for this book is available from the British Library

Library of Congress Cataloguing in Publication data
Lucas, P. (Peter W.)
Dental functional morphology: how teeth work / by Peter W. Lucas.
p.   cm.
Includes bibliographical references and index.
ISBN 0 521 56236 8
1. Teeth.   2. Teeth–Evolution.   3. Teeth–Anatomy.
4. Dental anthropology.   1. Title.
QM311.L83   2004
611′.314–dc22   2003063885

ISBN 0 521 56236 8 hardback

*To my wife Mariati and my daughters, Katherine and Diana,
with my everlasting love*

Contents

Preface

Teeth cause such dreadful problems in humans that interest in them by non-dentists would seem both unlikely and unhealthy. Who could get excited about tooth decay and gum disease? The physical reality of such apparently moribund structures is paralleled in our cultural perception of them. Diseased or not, the whole mouth is viewed as an unclean region of the body in most parts of the world, especially when it is crammed full of food. Parents, particularly in Western countries, often train children to keep their lips sealed when they are eating even though this is difficult to follow exactly and, indeed, little food seems to re-emerge if the instruction is disobeyed. It is debatable if this training is necessary. While it is possible to sit next to someone at a banquet and get sprayed with seafood, for example, from his or her mouth, the nutritive loss to the diner, represented by the sum of those fine particles, seems negligible compared to what is obviously going down their throat. This is a clear sign of the efficiency of the chewing process. The main reason, in fact, that food particles are expelled is that the person is talking while chewing. Talking involves the expiration of air and that is what pushes food particles forwards. This may seem a strange example but it makes a strong point: the thought of even catching sight of food that was, a moment previously, decorating a plate evokes visceral feelings (of a somewhat inside-out kind) rather than artistic ones. The plate, too, seems to lose its appeal after most food has disappeared and may be quickly consigned to the wash. In short, we appear often to be embarrassed, if not disgusted, by the major biological function of feeding and the need that underlies it, although we don't go as far in hiding it from social view as we do activities at the other end of the gut. Presumably, an intuitive understanding of hygiene explains that disparity, but restrictive practices about eating pervade most human societies and, according to an intriguing account by Visser (1991), have many cross-cultural features that are not easily explained on grounds of hygiene (the latter, of course, preoccupies developed societies: Lacey, 1994).

ix

So why write a book on how teeth work? I answer this, not by seeing any disguised elegance in the appearance of a good feed, or in the avocation of new table manners, but by admitting and attempting to transmit through this book a personal fascination with the fundamental role that teeth have had in our evolution and my dissatisfaction with current explanations on both how they work and why they evolved.

How do teeth work? One relatively uniform answer to this is already provided in numerous accounts in top journals, encyclopaedias and even school texts. It is that teeth variously crush, cut, shear or grind food. And with slight complications, that is more or less the prevailing wisdom in dentistry, zoology, palaeontology, anthropology and many other biological fields (with the eminent exception of food science). Despite apparent unanimity, these accounts are completely wrong. A genuine analysis of tooth action, one that could possess explanatory power rather than glib description, starts in the still somewhat obscure world of fracture mechanics – in the understanding of how food particles break. Such an analysis is not likely to be as edible as slogans like shearing or grinding: inevitably, analytical depth requires more than the coining of facile words and phrases.

The action of the teeth cannot be separated from that of the mouth, so an attempt is made consistently to understand oral processing as whole. However, this is the oral processing of solids, not liquids. The ingestion of liquids like nectar and honey is all tongue and no teeth, and drinking is actually not that common an activity in mammals – certainly not in primates – once they grow up. The process of growing up, of development, is not discussed, so those interested in suckling and weaning and how the young cope will not find anything on it. The largest body of information to be excluded here though is neurobiology: there is no space to include much of it here and some of its alleys seem currently to be very dark.

I am sensitive to the knowledge that the further the book sinks into a world requiring the learning of new terms, the more potential readers will be lost. Accordingly, I have tried to present my viewpoint in as simple a way as possible, deliberately seeking light generalization rather than long and dark specifics. One of the worst aspects of biology is the plethora of terms that it employs. If I added a full suite of terms from mechanics, then the book would be a slow read. Sensitive to this, I have made a deliberate effort to reduce the number of terms to a minimum. The overall intention is to provide a fundamental analysis of the feeding apparatus of mammals, based on the interface between outside and inside, i.e. the contact between foods and teeth. If this interface is properly understood, then I contend that the optimal design of the working surface of teeth, the organization of the structural support for this surface and the production of bite forces by muscles that move it – all should follow in a predictable way. If this analysis succeeds, then it should open the way to a fundamental understanding of the evolution of feeding adaptations in mammals. On the plus side, I hope that this book will be of value to anyone in bioscience with an interest in feeding. On the debit side, I will undoubtedly have made some dreadful mistakes and may sometimes appear uncharitable to those with other views. However, the book is meant to be constructive and, in this sense, research life has some resemblance to a game of chess: no one ever excels by just making moves that have been seen before. Unless I am mistaken, a lot of the 'moves' recorded here are new.

I hope that this book offers a cohesive framework on the function of teeth. For dentists and those basic scientists whose work might be covered by the term 'oral biology', this is an account of how teeth break foods down, untainted by the modern reverse trend. For food scientists and those concerned with food texture, it is about oral physics, on which psychophysical investigations of food texture can be superimposed (see Chapter 7). Common to both dentistry and food science has been a strong interest in 'applied science'. Whereas dentists look at patients, food scientists look at consumers. However, dental surgery exists as a discipline entirely separated from medicine because the dentition is the one area of the body requiring regular surgery. It can be argued then that in many parts of the world, 'the patient' is also 'the consumer'. Both dentists and food scientists might benefit from basic models of oral processing in order, on the one hand, to predict the outcome of surgery or, on the other, to provide a foundation for psychophysical investigations of food texture. Until recently, there was very little cross-talk between these disciplines in most countries (I exclude Japan from this), something possibly caused by lack of a sufficiently overarching viewpoint. For ecologists, I describe the actual mechanical properties of foods that could influence dietary niches and feeding rates. For materials scientists, it may provide some information on a restricted group of biomaterials – foods. For palaeontologists and those concerned with the evolution of the feeding apparatus in a wide group of organisms, the hope must be that enough is explained here to help in the generation of general theories for evolutionary change.

It seems a book tradition to tell people where they can find articles on its subject matter. Each of the above fields has scientific journals responsible for the vast literature on the structure and function of mammalian teeth, but if the definition of this book's scope is taken to encompass feeding, then the answer to this is really just about anywhere, even in physical sciences. Well-cited journals here include the *American Journal of Physical Anthropology*, *Archives of Oral Biology*, *Journal of Dental Research*, *Journal of Human Evolution* and *Journal of Texture Studies*. Some, such as the *Journal of Prosthetic Dentistry*, are underrepresented because of their clinical slant. Unfortunately, in recent years, there is another candidate for this roll of honour, one that likes to bestow its own credits. The *Annals of Improbable Research* has devoted much of its space to feeding research and several papers have won IgNobel awards. Sometimes, this recognition is truly deserved, but the frequency with which food research gets treated this way in the journal seems to suggest that its contributors are actually obsessed with the field too, although clearly from a different perspective.

The problem with being sure about scientific novelty is the need to wade through the mounds of information that modern biology accumulates so rapidly. Really, there is too little time to sit down with the enormous body of relevant literature to see if a theory fits well with the evidence or, alternatively, casts it into serious doubt. The purpose of this book is overtly to try to bring together some of these piles according to the overarching theoretical model that dominates the book. Anyone who reads it will probably know some of the areas that I have explored here, and in particular areas of expertise, may well know more than I do, but what I am banking on is that few will be acquainted with the full scope of this book.

One of the great benefits of writing this, afforded for the only time in my career, is the space that it provides in the preface to thank those who have been seminal influences. I am very grateful to some of the great men of materials science and biomechanics who sat down and just had a word (or several) with me. To give their names would be to suggest that they might support some of what is written here or even know about it at all. They don't necessarily know in either sense, so the temptation to thank them by name is resisted. Early influences are always the strongest. Of these, Bob Martin was extremely important in directing my thoughts as an undergraduate at University College London. As a postgraduate, the seminal influence was Jeffrey W. Osborn. Although we overlapped at Guy's Hospital for only 18 months, his influence was so powerful that it pervades the book. He was, and still is, the greatest of teachers and the most powerful thinker that I have ever met: I am grateful for the chance to thank him here. I should have learnt from Jeff just how hard it is to write a coherent book in uniform style, but I didn't. Although his research papers and a (now out of print) co-authored book called *Advanced Dental Histology* (many editions, published by Wright, Bristol) have probably been most influential, a much underrated (also out of print) book for dental undergraduates that he edited, entitled

Dental Anatomy and Embryology (Volume 2 of *A Companion to Dental Studies*, edited by A. H. R. Rowe & R. B. Johns, Blackwell, Oxford, 1981) has provided a magnificent general guide while writing this book. I would also like to thank Karen Hiiemae for getting me started in research, for persevering with me and for offering general advice on research directions. The subtitle of the book is adapted from an article by A. W. Crompton and K. Hiiemae, entitled 'How mammalian molar teeth work', published in *Discovery* (Yale University) **5**: 23–34 (1969). Douglas Luke helped me survive and turned me from a neophyte into, well . . . into something else. He also kept me in the business by pushing me to apply for a postdoctoral fellowship, which led to several years of joint publication.

Despite all this, if I had to pinpoint any particular period when I began to feel comfortable and capable in research, then it would have to be nine years spent in the Department of Anatomy at the National University of Singapore. I am eternally grateful to all the members of that department and to Professor Wong Wai Chow, its then head, for supporting me while I grew up. I miss that department and its staff very much and would like to take this opportunity to offer my greetings to them. In particular, I recall with gratitude, the advice offered in many conversations by Samuel Tay, Gurmit Singh and K. Rajendran (acknowledged for Fig. 2.17). To the current head, Professor E. A. Ling, I wish you all the best in your energetic leadership of the department. Mark Teaford was for a long time a co-author of this book. His sense of perspective and great friendship has helped sustain me through many crises of confidence. Mark very kindly read through much of the text and helped correct and clarify it in many areas. Through both Singapore and Hong Kong, I have collaborated at length with Richard Corlett, who taught me ecology. Brian Weatherhead brought me to the University of Hong Kong: thanks a million to you, Brian, and to the Anatomy staff here. I have had a warm working relationship with Brian Darvell in Dental Materials Science for most of my 10 years here, and also with his technicians, Paul Lee and Tony Yuen. Together, they designed the HKU Darvell field tester, described in Appendix A. A long-term collaboration with Iain Bruce in Physiology has also been very stimulating and fruitful. In recent years at the University of Hong Kong, I have been lucky enough to associate with several PhD students (Choong Mei Fun, Nicola Parillon, Jon Prinz, Kalpana Agrawal and Nathaniel Dominy) and postdoctoral fellows (David Hill and Nayuta Yamashita), all of whom have had great influence on me. Of these, Jon and Nate have been towering influences. P. Y. Cheng has worked with me for 11 years: I am deeply grateful for all his help. Recently, a medical student with strong palaeontological leanings,

Sham Wing Hang, has done a lot to focus my thoughts clearly. I am deeply grateful. In addition, I have benefited greatly from interaction with Kathryn Stoner and Pablo Riba during the 'Pantropical Primate Project'. For help during production of this book, I thank Eastman Ting, for several pieces of artwork that set the style, and Johnny Leung for photographs. Henrique Bernardo (rickybernardo@hotmail.com) did the cover and offered much helpful advice. Gavin Coates (gavincoa@netvigator.com) created the flick art. In addition, I would like to thank for either direct help or inspiration: Holger Preuschoft, Charles Peters, Roland Ennos, Walter Greaves, Patricia (Trish) Freeman, Mikael Fortelius (for lengthy correspondence as well as his papers, the influence of both of which permeates the book in many places), Josefina Diaz-Tay, Robin Heath, Jukka Jernvall, Michael LaBarbera, Mark Spencer, Peter Ungar and Chris Vinyard. I am also extremely grateful to Dr Rob Hamer (Wageningen Centre of Food Sciences) for inviting me to a food summit in Wageningen in November 1999, without the influence of which this book would not have been finished. I had the great privilege there of meeting many of the greats of food science, such as Drs Alina Scszesniak and Malcolm Bourne.

Specific acknowledgements for permissions to use figures go to the *Journal of Anatomy* (Fig. 3.8), *Food Quality and Preference* (Figs. 3.4, 3.5, 3.13, 3.14, 4.7, 4.8 and 4.10), *Archives of Oral Biology* (Fig. 3.12), Cambridge University Press (Fig. 4.12) and the *British Dental Journal* (Fig. 5.10).

Finally, to Tracey Sanderson and Cambridge University Press, thank you very much for sticking with this. To my mother and my extended family back in England, thank you for all your support. Lastly, and most importantly, to members of my immediate family, my wife Mariati and my daughters, Katherine and Diana: you have supported me loyally throughout my career and I simply do not have the words . . . emotions are better and mean much more.

Flickart

A flick-page animation starts here, illustrating the evolution of the human lower molar from the single-cusped tooth of an early synapsid that lived about 300 million years ago. The sequence is designed to 'morph' between existing fossils and may, inadvertently, involve variation off the main lineage. It is hardly possible to know for sure, but only basic features are shown anyway. Some partial restoration of cusp form has been necessary for fossils with worn teeth. The timescale is kept relatively even, so some changes are more rapid than others. Only the evolution of the lower molars is figured because that of the upper molars is more complex. The original names for their cusps, given in the nineteenth century (Osborn, 1888), make no evolutionary sense and tend to confuse.

Each diagram shows the jaw bone in grey with tooth crown evolution displayed above it. The tooth is viewed from the lingual aspect. The mesial side of the tooth is always to the left. Root development is indicated only by the bifurcation into two roots a little further on. Note though that the earliest teeth in the sequence did not have roots.

The trends in the animation can be noted as follows:

(1) A single-cusped reptiliomorph tooth, equivalent to the protoconid cusp of a mammalian molar, is the sole initial cusp. It always forms first, even in living mammals, and is generally very large.

(2) Two separate cusps form on either side of the protoconid. These are the mesial paraconid and the distal metaconid. Early mammaliaforms had this 'three-in-a-line' molar form.

(3) A small shelf develops low down on the crown. This is the cingulum.

(4) The three cusps then triangulate by relative movement of the paraconid and metaconid to form a trigonid. This tooth form is called a 'tritubercular' molar.

(5) The cingulum extends distally to start forming the talonid, a shelf that starts to develop three cusps. The resulting six-cusped tooth is a

'tribosphenic' molar from which those of all living therian mammals have adapted.

(6) On the primate lineage towards humans, the cingulum and paraconid are lost. The talonid evens up in height to match the remnants of the trigonid. Late on, all the cusps become blunter and reduce in height.

How to get excited about teeth

INTRODUCTION

Animals are destructive by nature. They do not build as plants do, by taking simple molecules and making themselves from them. Instead, animals operate by taking complex ready-made structures and breaking them down in their guts. Vertebrates are distinctive in that they are prone to attacking the largest structures. They are, and probably always were, very active organisms, with consequent big energy expenditures, and they evolved teeth and jaws early on to increase the rate of acquisition of these food items. Mammals descended from this line of biological warriors, but they evolved mechanical comminution (chewing) of food particles to precede the chemical comminution in their guts with so as to increase the rate of energy flow. They needed to turn an energy stream into a river so as to fund their all-weather activity cycle, attaining the latter at the immense cost of maintaining an internal body temperature well above their surroundings. We (making the assumption that the reader is human) are super-mammals, having extended this characteristic mechanical and chemical destruction beyond the realm of our bodies to our environment. Probably none of their devastating nature gives animals a good reputation among plants, which have developed an enormous array of defences to try to stop being eaten, and were any readers to be organisms other than human, then they would surely vouch for the exceptionally poor reputation that we currently have with every other species.

It is my set task in this book to try to dissect out facets of this general picture, to make them glint in the light and then claim that these features explain it all. Of course, this is a ridiculous remit, so I will go for a smaller assignment and just try to make something glint. That something is the mammalian dentition, set in the mammalian mouth. As with everything else in biology though, even an apparently manageable and limited undertaking can start to become awesome once you get into it. There is a lot to explain about just about anything and to be credible that explanation needs to start with the basics. Most scientists seem to have found the same thing in any field of investigation and it is as though they jump ship when they realize it, becoming honorary fellows of the plant kingdom, because explanations have to be built.

THE ORIGINS OF TEETH

Vertebrates have over 500 million years of ancestry, dental tissues being present in the very first forms. Until recently, it was thought that these tissues developed not in the mouth, because the first vertebrates were jaw-less and fed on very small organisms, but on the surface of the body as exoskeletal protection – 'dermal armour' as it is called (Smith & Sansom, 2000). The inclusion of dermal armour as 'teeth' definitely takes the very respectable ancestry of these organs back to the Silurian period (Janvier, 1996), and possibly to late Cambrian times, over 500 million years ago (Young *et al.*, 1996).[1] This view has teeth as the last remnants in a mammal of a vertebrate exoskeleton (an external rigid coat that not only provides support but also contributes protection as a form of armour). A mineralized exoskeleton (mineralized tissue being that which contains crystals of an inorganic compound – in vertebrates, always a compound containing calcium and phosphorus called hydroxyapatite) seems to have preceded the evolution of the bony endoskeleton that vertebrates now have. While an exoskeleton has great advantages in a small animal (that is why insects possess them), providing optimal stiffening and direct protection, a large animal would have to develop an extremely thick exoskeleton in order for it not to buckle (Currey, 1967). Such heavy, and probably very insensitive, animals would clearly be at a competitive disadvantage to those with endo-skeletons, which is presumably why the exoskeleton of vertebrates has been lost.

Enamel, the outermost tooth tissue (that which may contact a reader's fingernails in some circumstances), is unusual in that it is formed from the outermost layer of the body, the epidermis, in a process involving a very distinct set of proteins. Indirect evidence from molecular analysis has recently been interpreted as suggesting that some of these proteins might even have evolved back even in pre-Cambrian times (Delgado *et al.*, 2001). Dentine is the foundation of modern vertebrate teeth and derived from underlying connective tissue. It may have evolved alongside a bone-like tissue at a later date (Janvier, 1996; Smith & Coates, 2000), but the issue of what evolved first will probably only be resolved by examination of the genes that make these proteins (Kawasaki & Weiss, 2003). There is no need to provide a fixed viewpoint on this debate here: the science itself will evolve rapidly.

Whether 'teeth' evolved first on the surface of the body or actually in the mouth, it seems probable that they evolved in jawless forms. These early forms were very likely to be active feeders that sought large prey items. We know that conodonts (vertebrates off our direct line of descent and that lived over 440 million years ago) had muscle tissue somewhat like that of modern vertebrates (Gabbott *et al.*, 1995), which indicates that they were highly active organisms. They also had teeth set in a jawless mouth that show evidence of microscopic wear, indicating they were used in food acquisition (Purnell, 1995). Microwear, as this microscopic wear is called, is a clear indicator of the manner in which teeth are used and is strongly linked to diet, as shown in Chapter 6. The first jawed vertebrates were the placoderms. Curiously, the earliest members of this group did not have teeth. Later ones did, but may have evolved their teeth independently (Smith & Johanson, 2003).

Most vertebrates have teeth just for ingestion, the term given to the process by which food is taken into the mouth. The real breakdown of food is chemical. Although all animals are distinguished by the chemical breakdown that takes place within their guts, the most distinctive feature of a mammal is the mechanical breakdown that takes place in its mouth prior to this chemical activity. Although we tend to think of teeth as a chewing instrument, it is only mammals that really chew.[2] The earliest mammals added chewing (or mastication) to ingestion. Rather than just break food down chemically with their gut, they also comminuted food with their teeth. Why?

It is generally agreed that the need for mastication is related to the rate of energy requirement by mammals. Among vertebrates, only birds and mammals have exact control over their body temperatures. Other vertebrates have restricted activity cycles and limited ecological niches. Adaptation to cooler climes requires the development of a locomotor stamina that cold-blooded reptiles do not possess. The elevation of body temperature to a standard that is usually well above the ambient is very costly and demands a large increase in basal metabolic rate (which is the rate of energy consumption required to keep the body functioning). There are two basic methods by which such energy demands could be met in mammals. One involves simply ingesting much more food and letting a very large gut extract more nutrients per unit of time. This, however, works against mobility because the mammal would have a heavy weight of inert material in its gut that would simply slow it down. The alternative solution, to which mammals have resorted, is to reduce food particle size mechanically prior to chemical reduction in the gut. This adds new surface area on which enzymes in the gut can act more quickly, so increasing the rate of chemical breakdown (digestion). It is mastication, veiled by the lips and concealed by the cheeks, that provides this key advantage to mammals.

Birds are the other surviving group of vertebrates to possess high metabolic rates and it will not have escaped the attention of observant readers that they do not chew. It is unclear why this is and it should not be assumed that mammals are somehow superior: early birds appear to have had (simple) teeth but lost them. Weight was probably an important consideration for longevity of flight. Beaks are much lighter than teeth, but have disadvantages in being much more limited in their ability to break down foods and more importantly, are relatively insensitive compared to a dentition. The sensitivity of teeth is provided by receptors housed both inside them and in their sockets, which offer fine-scale detection of forces and also modulate salivary secretions. Both features signify a key asset of a complex mammalian mouth – the ability to discriminate foods by texture and taste at the front end of the gut so as to assess food quality prior to decisions about swallowing. Non-mammalian vertebrates have much less ability to do this because they generally only perforate large food particles and cannot make any detailed assessment of their contents.

Thus, birds may lack the ability to make many of the fine judgements that mammals can achieve with the plethora of sensory receptors in and around their mouths. Instead, they probably have to rely more on regurgitation or vomiting if they make errors during feeding. These reverse movements of the gut are part of the normal behaviour patterns of many birds when feeding their young. They are also seen in owls when they expel bones and fur through the mouth and in fruit-eating hornbills that regurgitate large seeds of the nutmeg family (Kemp, 1995).[3] The reliance of a mammal on oral sensitivity is made doubly clear by the rudimentary development of the sensory capabilities of the rest of the gut. In stark contrast to the mouth, the abdominal organs send very unclear messages to the brain about problems with their contents. It is extremely difficult, for example, to get a clear idea about where visceral pain is coming from and it often gets 'referred' to the skin. There is no reason to suppose that these human sensory dilemmas differ from the situation in other mammals. Clearly, all the major feeding decisions in a mammal are being made 'up front'. This provides them with a dietary diversity that would make it difficult to interpret signals from the gut very precisely anyway. For example, a monkey might feed on several food items within a 2-hour period. How could an individual animal with a diet of that breadth connect subsequent abdominal pain or diarrhoea to any one of these particular foods? It is thus possible to argue that a limitation on sensory discrimination within the mouth forces birds to adopt a more stereotyped diet than terrestrial mammals. They (and other non-mammalian vertebrates) probably use longer-distance cues to recognize food sources than mammals do. There is clearly a linkage here to the general diurnal (daytime) niche of birds versus the usual nocturnal niche of mammals. Lacking the visual ability to forage at distance due to nocturnal activity, early mammals probably had to develop shorter-range cues that included the ability to sense food texture very accurately in the mouth. This theme is developed further in Chapter 7 in relation to food selection in primates including humans.

Considerations like the above show the enormous timeframe over which teeth have evolved. Their beginning was a very long time ago. Given the wonders of modern dentistry and medicine, we sometimes forget the evolutionary basis for much of what goes on in the mouth. What is the connection between such problematic, easily infected, structures as our teeth with a noble lineage of 500 million years? The answer lies in our very recent history.

ARE TEETH JUST HANGERS ON?

Excepting humans, there is circumstantial evidence that mammals whose dental function is sufficiently impaired die and that the lifespan of the dentition could be the operative factor limiting the lifespan of many mammals (Chapter 6). In contrast, humans have in the last 10 000 years tried to control their food supply. No longer content with merely understanding where plants grew and when their parts were potentially edible or not, nor with simply tracking animals, our ancestors started to domesticate both plants and animals to produce reliable proximate food sources. And that is when problems started. Most of the dental decay (dental caries) and periodontal (gum) disease that we see started with the high consumption of cultivated (seed) grain products just less than 10 000 years ago. The nutritional content of these food items comes from contained starches that are usually bound in seeds into relatively insoluble, and thus indigestible, granules of micrometre size. Allied to cooking, a procedure recently supposed to define humanity better than tool use (Wrangham *et al.*, 1999), the starch is solubilized and its mechanical properties radically altered, such that it breaks up into very small particles (probably fragmented granules), which do not get cleared from the mouth very easily. An enzyme in the saliva starts to break the starch down into sugars and bacteria living in the mouth then convert those sugars into acid. Unfortunately, once acidity drops to below a certain level, tooth tissues start to dissolve and decay sets in. The major problem with the modern (Western) diet is not with these starches, but with sucrose. This forms an exceptionally adhesive layer to the teeth, more so than other common sugars. It may be thought that this effect should have some natural defence, but sucrose is not a common sugar in plants. It is the transport medium for energy but is rarely concentrated, being rapidly converted to other sugars within plant cells – e.g. for building up the cell wall. Many fruits do not contain sucrose and, indeed, a large group of fruit-eating birds lacks the enzyme sucrase needed to break this sugar down, developing diarrhoea if they consume it (Martinez del Rio, 1990). Our mammalian ancestors could process sucrose, but the effects of high concentrations of this sugar on the teeth in recent human evolution have been profound.

So teeth are not intrinsically dirty structures in an unclean cavity. Far from being unclean, the mouth of mammals is a very efficient self-cleaning, self-clearing, system for natural diets. It has to be because food is certainly not sterile and many microorganisms find the mouth an acceptable en-vironment in which to survive, if not thrive. The epithelial lining of the body provides protection, of course, but only so long as there is no break in it. The teeth, however, provide that break. Besides the risk of infection around their roots, the vulnerability of tooth material itself is great. The greatest danger facing dental tissues is acid. Any drop in the oral pH below about 5.5 (which is really only very mildly acidic) and tooth tissues start to dissolve. What prevents this? The major factor is oral fluid. Saliva jets out into the mouth from four major orifices, and many smaller ones, pro-viding a mildly alkaline bicarbonate spray to prevent dissolution (Edgar & O'Mullane, 1996). For further detail, see Chapter 3.

I believe that that the presence of teeth in every major vertebrate lineage except birds makes it obvious that there was very high selective pressure for the development of sturdy teeth in strong jaws. This pressure is also responsible for the diversity of tooth form in mammals in relation to diet. The strongest evidence for this comes from the frequency of convergent (in-dependent) evolution of dental features in mammals. The most striking of these is the suggestion that the basal form of the mammalian cheek tooth, the tribosphenic molar (the evolution of which can be flicked between pages 1 and 159), may have evolved twice, quite independently, about 200–150 million years ago (Luo *et al.*, 2001).[4] More recently, a prominent feature of upper molars called the hypocone, not present in the tribosphenic form, has arisen independently (convergently) in many mammalian lineages (Hunter & Jernvall, 1995). In fact, the adaptability of hard tissues like the teeth has recently been termed too great to rely on for characters by which to judge the evolutionary separation of mammalian groups: some aspects of soft-tissue arrangements may be more conservative (Gibbs *et al.*, 2000). This is a complete reversal of the predominant view just 20 or so years ago. This pliability of the teeth is despite these structures being the complex product of the action of more than 200 genes (Jernvall & Thesleff, 2000), and the fact that a dental tissue like enamel (Chapter 2) possesses a sophistication far beyond that which can be produced artificially.

BASIC FUNCTIONAL CONSIDERATIONS

The teeth are set as close as possible to the sense organs (the ears, eyes, nose and tongue) so that they can aimed accurately, and set close to their central processing unit (brain) so that signals can be sent to and fro with minimal delay.[5] All this apparatus is housed in a light appendage (the head), set on an agile muscular stalk (the neck). The following is an attempt to sketch the arrangement of the functions of the mammalian head and neck and the compartments that they occupy (Fig. 1.1), emphasizing just how distinct mammals are from their reptiliomorph ancestors and other vertebrates.

Chewing and swallowing

The major function of the head is to ingest (take in) food. Much of the anterior (front) part of the head contains the jaws and teeth that ingest food particles. Mammals chew (masticate) food whereas virtually all other vertebrates do not. The evolution of mastication has required many evolutionary modifications to the head from the reptilian state. Moving the jaws so as to break food particles requires large muscles. These masticatory muscles, the temporalis, masseter and pterygoids (Chapter 2), have differentiated in mammals so as to bite food without sending this force pointlessly through their jaw joint. This keeps head weight down because the food should absorb a considerable amount of the energy imparted by muscular work, the optimum being to channel all this work eventually into its fracture.

In most vertebrates, ingestion is followed immediately by swallowing, with a very sensitive muscular bag, the tongue, organized to direct food particles backwards to the next part of the gut. In mammals, the tongue

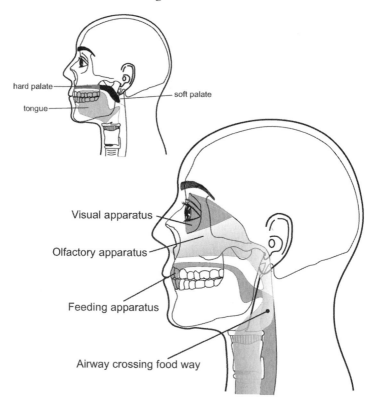

hard palate

soft palate

tongue

Visual apparatus

Olfactory apparatus

Feeding apparatus

Airway crossing food way

Fig. 1.1 The main diagram shows functional compartments of the anterior part of the head and neck in the human. The diagram above left shows how small the oral cavity really is, being the unshaded space lying between the tongue and hard palate. The rigid hard palate extends backwards into the pharynx as the soft palate, which is a mobile flap.

has developed a capacity to throw food sideways onto the teeth for chewing as well as backwards for swallowing. Assessment of whether to swallow or to chew further requires the tongue to have a surface positioned directly above it against which it can manipulate the food. The palate, another mammalian innovation, acts as this template.

Throwing food sideways requires a wall on the other side of the dentition to prevent food from escaping. Muscular cheeks provide this. When food particles have been chewed sufficiently, they tend to aggregate into a sticky mass called a bolus. The tongue can then propel this bolus backwards for swallowing. The tongue, palate and pharynx (the muscular tube directly behind the mouth), all act together during swallowing. The pharynx is uniquely mammalian possessing a musculature that appears to have no equivalent in other vertebrates (Smith, 1992).

Sensing food

The senses are mounted largely as paired organs on the head and are generally separated widely so as to allow the location of food to be fixed. The ears sit on the sides of the head while the eyes sit in the orbits just above the jaws. Both are admirably equipped to localize food direction. However, the nostrils (not the eyes) are too close together. Thus, the nose probably has little directional ability to recognize smells, which have to be determined by the animal following the concentration gradient of a particular chemical until it reaches the source (or the source reaches it). Airflow is needed for smell, so this sense is built into the air intake for the lungs. The nostrils lead to large nasal cavities lying above the mouth and separated from it by the hard palate. However, smell is not just important for leading mammals to food. When food is being chewed, the soft palate does not seal off the foodway from the airway completely. It is probable that food vapours pass from the mouth back around the soft palate, and up into the nasal cavity (Fig. 1.1). In the part of the pharynx behind the mouth, the so-called oropharynx, the airway and foodway are a single structure to allow this. Lower down, the airway and foodway separate with the larynx and trachea lying in front of the oesophagus.

The last sense is that of taste, which is largely embedded in the surface of the tongue.[6] Potentially, anything in solution in the mouth can be tasted, but it appears that many vital substances are not, key among these being proteins, calcium and phosphate ions. Recent developments in taste research make pronouncements uncertain though, and I postpone more consideration until Chapter 3. Taste and salivation combine as detection mechanisms.

Much of the head is dedicated to housing and protecting the central processing unit of this sensory information, the brain. This needs to be set close to the jaws because timing losses due to the speed of nervous impulses are then minimized.

Communication

The crossing of the foodway and airway not only allows the ability to smell food in the mouth, but also the possibility of vocalizing through it. Most mammals make noises. These noises can seem rather limited to us because these animals seem to communicate better by smell and body language (perceived by sight). Wilson (1975) sets communication via sound on the highest grade. In order to speak, humans use the larynx to break up expired air into bursts by opening and closing the airway. Our larynx lies very low so that these bursts of air pass through the oral apparatus, with the tongue moving against the palate and teeth, not to chew, but to produce speech. The consequence is a potential for choking. However, unless it can be demonstrated that this need to communicate has influenced the design of the oral cavity, this is all really beyond the scope of this book. There are other candidates for evolutionary modification to the chewing apparatus than sound. Higher primates (monkeys and apes, including ourselves) can make faces, adding these expressions to behavioural gestures and vocalizations. These movements are mostly a function of our facial muscles. However, some of these have important actions during food ingestion and chewing, so there is the potential for conflict and compromise of function here too.

THE BASIC MODEL

This book considers the function of teeth in relation to the ingestion, chewing and swallowing of food. It attempts to elucidate the principles that underlie the evolution of tooth shape and size and those mechanisms by which the dentition can be maintained. To do this, I need to understand the properties of foods that drive evolutionary change. The following model has its antecedents in food science, most particularly in a paper by Szczesniak (1963), but it differs slightly in emphasizing a food surface versus non-surface dichotomy and by viewing the action of the teeth as to change the boundary between the two by the process of fracture.

When a mammal chews food, it usually breaks particles into fragments, producing new surface area by cracking. This extra area increases the rate at which digestive enzymes in the gut act. This is the only general explanation advanced for the evolution of mastication in mammals (though more specific explanations will be offered in Chapter 3). I assume in this book that the rate of food breakdown is optimized to an animal's needs and, therefore, that features of its anatomy and physiology can be interpreted in this light.

The effectiveness of the forces that are produced when teeth contact food depends on the characteristics of the food surface that they act on. The form and extent of the external food surface is referred to as the *external physical attributes* of the food. These characteristics include food particle size and shape, the total volume of particles in the mouth and attributes of surface quality such as surface roughness and stickiness. These surface attributes can be contrasted with the resistance of food to the formation of new surface, this being the object of the chewing process. This resistance is embodied in the *internal mechanical properties* of the food, including the Young's modulus, yield strength and toughness.

The rate at which food particles break down can be subdivided into many anatomical and physiological variables. However, at a conceptual level, it can be subdivided simply into the product of two discrete events: the chance of hitting a food particle on the one hand and the fragmentation that is produced when any particle is hit on the other. From the food side of the problem, the chance of hitting a food particle depends only on its external physical attributes. For example, small particles are less likely to be hit than large ones. A smooth particle may slip from under the load whereas a rough one might be trapped and so on. I will deal with these attributes in Chapter 3. Once the particle is hit, then its resistance to fragmenting is defined by its internal mechanical characteristics. These characteristics include major mechanical properties such as toughness, Young's modulus and strength (defined in Appendix A). From the mammal's side, the chance

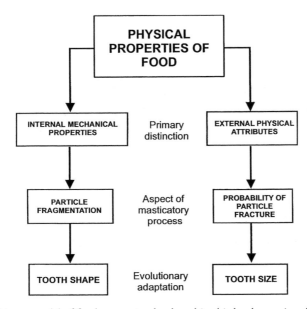

Fig. 1.2 A binary model of food properties developed in this book stressing the difference between surface attributes (right column) and those properties that act variably to prevent the surface from being extended by food breakdown (left column).

of hitting a food particle with the teeth is enhanced by making the tooth bigger – i.e. by changing tooth size. In contrast, the effect of the force that the tooth exerts on that particle depends on the contours of its working surface – i.e. on its tooth shape.

Figure 1.2 shows how this binary model is developing into two separate streams of reasoning. A fundamental aspect of this book will be pushing this model to the limit, progressively examining this classification of food properties so as to understand how teeth work. To do this, I rely a lot on optimization theory, even when the relevant numerate models are not yet constructed. For those averse to this method of investigation, I refer them to a book on this subject by R. M. Alexander (1996) and his recent short defence of it (Alexander, 2001). Achieving my objective requires some descriptive detail and literature citation. Vital though it is, citation interferes with textual rhythm and seems somehow to make it all rather sombre. Heraclitus said that everything flows (Barnes, 1987), but perhaps he might have exempted much of this book.

The basic structure of the mammalian mouth

OVERVIEW

Before processes of ingestion, mastication and swallowing can be considered, the structure of teeth and their general arrangement in the face have to be sketched. Function creeps in, but only so that the sense in certain structural arrangements is clarified.

WHAT ARE TEETH?

The teeth of a mammal are stones, anchored in tight-fitting holes in the bones of the upper and lower jaws and projecting through the lining tissue of the jaw into the mouth (Fig. 2.1). The part of the tooth that projects into the mouth is called the tooth crown while that part set into the jaw is called its root. The working surface of the crown is described by its most prominent features: pointed elevations are called cusps if they are large, tubercles if they are small. Roughly circular depressions are called fossae (singular, fossa). Raised folds can be called ridges or crests (sometimes written as cristae; singular, crista), but I will sometimes refer to these features later as blades. The creases that run between the bases of cusps are called fissures. Figure 2.2 shows some of these features diagrammatically on a typical mammalian molar and gives compass bearings needed to identify them unequivocally in a skull. Many features, such as fissures (one is indicated on Fig. 2.1), are depicted best in photographs and reflect the fact that the cusps of most molars rise like mountains alongside V-shaped valleys. Fissures are like the rivers running at the base of those valleys.

A typical mammalian tooth contains three mineralized tissue layers that are firmly bonded to each other in spite of the fact that few structures traverse their junctions. The outermost layer of the tooth crown is the enamel (Fig. 2.1). This tissue is developed from the epithelium of the mouth and is very heavily mineralized. It is backed by dentine, a tissue that forms

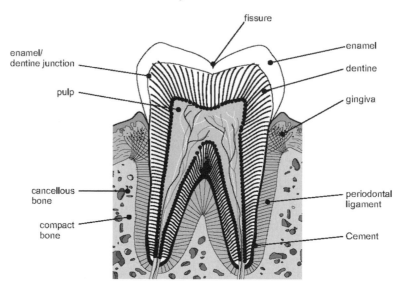

Fig. 2.1 The basic form of a mammalian tooth, illustrated by a human molar. The crown is covered by enamel and usually projects entirely into the oral cavity, while the enamel-less roots are buried in a socket in the jaw bone. The junction between the two is called the cervical margin (the 'neck') of the tooth. The gingiva (gum) forms a roof for the periodontal ligament, the soft tissue of the tooth socket. The ligament (much narrower than shown) is crossed by collagen fibres that are anchored in the cement of the tooth and in the bone of the socket wall. The tooth is kept alive by blood vessels that enter through holes in the roots and which feed the cells that form the dentine (odontoblasts – cell bodies shown as circles). Nerves that supply the periodontal ligament also pass into the tooth.

the foundation of all teeth, and which derives from underlying mesodermal (or, more strictly, ectomesodermal) tissue. Inside the dentine is a pulp cavity containing living tissue, which can repair dentine as a response to tooth wear. A thin tissue called cement (or cementum) lines the outer surface of the dentine in the tooth root. The cement and dentine, together with the soft tissue of the tooth socket, the periodontal ligament, and also the

Caption for Fig. 2.2 (*cont.*) The equivalent posterior extension in the upper molars, the talon, was a late evolutionary development (evolving independently in many mammalian lineages) and usually has one major cusp. Additional smaller cusps can be present and any or all of these can be connected by ridges/crests. The named basins (fossae) lie below cusps: the protocone is aligned over the talonid, while the hypoconid is lined up under the trigon. There is a lot of variability on this basic form (for example, the flickart on pages 1–159 shows how the paraconid was lost in the evolution of higher primates). Everything possible on a tooth crown is named and the 'wealth' of terminology is truly awesome. In line with the advice of Butler (1978), I have strived to keep things simple.

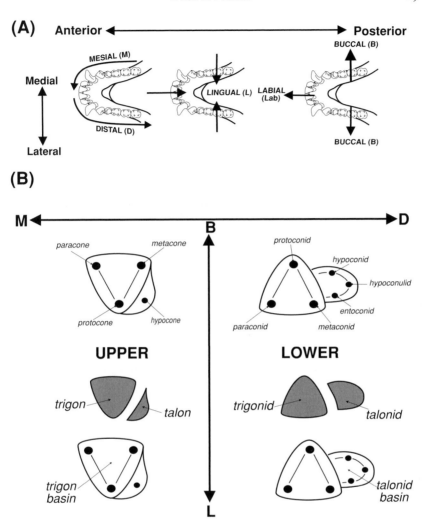

Fig. 2.2 (a) Anatomists use a straight compass to describe directions, but the mouth has its own bent version. Anatomical directions are shown above, where anterior (to the front) is the opposite of posterior, and medial (towards to the midline) is contrasted with lateral (away from it). In the mouth, mesial is the opposite of distal, while buccal is opposed to lingual. Buccal (meaning 'towards the cheek') is replaced by labial (towards to the lips) for the anterior teeth. (b) The basic mammalian molar form. Cusps are indicated by circles, while ridges or crests are marked by lines. Both upper and lower molars have a triangular region. This is called the trigon in the upper molar (the suffix '-id' being added for lowers) and is bounded by three major cusps. The lower molars have a platform (posterior to the trigonid in all living forms) called the talonid, which is bounded typically by three cusps.

(*cont. on previous page*)

surrounding jaw bone all seem to have a common developmental origin and possess structural similarities, being built on a collagenous framework. Cement forms an anchorage for the thick bundles of collagen fibres that pass through the living tissue of the periodontal ligament to the densely mineralized bone of the socket wall. However, it can extend over the crown of a tooth as well as the root. The pulp communicates with the periodontal ligament via a hole or holes in the tip (apex) of the tooth root. The teeth at the back of the mouth tend to have more than one root (a two-rooted tooth is shown in Fig. 2.1) for greater stability under load. The pulp and periodontal ligament are well supplied with blood and there are also nerves that supply the walls of the blood vessels and conduct impulses from sensory receptors. The collagen fibres of the periodontal ligament have various orientations and are separated by cells of both epithelial and ectomesenchymal origin. There are also fibres in the elastin family called oxytalan fibres and much extracellular gel-like material. The ligament cushions the tooth and controls small movements within the socket, as will be discussed a little later.

TYPES OF TEETH AND THEIR DEFINITIONS

There are three classes of teeth in a mammal – incisors, canines and post-canines. The incisors and canines will often be referred to as the anterior teeth, while those behind the canine will be called posterior teeth. All develop inside the jaws and erupt through the jaw sequentially so as to appear in the mouth. Each tooth class develops in its own series over two distinct tooth generations. Figure 2.3 shows a sketch of the permanent dentition of the lower jaw of two Old World higher primates: a male adult long-tailed macaque (the 'monkey' of much medical literature) and a male gibbon (an ape). The normal dentition has bilateral symmetry, and the lower teeth are generally aligned fairly precisely against the upper teeth, which are embedded in the upper jaw.

Set at the front of the jaws, the incisors generally have a simple shape. In most mammals, these are short pointed teeth. However, in higher primates (New World monkeys, Old World Monkeys, apes and humans), incisors have a bladed working surface. In rodents, lagomorphs (rabbits, hares, etc.) and one primate (the aye-aye), the central incisors are wide bladed teeth that grow throughout life. The canine teeth, when they are present, are set behind the incisors. Canines nearly always have pointed crowns and generally project beyond the rest of the dentition. Behind these are the postcanines, which are responsible for mastication. There are two names

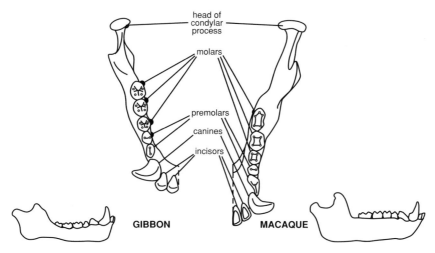

Fig. 2.3 The lower jaws of two male primates, an ape (the Bornean gibbon, *Hylobates muelleri*) and an Old World monkey (the long-tailed macaque, *Macaca fascicularis*), to illustrate basic features of the arrangement of the jaw and teeth of mammals. The broad lower incisors are called 'spatulate' in the literature.

generally given to postcanine teeth: molars and premolars. Molars develop in the first generation, but the most posterior members of the series are retained throughout life. However, most of the more anterior molars are replaced by premolars. These are the descendants of molars: by definition, they are 'postcanines of the second generation'. Deciduous molars are those that are shed (fall out) so as to provide space for their descendent premolars, whereas permanent molars are not succeeded and remain in function until death.

It is thought probable that early mammals had at least 12–13 permanent teeth in each half of each upper and lower jaw: up to five incisors, a canine, four premolars and three molars. In some mammalian lineages, there has been a marked tendency to reduce this number (e.g. in primates: Martin, 1990), while in others like the cetaceans (whales/dolphins), there has been an increase. Either an increase or a decrease creates a problem: in order to describe teeth, we need to identify them.

When a common name is given to structures in different mammalian species, we really need to establish their genetic identity. Structures in different species of organisms formed by the same set of genes are called homologous structures. The problem arises because, for most mammalian structures beyond the simplest, the developmental mechanisms that create these structures have not been tracked back to the complicated interaction of gene products on which their existence depends. Far too many genes act together to produce the dentition to be clear about what creates the identity of a tooth class. There are clues (Yucker *et al.*, 1998), but no certainty. So until the homology of structures is known for certain, arbitrary indicators must be used.

The homologies of the teeth of mammals are currently determined in a very arbitrary way, being identified by their position with respect to a bony suture lying between two bones of the upper jaw, the premaxillary bone in front and the maxilla behind it. The first tooth behind this suture is named the upper canine. There is, for some obscure reason, only allowed to be one such canine on each side of the upper jaw. Incisors lie in front of this canine, while postcanines lie behind it. The incisors and postcanines (deciduous molars, premolars and molars) are each numbered from mesial to distal. However, this procedure only defines upper teeth. Names for the lowers are organized according to a single criterion: the tooth fitting just in front of the upper canine is the lower canine. Everything follows from this.[1]

Simple though this nomenclature appears to be, it is just deeply entrenched convention. There have been surprisingly few attacks on it, partly because of the tenacity of the conservative response. Yet a logical argument has been advanced which casts great doubt on the relationship of this convention to actual homology (Osborn, 1978). Osborn's theory, the developmental details of which are not of concern here, suggests that tooth classes can be recognized by a commonality of shape. Within any tooth-class series, there will be a clear gradient of size and shape, the form of any individual tooth depending on its position within one of these series. The boundaries between tooth classes can then be recognized by non-sequiturs of form, i.e. by recognizing abrupt breaks in shape and size. Osborn further suggested that the numbering of teeth should follow the order in which teeth develop rather than their position in the jaw. This is based on the assumption that teeth tend to be lost at the ends of series (where the end of any series can be marked by identifying the last tooth to develop in either the first or second generation). Thus, the first teeth to develop in a particular tooth class in any mammal would be homologous across all mammals with the others numbered in order of their subsequent appearance. Using these simple rules, Osborn showed that the application of this system sometimes leads to very different tooth identities to those of conventional wisdom. For example, his system has no need to postulate that there is only one canine on one side of an upper or lower jaw and there is no reason to identify canines by their position in that jaw, either in relation to a bony suture or to a tooth lying above and behind it. Just a little thought by anyone who has not been taught the prevailing dogma is perhaps enough to see that the evolution of the dentition is unlikely to have been constrained by a fixed relationship to bone. Without dwelling on this further (because it will only be returned to briefly in Chapter 7), it seems quite likely that this novel system of homology will be shown to be correct. The ramifications could be considerable. The commonest reason for ignoring it – that it would destabilize long-held nomenclature for the dentition – should not be a barrier to its investigation. Even Osborn (1978) felt that the two issues, taxonomic nomenclature and biological identity, should best be separated.

The development of the dentition is the subject of a lot of current research, but it is far from generating clear answers. Among the general facts that developmental biologists have to contend with is that teeth can develop perfectly normally outside the mouth, e.g. the ovarian cyst pictured in Sharpe (2000) that looks to possess most of the tooth types in the lower jaw. The crowns of these teeth look completely normal but are located in an utterly bizarre setting. (Such a structure has, on occasion, been reported to have broken through into the vagina of a patient.) Yet location must be important in dental development or else upper and lower teeth could not develop in approximate alignment, a matter for the next section.

ALIGNMENT OF TEETH

The upper and lower teeth of most vertebrates other than mammals never contact and wear only because they contact food. When the jaw is closed, the simple conical crowns of the upper and lower teeth of a typical reptile, like a lizard, interdigitate with each other, but are spaced such that there cannot be contact. Mostly, this is achieved by having the lowers lie inside the uppers forming a zigzag arrangement (Fig. 2.4), although in some living crocodilians, the upper and lower teeth lie along a line (Brochu, 1999).

In contrast, the need of mammals to fragment food particles requires complex postcanine teeth that have to touch when the jaw is closed. There is obviously great selective pressure against adding tooth-on-tooth contacts to those produced by food, because this is potentially very dangerous. The principal mode of wear of teeth is abrasive wear, in which hardness is a major factor (Chapter 6). Teeth are generally the hardest items in the mouth and are therefore likely to do much more rapid damage to each other

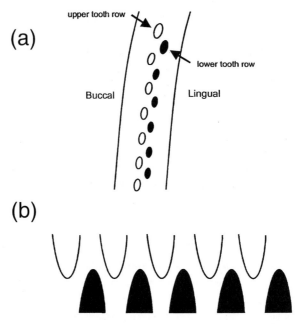

Fig. 2.4 The arrangement of the upper (unfilled) and lower (black) single-cusped teeth of a reptile. (a) A view from below, the outline being that of the maxilla. The lower dental arch fits inside the upper and the teeth are spaced so as to avoid tooth–tooth contact and wear. (b) A side view with interdigitating cusps. In some living crocodilians, the upper and lower teeth lie 'in line' rather than one inside the other.

than food is capable of. Forceful contacts are made during chewing (just after breaking food particles) and swallowing. Wear potential from these contacts is minimized by various anatomical and physiological variables. The first is precise alignment of upper and lower teeth (called occlusion by dentists). Opposing teeth have surfaces that match very well and the contacts that are made as they glide together are used as a physiological guide to bring the teeth together in a standardized fashion. The actual precision of this alignment depends on the type of tooth. There are fewer misalignments (called malocclusions by dentists) in mammalian species where the postcanines have tall cusps with more prominent ridges than in those with low cusps where malocclusions are common (Mills, 1955).

The alignment of teeth cannot just depend on the position of a tooth after it erupts because this is unlikely to be accurate enough. It depends also on the ability of mammals to move teeth very subtly in the jaw so as to adjust their position. This movement is possible because the bone surrounding the periodontal ligament can remodel under stress.

Spreading the force of tooth contacts across the dentition by precisely aligning contacts is not necessarily sufficient to avert wear at the end of chewing strokes. It is also necessary to decrease the time of contacts. This is achieved by 'switching off' the activity of the key jaw closing muscles about 12 ms after tooth–tooth contact (Gibbs *et al.*, 1981).

Just as in other vertebrates, there is a tendency for the mammalian lower dentition to lie medial to the upper dentition. It follows that, in order for the teeth to be aligned for conjoint action, the lower jaw must be moved laterally at some point. In most mammals, this is the only other movement besides simple open and close, but some mammals, like rodents and elephants, have an anteroposterior component to jaw movement during chewing (Maglio, 1973; Hiiemae, 1978).

INDIVIDUAL TOOTH TISSUES

Each hard tissue is a composite, i.e. an organized mixture, in which minute ceramic particles (inorganic crystals) are set in organic glue containing plenty of water. The protein in the glue acts like a scaffolding while the tissue is being built. This protein may be fibrous, as in dentine, or mostly globular as in enamel (Fincham *et al.*, 2000). Normally, this scaffolding remains after formation, but to get the very high mineral content of enamel, it is largely removed during a process called maturation. The orientation of the protein framework (called a protein matrix by biologists) seems to determine how the mineral is laid down.

The inorganic crystals in each tissue are made of hydroxyapatite, a calcium compound with the formula $Ca_{10}(PO_4)_6(OH)_2$. In large blocks, this stony material is stiff and hard, but it lacks fracture resistance in tension or bending due to its low toughness. However, as outlined in detail in Chapter 5, the smaller a solid object is, the stronger it usually is. The crystals in hard tissues are nanometre-sized and are actually so strong that, when these tissues crack, the cracks run between crystals in the surrounding protein matrix and not through them.

The protein matrix surrounds the crystals and appears to contain narrow channels through which water can flow. When dental tissues are compressed, they reduce their volume slightly, densifying because their Poisson's ratios are less than 0.5 (see Appendix A). Assuming these tissues are saturated with water (and the tooth crown is always coated with saliva), this loading must push water out of the tissue because water, like other fluids, has a low compressibility. The results of Haines *et al.* (1963) and Fox (1980) are both consistent with water flowing through pores in dental tissue. In part, this flow must be through enamel, but Paphangkorakit & Osborn (2000) have actually demonstrated this fluid flow for dentine under load.

The two-phase liquid–solid structural organization of dental hard tissues gives them very different properties to a slab of plain ceramic, as in an unglazed teacup, a slab of cement or a block of limestone, which they might superficially appear to resemble. The mechanical advantages of their intricate make-up are clear-cut: at the cost of a lower stiffness and the loss of some compressive 'strength', dental tissues have much greater toughness than plain ceramics. As Gordon (1991) pointed out, the greatest sin in engineering materials is a lack of toughness. Selective pressures on structures like dental tissues prove the point: teeth do not break down themselves, even after breaking down food particles tens of thousands of times.

Enamel

Enamel forms a hypermineralized cap for the tooth, never more than a few millimetres in thickness, that coats at least part of the dentine so as to provide a hard working surface to a tooth crown.[2] It is the most highly mineralized of all vertebrate tissues, being between 85% and 92% mineral by volume in the permanent teeth of humans. Its other constituents are mostly water, with some organic matter that is largely protein. The mineral crystals of enamel have the same constitution as in bone or dentine but grow much bigger during the final stage of enamel formation, called enamel maturation. At final size, these crystals, which resemble long bars with a hexagonal cross-section, measure about 30–40 nm across. They are much longer than these cross-sectional values, but the actual length seems contentious.

The first-formed enamel, right at the enamel–dentine junction, has not been much studied, but appears structureless under the light microscope and its crystals appear to have a random orientation. The rest of the enamel is highly organized with a very definite pattern of crystal orientation reflecting that of the pre-existing protein matrix (Osborn, 1981). Neighbouring crystals are usually roughly parallel, but there are tiny areas where crystals are strongly angled to each other. The crystals pack less well in these areas, resulting in a 0.1 μm wide strip of tissue that is relatively deficient in mineral. This crystal-free region is circular in form (Fig. 2.5). In some species the circle is complete, whereas in others, such as in human enamel, it is horseshoe-shaped. Complete or not, this crystal-free region bounds a roughly circular patch of enamel, usually around 5 μm in diameter, called an enamel rod or enamel prism (Fig. 2.5). These rods are visible under

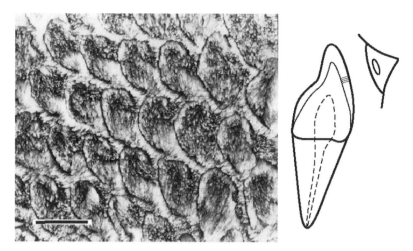

Fig. 2.5 Scanning electron micrograph of acid-etched enamel showing the form of enamel rods in human tissue. The multi-crystal rods are viewed end on, each being defined by a semicircular hood where crystals are deficient. This hood, sometimes called a rod sheath, lies at the boundary between the territories of secreting cells. At these points, crystals lie at large angles to each other. This is no developmental blunder, but central to the mechanical properties of enamel. Rods have a variety of shapes in different mammalian species. In human tissue, the rod is bounded on three sides by this crystal-free sheath but, at the lower border of the rod, there is no boundary because cellular territory is continuous and crystal disjunction does not result. Scale bar, 5 μm.

a light microscope (Fig. 2.6) and extend through the enamel from close to the dentine border nearly to the enamel surface. Very often, the rods curve.

Human enamel has been particularly well studied. The path of each rod is often slightly out of phase with that of its neighbours above (i.e. towards the cusps) and below (towards the root). This phase change is regular in the inner enamel and sufficient in magnitude to make rods that are 10 units apart appear to be running at a large angle to each other (Osborn, 1974). These periodic shifts, called decussation, give rise to optical effects called Hunter–Schreger bands. Many mammals have these bands (Fortelius, 1985; Boyde & Fortelius, 1986), which probably help to prevent a straight crack from cleaving the entire thickness of enamel (Rensberger, 2000). On exposure at the working face of the tooth, they also roughen its surface, so helping grip. At the cusp tips, rods twirl around each other at very tight angles and the crack-stopping effect of this arrangement has been demonstrated experimentally in the cusps of human teeth (Popovics *et al.*, 2002).

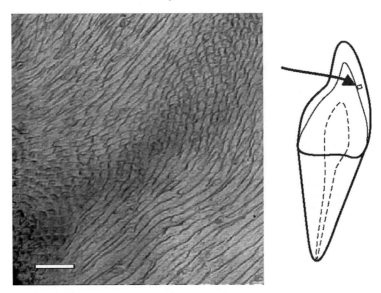

Fig. 2.6 A low-power light micrograph of inner human enamel. The rectangular box on the right indicates where the section belongs within a tooth. In the centre of the picture, enamel rods can be seen end-on with their characteristic semicircular hoods. To either side of this band, rods are seen in side view. This banded appearance is due to the wavy arrangement of enamel rods, with the path of each rod being slightly out of phase with that of its neighbours. This phase change is sufficient to produce the changes in orientation seen in the picture. Viewed from a distance, rod directions appear to change abruptly, but closer scrutiny (e.g. to the upper right side of the central band) shows that the change is gradual. These 'side view/end-on/side view' bands have been known for about 200 years and are called Hunter–Schreger bands. The dark colour of the central band is an optical effect and has no significance. Ground longitudinal section stained with Bohmer's Haematoxylin. Scale bar, about 50 μm.

Enamel formation by ameloblasts is continuous, but not completely regular. The quantity of matrix being secreted varies on a daily basis. Daily increments can be visualized as cross-striations (Fig. 2.7a), the measurement of which in apes and humans shows that 2.5–7 μm of this tissue forms each day (Dean, 2000). Somewhat staggeringly, there is also a periodic change in formation on a 4–10-day schedule (Dean, 2000) when, for no known reason, it appears that the entire population of ameloblasts suddenly shifts slightly in the tranverse plane (Osborn, 1971a). This shift leads to lines called striae of Retzius (Fig. 2.7b). Integrated measurement of these incremental lines gives accurate information about the timing of tooth formation in living and fossil mammals (Dean, 2000).

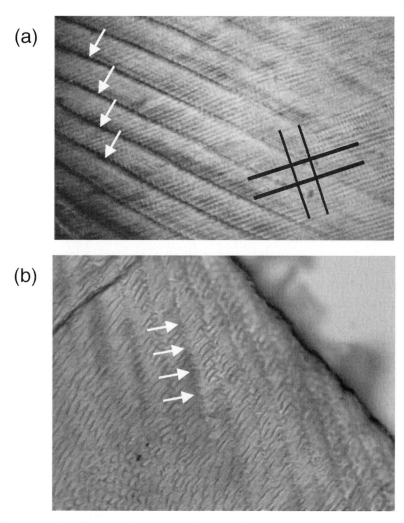

Fig. 2.7 Optical effects called incremental lines that form in human enamel on a regular basis. In both (a) and (b), the white arrows are directed at the striae of Retzius, a fancy name for the results of an event without parallel in any other mammalian tissue. At an interval that hovers around a mean of 1 week (depending on the species), ameloblasts suddenly move transversely, causing a wriggle in the subsequently formed enamel rod, which when viewed at sufficiently low magnification (as in (b)), looks like a dark line. (a) also shows rod direction (parallel to the thick black lines) and cross-striations (perpendicular to rod direction, indicated by the thin black lines). The latter represent daily increments of enamel formation.

Unlike the mineral, the protein matrix of enamel has been difficult to study because the soft tissue tends to collapse after demineralization because there is so little of it. Some proteins have now been identified, but detailed understanding is still not there. The amelogenins, of which there are at least seven, initially make up about 90% of the protein and are most important. They are hydrophobic and probably control crystal size and orientation by laying down the scaffolding on which crystals develop. This scaffolding appears to consist, rather bizarrely, of a mass of macromolecular spheres, each about 20 nm in diameter (Diekwisch, 1998; Fincham *et al.*, 2000). Some of the remaining 10% of proteins, called enamelins, are acidic, proline-rich, and may nucleate crystals. They are concentrated in the region around the enamel–dentine junction creating a mineralizing front. There are also tuftelins (Deutsch *et al.*, 1995), so called because they were identified in hypomineralized areas called enamel tufts. Amelogenins are removed during maturation by enzymes released from ameloblasts, which then suck out vast amounts of protein (degraded with proteases) and water. The other proteins remain in some form (Robinson *et al.*, 1997), particularly close to the enamel–dentine junction. However, unlike mesodermal (or, more accurately, ectomesodermal) mineralized tissues such as dentine, cement or bone, these proteins do not appear to be fibrous. A better description might be fibrillar – and perhaps, in combination with the water, they form a thick gel.[3]

The maturation process must clearly rely on flow channels in the matrix. The crystals of enamel, which are initially small, grow to fill much of the space left by the withdrawn matrix. However, channels remain in mature enamel and appear to have a bimodal size distribution, being very narrow except in the crystal-free edges of rods (Zahradnik & Moreno, 1975; Shellis & Dibden, 2000). Fox (1980) hypothesized that the narrowest of these channels are extremely important in absorbing energy under load. He supposed that fluid is squeezed slowly out of the gel-like matrix when the tooth is loaded, but it returns afterwards because this matrix attracts water, being hydrophilic. In itself, this model does not provide much resistance to flow, but when channels are as narrow as in enamel (in the low nanometre range), the electrical layers on the walls of the channels interact significantly with ions in the water to raise overall resistance to their flow. Fox predicted, and found, that soaking enamel in solutions of greater ionic strength (such as stannous fluoride) would increase resistance to flow, effectively toughening the enamel because the energy absorption in viscous flow cannot be fed into fracture in the way that elastic energy absorption can. In some ways then, enamel may resemble the stiffest of sponges.[4]

Dentine and pulp

Dentine actually forms most of the tooth. It is rather like bone and cement, with about 48% of its volume being mineral and an organic matrix based on collagen fibres. The cells that form it are called odontoblasts. Unlike in bone, where the formative cells become embedded in the tissue, the odontoblasts retreat in a row away from the tissue as they develop it towards the pulp. They take their cell bodies (their nuclei and most cellular components) with them but lay out long cellular processes behind them as they go. These processes extend initially through the entire thickness of the dentine, remaining in unmineralized spaces called dentinal tubules (Fig. 2.8). The cell bodies of the odontoblasts end up lining the pulp cavity. The very first formed dentine, in humans about 20 μm or so, is called mantle dentine and is less mineralized than that which follows.

The hydroxyapatite crystals are plate-like in dentine, being very long as in enamel, but only 2–3 nm in thickness. The organic matrix of dentine is based on collagen fibres, whose orientation seems to determine that of the crystals that initiate within and around them. Scattered throughout the dentine, there are spherical regions of crystals called calcospherites. Calcospherites are obvious structures when dentine is forming, but they generally fuse together as mineralization proceeds. However, their margins can be recognized in several sites after formation where mineralization remains incomplete (Fig. 2.8a).

The pulp cavity is the living centre of the tooth, existing just to service the nutritional and functional requirements of odontoblasts on its periphery. Odontoblasts live on for most of the life of a tooth, regularly forming dentine at a slow pace after tooth eruption. As a result, the size of the pulp cavity gradually reduces as the dentine is thickened. The inside of the tubules gradually mineralize (Fig. 2.8b) and are slowly obliterated, particularly in root dentine. All this provides a physiological adjustment for gradual tooth wear. However, this leisurely secretion is not likely to be the main reason why teeth have a vulnerable soft centre. A much more likely reason is the potential that it gives for a local response to more rapid wear of the tooth crown. The mechanism for this response is not properly established, but the long processes of odontoblasts appear to detect the movement of fluid inside the dentinal tubules following tooth loading (Paphangkorakit & Osborn, 2000). The pulp is well supplied with nerves, most of which head towards its periphery where they form a plexus near the odontoblasts. Some nerve fibres even pass for a short distance into the

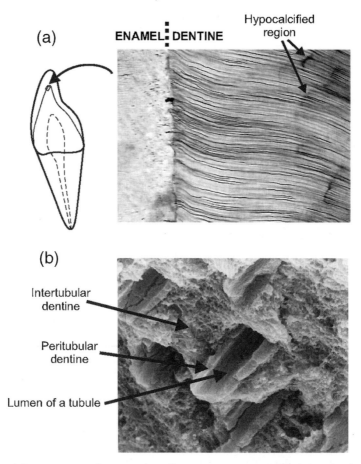

Fig. 2.8 The structure of mammalian dentine, again exemplified by the human. (a) Dentine contains cylindrical cavities called dentinal tubules that wind from the junction with the enamel back to the pulp cavity. When the tooth is first formed, they contain elongated processes of the odontoblasts and also fluid. Initially, the odontoblastic processes extend right to the enamel and can even be squeezed into it, but they later retract away. Note the hypocalcified region in the dentine, marked by the semicircular contours of calcospherites that have failed to fuse in this region (see Fig. 2.10). For some reason, this appears to have no functional consequences. The hydroxyapatite crystals in dentine are completely separate to those in enamel (Diekwisch *et al.*, 1995), again without consequence. (b) The tubules gradually infill with peritubular dentine, which is much more heavily mineralized than the dentine between tubules (called intertubular dentine).

PULP ⋮ DENTINE

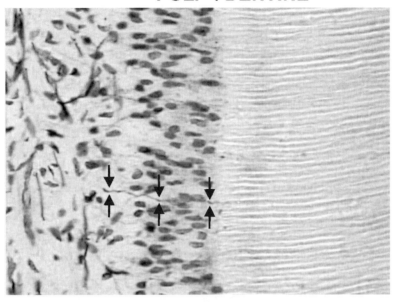

Fig. 2.9 The pulp exists to service odontoblasts, cells whose nuclei can be seen as dark oval masses next to the dentine. Most of the nerve fibres in the pulp form a plexus near the odontoblasts and some, including that which is arrowed in the figure, pass between them to enter a dentinal tubule. One way or another, this makes dentine sensitive to load, triggering the commissioning of extra tissue so as to thicken it.

tubules (Fig. 2.9), but none appears to communicate with the odontoblasts via synapses (Holland, 1994).

It is thought that the cell body of the odontoblast in the pulp may relay this signal to nearby nerves. Quite how this signal is organized is unclear, but odontoblasts start to produce dentine at an increased rate as soon as a sufficiently intense signal is received. There appears to be no need for the dentine itself to be exposed in the mouth for this response to begin.

As with other highly structured mineralized tissues (e.g. mollusc shell: Currey, 1980), the organization of dentine suffers when it is laid down rapidly. Reparative dentine does not contain ordered tubules and has a lot less mineral than the dentine that is formed before a tooth is in function. This difference in mineral content between regular and reparative dentine is important in understanding the wear patterns of exposed dentine.

The properties of enamel and dentine

Appendix B lists the principal mechanical properties of these crucial tissues. Unsurprisingly, enamel is generally much stiffer and harder than dentine. However, it is becoming clearer that there is some overlap. Peritubular dentine, for example, has close to the hardness of enamel (Kinney *et al.*, 1996) and there is probably nothing difficult in developing very high mineralization levels in vertebrates. Lungfish, for example, long separate on their own evolutionary path, have a form of dentine (petrodentine) within their massive tooth plates that has the hardness of mammalian enamel (Bemis, 1984). What seems special about mammalian enamel, however, is its mechanical anisotropy (difference in property values in different directions).

Microstructurally, enamel and dentine are both very heterogeneous tissues and it might be thought likely that this is reflected in their properties. The surprise though is that enamel seems to show more anisotropy than dentine, e.g. in toughness (Rasmussen *et al.*, 1976). The reason for this seems to lie in the avoidance of crown fracture. The theoretical work of Spears *et al.* (1993) suggests also that some anisotropy of the Young's modulus of enamel is essential if stresses are to be passed from the enamel through to the dentine. Their specific prediction, that enamel should be considerably stiffer along rods than across them, remains unproven (White *et al.*, 2001), yet it makes much sense out of tooth structure. Many prior studies have shown that if enamel were isotropic, then stresses would be taken mainly by the stiff enamel cap and only transferred to the dentine around the cervical region, the neck of a tooth. Spears *et al.*'s suggestions would result in the dentine being loaded much more evenly and make a lot of structural sense. How, for example, could odontoblasts in the pulp respond to wear if the dentine were not under significant load?

Inner enamel behaves differently to the bulk of the tissue. The first-formed enamel, that closest to the dentine and which appears the least organized region, resists cracking under indentational loads much better than the rest and seems to protect the enamel–dentine junction (Habelitz *et al.*, 2001), something that also only makes sense if this region is heavily loaded. One of the best-documented features of enamel, and one that is clinically important, is the effect that enamel structure has on controlling cracks. These fracture paths mirror toughness values in that cracks avoid crossing the rods and tend to travel only in the rod margins. Thus, variation in rod directions and in the amount of decussation of the enamels of different mammals could and should have a very strong effect on wear resistance and on the maintenance of the sharpness of enamel ridges. Work on this is now progressing (Rensberger, 2000).

Dentine is structurally just as heterogeneous as enamel, but it has nothing like this mechanical anisotropy. Fracture paths show no definite relationship to tubule directions (Andreasen, 1972) and, despite differences in toughness, there is little evidence for any crack deflection at the peritubular–intertubular dentine boundary. Yet, there are clearly important regional variations in dentine within a tooth that have important consequences for wear when this tissue is exposed on the crown surface. The dentine that lies just under the enamel, called mantle dentine in humans, is at least 10% less hard than the rest of the first-formed (or primary) dentine (beavers: Osborn, 1969; gibbons: pers. obs.; humans: Renson & Braden, 1971) and overall, the microhardness of primary dentine is generally twice that of secondary dentine (Baker *et al.*, 1959; Habelitz *et al.*, 2001). A careful study of the wear patterns of the dentine of the incisors of the beaver (Osborn, 1969) shows a close correspondence with the microhardness of the dentine and there is circumstantial evidence for this in virtually all mammals where both primary and secondary dentine are exposed.

Yet, it should not be thought that enamel and dentine are faultless and impeccably nano-engineered. In fact, flaws are obvious in both tissues by observing them at low magnification in a light microscope. Figure 2.10 shows one typical flaw in both tissues. Relatively massive spaces (massive compared to processes involved in crack formation anyway) called interglobular dentine are commonly found and represent regions of very low mineral density. Why do these not just collapse under load and send cracks spraying out to the enamel–dentine junction? The answer does not seem obvious. Hypomineralized regions called tufts, flattened irregular sheets, also characterize the inner enamel (Osborn, 1981). The flaws may run considerable distances towards the crown surface, without apparent consequence. More rarely, 'normal' enamel contains what look to all intents and purposes to be cracks. These lamellae are thought to represent cracks that developed early in enamel formation and which are filled by protein.

Much has been made of the hierarchical complexity of mineralized tissues like enamel and dentine and of the contribution that an understanding of their structure could make to improving the properties of artificial composite materials. The aim of some is to produce similar materials that, like enamel and dentine, are structured as a nanometre scale. However, perhaps when these artificial nanomaterials can be designed to survive the scale of flaws that real dental materials possess without apparent consequence, this will signal true understanding of these amazingly complex dental tissues.

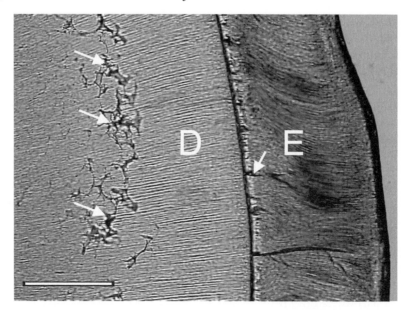

Fig. 2.10 Flaws in human enamel (E) and dentine (D) are obvious in this micrograph. The large spaces (arrowed in D) are called interglobular dentine, while the lines in enamel extending from the tissue junction are called tufts (arrowed in E). Both are very low in mineral density. The multiple semicircular structures bounding interglobular dentine are called calcospherites and represent the manner in which dentine mineralizes. Scale bar, ~1 mm.

JAW BONES

The upper teeth are housed on the maxilla and premaxilla, the lowers on the mandible. The teeth in both jaws form arches that follow the margins of the tongue. Normally, the lower arch is narrower than the upper arch, which means that lateral movements of the mandible are necessary in order to align upper and lower postcanine teeth properly for chewing.

In most mammals, dependent on their age, there is a pattern of sutures (stiff fibrous joints) between the bones in the upper part of the face allowing for very limited movement between them. The lower teeth are mounted in the mandible that may or may not be a single bone. Anteriorly, at the midline, the two dentaries (as the bones of the mandible are called) are often connected much less tightly and there can be a substantial amount of movement between them (e.g. in the dentally primitive Madagascan mammal, the tenrec: Oron & Crompton, 1985). Scapino (1965) even referred to this anterior symphyseal region[5] in carnivores as akin to an extra jaw joint. Studies of the anatomy of the tissues in primates with unfused symphyses suggest that its flexibility is variable, with some species having very tight connections and others much looser ones (Beecher, 1977, 1979). Quite independently, a number of mammals, including the higher primates and the giant panda, have fused their symphysis (i.e. converted it to bone), so that the whole lower jaw is just one unit. A large amount of work by Hylander and colleagues (Hylander *et al.*, 2000) suggests that symphyseal fusion can be explained by the need of some mammals to recruit jaw-closing muscles of both sides of the body to elevate the force in a bite (although this is not without counter suggestion; Lieberman & Crompton, 2000).

Bone differs from tooth tissue in being alive. It contains blood vessels that run in a network of channels throughout the tissue, feeding the bone-forming cells and also transporting bone-eating cells rapidly to any location. The result is that bone is constantly in flux both internally and externally, allowing it to respond homeostatically both to changes in mechanical strain within the bone itself and to whole-body equilibrium. An example of this is provided by tooth loss in humans. The mandible and maxilla have a basal part, which forms the basic support of the face and which remains even if the dentition degenerates. The roots of the teeth are housed in what is called the alveolar process of these bones. If the teeth are lost, this bone is gradually resorbed (eaten away) and over a period of 10 years or more can be completely lost. Heath (1982) studied human denture wearers and found a correlation between maximum bite force and the height of the mandibular body.[6]

The local signal for bone remodelling has been thought for more than 100 years to be related to its mechanical environment and there is a quite massive body of literature that has attempted to define the biological rules that control this. Whenever food is loaded between the teeth, the teeth and surrounding bone are deformed. It is thought that bone tissue remodels in response to the strain levels that it sustains. Biewener (1992) describes how this has been investigated, but the strain threshold level for either bone loss (resorption) or deposition has not been established clearly and it has not even been proved that strain provides the trigger for tissue changes. Having a cellular response purely based on a strain criterion will not guarantee freedom from fracture and microcracks in bone are in fact quite common (Lanyon & Rubin, 1985). Local compressive strain level within bone has been linked with the onset of bone formation, but the rhythm of loading patterns is also implicated (Rubin *et al.*, 2001; Mao, 2002). It is entirely possible that local strain energy density, i.e. the energy absorbed per unit volume of tissue, plays an important role. The skull may behave differently from the limbs and one study found that no local stimulus was necessary for skull bone deposition: that of the armadillo thickens substantially in response to general exercise, apparently quite independently of workouts of its jaws during feeding (Lieberman, 1996).

The stress that corresponds to a particular strain will depend on the stiffness (Young's modulus) of bone (Appendix A). The outer bone appears completely dense to the naked eye and is called compact bone. Its stiffness varies subtly with location (Dechow & Hylander, 2000), possibly related to the direction of muscular pull. Inside this is a network of bony struts getting the name 'trabecular bone'. The stiffness of trabecular bone depends not only on the density of the struts, but also on their geometrical arrangement and orientation (Gibson, 1985).

Sadly, I conclude this section by confessing that my efforts to establish overall generalities relevant to the control of bone formation or loss in the face have failed.

THE TOOTH SOCKET

Non-mammalian vertebrates generally have their teeth fixed to the skeleton of the face either by fibrous tissue or direct to bone (this bone being termed bone of attachment). In contrast, mammalian teeth nestle in sockets: a small fact with great implications. It is the socket that allows a tooth to drift slowly in the jaw and come to rest in a position where it is accurately aligned to its counterparts in the other jaw. The movement involved is not rapid, depending on the rate at which the body can remodel bone. As with so many aspects of dental function, the mechanisms involved are not well understood, but without this ability to move, mastication would be impossible because the teeth would not match up. The structure of the lining tissue of the tooth socket, the periodontal ligament, is therefore worthy of some attention.

Most of the ligament, three-quarters of it or so, is fibrous tissue in the form of long bundles passing from the cement to the bone lining the socket (Fig. 2.11).[7] These fibre bundles form an open network in the periodontal ligament, running in a variety of directions that enclose cells, interstitial fluid, blood vessels, lymphatics and nerves. There are also elastin fibres, although their role is not yet very clear (Kirkham & Robinson, 1995). Nerves and blood vessels cross the ligament to enter the apices of the tooth roots but the ligament itself is also very richly supplied. The collagen fibres of the ligament have a rapid metabolic turnover, which requires a good blood supply. Thus, copious blood vessels enter the ligament through numerous small holes in the bone of the socket. The ligament is also well innervated

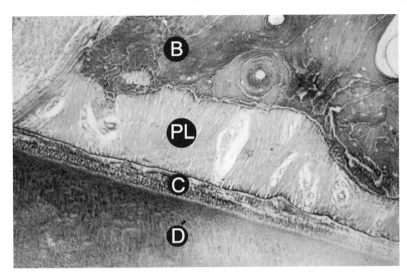

Fig. 2.11 The periodontal ligament (PL), lying between the cement (C) and bone of the socket (B), is largely composed of collagen fibres set in a gel-like matrix. Both are important in explaining why teeth do not abrade against the socket walls, given that the ligament is so narrow. The dentine (D) underlies the cement. The periodontal ligament is generally between 0.1 and 0.3 mm thick.

and contains mechanoreceptors that relay information about the direction of loading on a tooth (Linden, 1990).

Cement starts as mineral glue that forms around the incoming collagen fibres of the periodontal ligament. However, unlike enamel, it can be added to throughout the life of a tooth so as to compensate for the loss of tooth height. Cement that is laid down later, after tooth eruption, develops its own, much finer, collagen framework (Jones, 1981). If it is formed slowly, then the cementoblasts that form it can retreat away, but if it is formed rapidly, cells get trapped in the matrix and eventually die. The rate at which cement has formed is evident from its 'cellularity'. In humans, rapid cement formation around the apices of the roots is a response to rapid wear of the crown. In some herbivores, the crown of the tooth is also coated in cement (Jones & Boyde, 1974).

The periodontal ligament is generally very narrow, generally being 0.1–0.3 mm in width in the human. Large transverse (lateral) forces will widen the ligament. For example, a very wide socket is normal for the upper canines of Chinese water deer (*Hydropotes inermis*), which are presumably continually being knocked by lateral movements of the jaw (Aitchison, 1946). However, the teeth of most mammalian species tend to fit very tightly into their sockets. Synge (1933) suggested that the reason for this tight fit is that the displacement of the tooth crown in a lateral direction at any point on its surface is proportional to the cube of the width of the ligament. In other words, doubling the width of periodontal ligament would increase tooth movement eight-fold. A snug fit of a tooth in its socket is thus absolutely essential or it will waggle like a water deer's.

The assumptions that were necessary for Synge's result do not sit very well with its biology. He assumed that the ligament was elastic and incompressible (rather like a rubber) and sandwiched between an infinitely rigid wedge-shaped tooth and bony socket. However, the critical assumption is probably that of incompressibility, which is likely to be provided by the large quantity of intra- and extracellular fluid within the ligament, much of it 'bound' in a gel-like matrix. Incompressibility is supported by the observation that when a tooth is pressed into its socket, the bony crests standing just beneath the gum move away from the tooth (Picton, 1965). Whatever, the ligament is far from ideally elastic, taking time to recover after being pushed into its socket. Synge knew this from investigations done with Dyment (Dyment & Synge, 1935) and he made a number of revisions to his theory, including the effect of a small compressibility. However, his theory did not offer a specific role for the collagen fibres in the periodontal ligament, stating that if these alone were responsible for resisting tooth movement, then the displacement of the tooth crown in a lateral direction at any point on its surface would be directly proportional to the width of the ligament. Without the 'cube' relationship, an argument for the tightness of the socket is lost. Rather like tendon fibres, the collagen fibres of the ligament possess an undulating form (called a planar crimp). This probably pulls straight at a ~4% elongation (Kastelic & Baer, 1980), but prior to that, these fibres offer no resistance to tooth movement. Once straightened, they are predominantly elastic and must be very important in preventing a tooth from bottoming out in its socket. There is experimental evidence in goats that collagen fibre direction adapts to the degree of tooth depression (Lieberman, 1993).

If a tooth goes out of function, then the ligament narrows. It is accordingly likely that most teeth are loaded mostly in a direction close to their long axis. If they were not, then they would realign until this were so or else quickly work very loose. Resistance of the ligament to axial forces that press teeth directly into their socket is obviously extremely important. Synge (1933) also gave an estimate of this for a human upper central incisor with a conical root. The pressure, P, in the ligament is highest at the apex where

$$P = P_A - (2F/A) \qquad (2.1)$$

where P_A is atmospheric pressure, F is the axial force and A, the area of the cross-section of the tooth root at the top of the socket.[8] If the ligament is assumed to possess the properties of rubber, then Synge worked out that an axial force of 1.7 N moves the tooth only 7 μm into its socket, while a transverse force of 0.85 N applied to the incisal edge moves it 226 μm. Real data somewhat exceed this for axial forces but not for transverse ones. For example, Parfitt (1960) found experimentally that an axial force of 1.7 N actually pushed an upper incisor 25 μm into its socket. (Anderson (1976) states that human teeth probably only depress about 50 μm at most into their sockets.) This is much greater than Synge's predictions, but does not (in my opinion) detract from the value of his work. Mühlemann (1951) found that the same horizontal force as Parfitt's would move the tooth 150 μm. However, the ligament is less stiff at low forces than higher ones, which suggests collagen fibre involvement that Synge would not have anticipated. Much experimental work has followed these early studies and is summarized in Berkovitz *et al.* (1995), but fundamental explanations of their results seem far away.

MOVEMENT OF TEETH WITHIN THE JAWS

Teeth develop inside the jaw, emerging as and when needed by a process called tooth eruption.[9] A soft lining to the tooth socket allows the erupted teeth of mammals to 'drift' in the jaws so as to make minor adjustments in alignment. This drift can be either mesial or distal in direction (Berkovitz *et al.*, 1995). Very likely in ungulates that have a diastema (shown in Fig. 3.10) between the cheek teeth and incisors, the anterior cheek teeth drift distally while the posteriors drift mesially, thus keeping the cheek tooth battery together in an unchanged position. Unless the tooth is completely stable, areas of both bone formation and resorption can be found in and around the socket so as to facilitate this tooth movement.

The principal reason for this drift must be the direction of the muscular force. This will determine the effective direction of the bite force on all food particles except the most rigid. While the resultant of the force must be directed along the long axis of the tooth in order for the position of the tooth not to shift, the direction of the force depends on gape. In humans, the muscle closing force has an anterior component when the jaws are almost closed, and a slightly posterior component when the jaws are wide

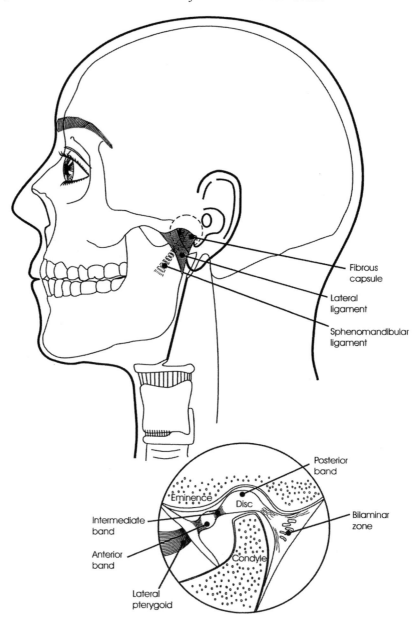

Fig. 2.12 The structure of the human temporomandibular joint. This has been researched in great detail, but some aspects of its structure and the explanation of these remain controversial. In mammals, a fibrocartilaginous disc separates bony surfaces lined by fibrous tissue, effectively forming two joint spaces, one above and one below this disc.

apart (Paphangkorakit & Osborn, 1997). The force direction at jaw closure is consistent with the results of Osborn (1961) who found that teeth push forwards (mesially) in the jaw during an empty clench. The approximal wear that develops from this is described in Chapter 6.

THE TEMPOROMANDIBULAR JOINT

The upper and lower jaws are connected together by a synovial joint, called the temporomandibular (or jaw) joint, which ensures their stable alignment. The joint lies between the head of the mandibular condyle and the squamous part of the temporal bone of the skull and is unique to mammals. The joint has been investigated in great detail because of its complexity and importance (Rees, 1954; Hylander, 1992). Like all synovial joints, it has a fibrous capsule enclosing the joint surfaces. Viscoelastic synovial fluid bathes the joint surfaces, so reducing friction to very low levels (the coefficient of friction in joints varies, but is generally <0.04; Swanson, 1980).

In most mammals, the joint is basically a hinge (capable of rotatory motion), but it can also glide (translate) at wide openings. In some mammals with fused symphyses, any movement of the lower jaw necessarily involves movements of both the left and right temporomandibular joints because the mandible is a single bone. The limits of jaw movement in humans are mainly due to two prominent ligaments, the lateral ligament that lies on the capsule and the sphenomandibular ligament, which is separate from it (Fig. 2.12) (Osborn, 1993). In humans, the mandible appears to translate under the control of the upper end of the lateral ligament, but rotate around its lower end, rather like an upside-down child's swing (Osborn, 1995a).

In humans, the articular surface of the mandible sits proud as a condyle, which nestles behind a projection of bone on the posterior aspect of the zygomatic part of the squamous temporal bone of the skull called the articular eminence (Fig. 2.12). These bony surfaces are covered in fibrocartilage rather than the hyaline cartilage that covers the surfaces of bones in most other synovial joints. This is partially a consequence of these bones developing in fibrous membrane rather than from cartilaginous precursors

---

Caption for Fig. 2.12 (*cont.*) This is a unique arrangement for a synovial joint. According to Osborn (1985), this is to allow the head of the mandibular condyle to bed into the disc at low stresses, due to the J-shaped stress–strain curve of this soft tissue (see Appendix A and Fig. 5.7). The two important ligaments that help to control jaw movement, the lateral and sphenomandibular ligaments, are shown.

(Hylander, 1992), but the distinction is important mechanically too. Unlike hyaline cartilage, fibrocartilage has a J-shaped stress–strain curve (Fig. 4.13) (Tanne *et al.*, 1991; Teng *et al.*, 1991). At low stresses, it is very pliant which means that the head of the condyle tends to sink into the fibrous tissue under the skull, allowing the condyle to 'bed down' (Osborn, 1985). At high stress, this tissue is much firmer, so allowing it to slide (see Fig. 5.7).

The joint surfaces do not contact because a fibrocartilaginous disc is placed between them. It is attached to the capsule all round, so dividing the joint cavity into separate upper and lower spaces. Anteriorly, the disc attaches both to the squamous temporal bone and the neck of the condyle. It also appears to receive fibres of the lateral pterygoid muscle, although whether these fibres embed only in the capsule of the joint or only pass into the front of the disc instead is not clear. Essentially, the disc moves with the mandibular condyle, but not in complete synchrony, something that is a cause of trouble with the joint in humans.

SOFT TISSUES

The mammalian mouth (or oral cavity) is designed around the need for mastication. It is bounded above by the hard palate, which separates it from the upper respiratory tract lying above. Reptiles, other than some crocodilians (Smith, 1992), do not have a hard palate because they do not chew. They swallow large particles, which will not fit into the airway. In contrast, the small (sometimes minute) fragments that can be produced by chewing can be inhaled if the airway is not protected, so particle size reduction of foods by mastication requires the airway be as disconnected as possible from the foodway.

Food that enters the mouth encounters a cavity lined with a tissue called oral mucosa that is somewhat like thin skin (Fig. 2.13). As with skin, the lining of the mouth is keratinized in all places where there is considerable friction. Contact with food particles produces this friction and the high toughness of keratin (e.g. in horse hoof: Appendix B) prevents any subsequent wear. The frictional coefficient of dry oral mucosa against dry food is high (see below). To avoid this, the mouth needs to be constantly lubricated by saliva.

The floor of the mouth houses the tongue, which is a large muscular bag capable of a wide range of movement. The oral cavity is actually mostly tongue. It occupies virtually all the space inside the dental arches, leaving only 10–20% of cavity volume vacant for food particles. Tongue movements do not impinge on this space because the tongue is thought to have a

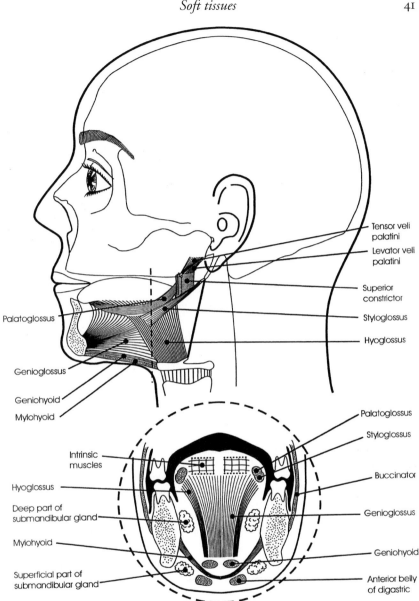

Fig. 2.13 Above, features of the musculature of the tongue, soft palate and pharynx in the human, exposed from the side. Below, a cut across the face in the coronal plane to show details of the floor of the mouth.

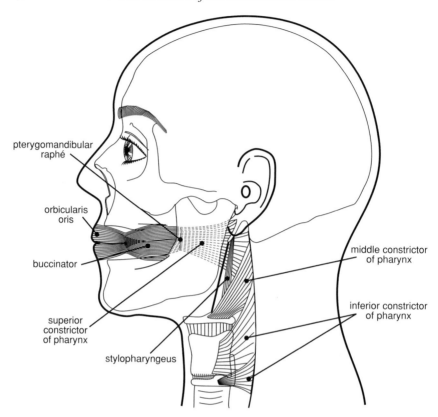

Fig. 2.14 The major muscles of the pharynx, the constrictor muscles, are a mammalian novelty. They form a semicircle of muscle around the back of the oral cavity and larynx and contract in a peristaltic wave during swallowing. Curiously, the superior constrictor is attached to the cheek muscle, the buccinator, another novelty, even though these have different functions, developmental origins and nerve supplies. The vertical line at which the buccinator and superior constrictor fibres interdigitate is called the pterygomandibular raphé. Anteriorly, the buccinator appears to send many of its fibres into the orbicularis oris, the muscle of the lips. These fibres decussate (cross) at the corner of the mouth.

constant volume whatever its position (Kier & Smith, 1985). The oral surface of the tongue is purposefully rough. Its upper surface is roughened with keratinous spikes called filiform papillae in order to grip food particles. Taste buds on the tongue surface are associated with other types of papillae. All taste buds are anatomically similar, but their component cells are sensitive to different chemicals depending on their membrane receptors (see Chapter 3).

Below, the mouth is separated from the neck by a thin diaphragm called the mylohyoid muscle (Fig. 2.13). This is connected at the front and sides to the lower jaw. Behind, it is attached to the hyoid bone, a small bone that 'floats' below the tongue suspended by ligaments and the tone of its muscles. Normally, the mylohyoid muscle hangs loosely downwards, but it contracts strongly during swallowing (Hrycyshyn & Basmajian, 1972).

Posterior to the mouth lies the opening to the pharynx. The tongue bridges the two spaces with its posterior one-third lying in the pharynx. In contrast to the roughened oral surface of the tongue, the posterior part is smooth so as to allow food particles to slip over it easily. In the pharynx, the airway crosses the foodway (Fig. 1.1). The mammalian pharynx possesses a series of semicircular constrictor muscles attached to the back of the mouth and larynx, which reptiles do not possess (Smith, 1992). Rapid peristaltic contraction of these muscles assists in pushing the food bolus down the pharynx to the oesophagus as rapidly as possible (Fig. 2.14).

SALIVARY GLANDS

Lubrication of the mouth is achieved by the output of an oral fluid called saliva secreted from a large number of glands located around the mouth. The largest of these glands in most mammals are the parotid, submandibular (sometimes called the submaxillary) and sublingual glands (Fig. 2.15). The first two are much more important than the last, which is actually a mass of very small glands lying close together. Saliva jets from four major orifices, two on either side of the midline, and from many small openings. The largest glands are the parotid glands, which are housed just behind the lower jaw and whose ducts open into the vestibule of the mouth opposite the molar teeth, the most important postcanines (Fig. 2.15). The ducts spray these teeth directly. The submandibular glands lie partly in the floor of the mouth and partly in the neck, wrapped around the posterior edge of the mylohyoid muscle (Fig. 2.15). Their ducts open close together in the floor of the mouth just behind the lower incisor teeth.

Saliva itself is a dilute solution of glycoproteins with four very important functions from the viewpoint of this book:

(1) It reduces friction between food and mucosa to low levels. This requires fairly high glycoprotein concentrations.

(2) It wets the new surface of the food produced by particle fragmentation so that these will be encouraged to bind together to form a bolus for swallowing.

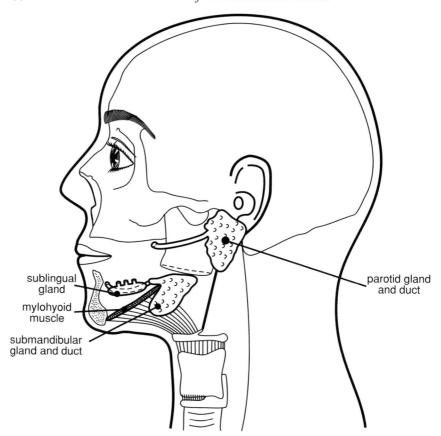

Fig. 2.15 A sketch of the major salivary glands in the human. Both the parotid and sub-mandibular glands have grown so large that their bulk, or part of it, lies outside the oral cavity.

(3) It helps to solubilize potential taste compounds in foods such that they can be sensed on the tongue.

(4) It buffers food acidity, so preventing acid erosion of the mineralized tooth tissues.

The composition of saliva depends substantially depending on the gland that produces it. Minor glands, including the sublingual, mostly produce viscous secretions that cover the mucous membrane. This thin covering is not lost at swallowing and is needed to keep intra-oral friction low. J. F. Prinz (pers. comm.) shows that surface-dry mucosal surfaces can have a coefficient of friction between them as high as 1.5, while saliva brings this down 10-fold or more. The secretions of the parotid gland are much less viscous, with submandibular saliva being intermediate.

Physically, saliva has many functions. It is the anti-stick lining of the mouth – only fatty films from foods usually evade the clearance of food by swallowing. Even chewing gum does not stick either to the teeth or mucosa. Nevertheless, saliva (particularly parotid saliva) is good at wetting food surfaces, due to a surface tension well below that of water (Glantz, 1970), and good at sticking food particles together into a bolus by virtue of a viscosity that is well above it (Prinz & Lucas, 2000).

The viscosity of saliva has always been measured by conventional viscometry, i.e. by shear. Currently accepted values, which vary with the rate of shear, are probably underestimates. There is circumstantial evidence during critical events in the mouth, such as during swallowing or food bolus evaluation, when the tongue thrusts up against the hard palate, that saliva is subject to extensional flow, probably biaxially (de Bruijne *et al.*, 1993; van Vliet, 2002). This type of flow changes the volume of saliva slightly by 'stretching' it. Volume change may seem counter-intuitive to those who think in terms of the action of hydraulic pistons, but it is perfectly possible to change the volume of liquids in some circumstances, and this is the only way in which a film of oil in an immiscible fluid can be emulsified (i.e. can break down into small droplets). Such emulsification does happen in the mouth, as also the production of many small bubbles in the saliva, which is somewhat analogous (de Bruijne *et al.*, 1993). The point is that resistance to extensional flow appears much greater than in 'normal' shear, requiring a rethink of salivary viscosity.

THE MUSCLES

There are a very large number of muscles involved in oral function and their general morphology and fibre direction varies substantially between mammalian species. The muscles can be grouped into (1) jaw movers, (2) tongue muscles, (3) facial muscles that move or fix the mouth slit and (4) neck muscles that adjust the position of the cranium.

Jaw movers

The major work input for fracturing food particles comes from the large muscles that elevate the lower jaw. These are the masseter, temporalis and medial pterygoid muscles (Fig. 2.16). The temporalis is a fan-shaped muscle, attached to the side of the vault of the skull and fascia overlying it. Its fibres

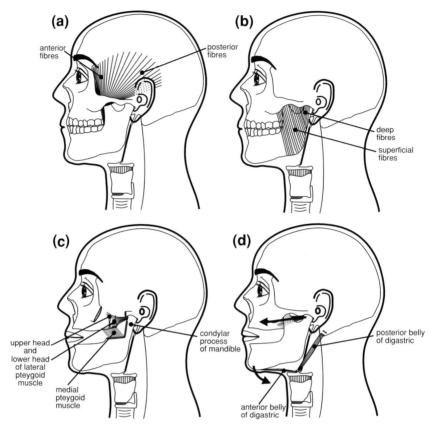

Fig. 2.16 The orientation of (a) masseter, (b) temporalis and (c) the medial and lateral pterygoid muscles in humans. All except the last muscle provide power to the bite. The lateral pterygoid is also shown in (d) assisting the digastric muscle in opening the jaw.

descend medial to the zygomatic arch and attach to the coronoid process of the mandible. The masseter muscles run from the whole length of the zygomatic arch, both from the undersurface and medial side, down to the lateral side of the ramus of the mandible. In most mammals, it is a complex multipennate muscle (Anapol & Herring, 2000). The medial pterygoid, which has fibres that are roughly parallel to those of the masseter, but which is on the inside of the mandible rather than the outside, adds weight to the bite.

The *jaw-closing* muscles (the elevator or adductor muscles) of lower vertebrates are critical for feeding, but they are relatively undifferentiated, forming a common muscle block roughly where the masseter and medial pterygoid lie in mammals. When this jaw-closing muscle acts, it probably presses the jaw joint into the base of the cranium. This does not matter too much if the forces are small, but mastication can involve very large forces. Crompton (1963) was the first to work out a potentially optimal solution, showing how the development of a discrete masseter, medial pterygoid and a posterior temporalis musculature considerably eases the elevation of a jaw. If the masseters acted alone, they could only close the jaw by exerting a very high compressive load on the jaw joint. The posterior temporalis muscle is important in continuing the movement without driving the condylar head into the skull. Figure 2.17 sketches the logic of the arrangement of jaw-closing muscles in mammals, but, since it can be difficult to grasp geometrical ideas like this from paper descriptions, Fig. 2.18 suggests a demonstration with a human mandible. Despite the logic of this, Osborn & Baragar (1985) showed clearly that it is impossible to unload the jaw joint by muscle arrangements such as this. Osborn (unpubl. data) suggests that it is far more probable that evolutionary changes favoured muscular efficiency.

The *jaw-opening* muscles (also called jaw depressors or abductors) include the lateral pterygoid, the digastric (Fig. 2.16d) and other suprahyoid muscles (such as mylohyoid, geniohyoid and sternohyoid – Fig. 2.19). The anterior digastric and the geniohyoid are probably the most important. In humans, the anterior digastric and the lateral pterygoid muscles act as a couple to open the jaw by both rotation of the joint (at any angle of opening) and anterior translation of the condyle (pronounced at wider gapes), but this is not true in many other mammals. The digastric muscle, for example, is variable even in closely related primates like humans and great apes (Aiello & Dean, 1990).

The mammalian mandible can be moved from side to side and forwards and backwards. Side-to-side movements are more pronounced in most mammals. In some ungulates, this is extensive, while in carnivores it hardly exists at all (Hiiemae, 1978). The lateral pterygoid muscle in particular, acting just on one side of the jaw, drags the lower jaw towards the opposite side, but other jaw muscles are potentially capable of assisting it depending on whether they are situated so as to produce mediolateral pull. There can be considerable variation in lateral movement even within a masticatory sequence, when processing one mouthful of food. In contrast, when jaw movements are viewed from the side, they appear stereotyped, varying only in amplitude. This pattern is reversed in rodents where the typical chewing orbit (see Fig. 3.6 for this in humans), with different opening and closing jaw

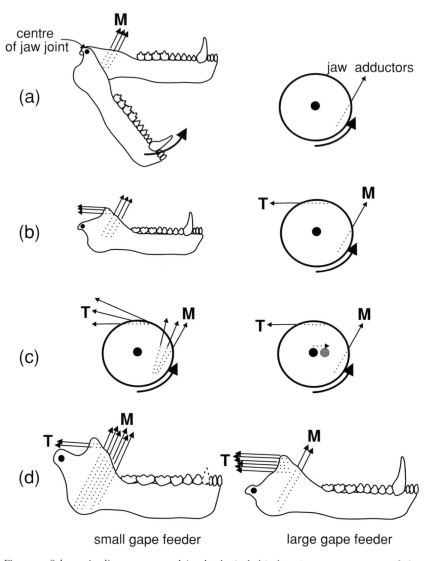

Fig. 2.17 Schematic diagrams to explain the logic behind various arrangements of the jaw joint and muscles in mammals. (a) A pseudo-mammalian jaw given one jaw-closing muscle – the equivalent of the pterygoideus part of the jaw adductor mass (called M) of a reptile – moving the jaw through a wide gape. At right, this jaw is modelled as a disc with a central 'joint'. The jaw muscle can rotate the disc through a given angle, close to where it forms a perfect tangent, but at other angles will tend to pull on the joint itself. (b) This mammal has now developed a process in front of the joint (the coronoid process) and differentiated a muscle mass (T, standing for temporalis) that pulls at a very different angle to M

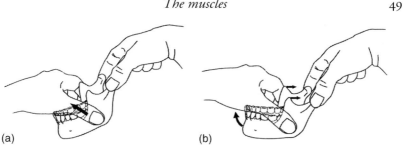

(a)   (b)

Fig. 2.18 A game for two players to demonstrate Crompton's (1963) explanation for the organization of jaw muscles in mammals. If you have access to skeletal material or a plastic model, you could try the following on a mandible with a friend. (a) Grip the outside of the mandibular condyles with the thumb and index finger of one hand and attempt to swing the jaw so as to close it by pulling upwards on the angles of the jaw with the other hand. This will simulate the action of the masseter and medial pterygoid muscles. The jaw will move but with difficulty. You will feel the force through the mandibular condyles that would, in life, push them through the floor of the skull. (b) Now ask a friend to pull back lightly on the coronoid processes at the same time. This simulates the action of the posterior temporalis muscle and causes the jaw to elevate with ease. The point is that, in order to rotate the jaw upwards, no one pair of muscles with a fixed line of action can produce an even bite force at all angles of jaw opening. (Courtesy of Dr K. Rajendran. Note that a human jaw is depicted: all human material should be treated with very great respect.)

trajectories, is seen not from in front, but from the side (Hiiemae & Ardran, 1968; Weijs, 1975).[10] The explanation for this anterior–posterior movement seems to lie in the mastication of very large mouthfuls of food, such that they can be processed bilaterally (i.e. on both cheek tooth rows at once).

Caption for Fig. 2.17 (*cont.*) (drawn for the medial pterygoid muscle, but which could also represent the masseter). This muscle arrangement closes the jaw through a far wider angle without loading the joint, as shown by the disc. (c) Many models of jaw-muscle action (too complex to describe here) emphasize the optimization of muscle activity so to produce the largest bite force for either the least effort or the least loading of the joint. A mammal that feeds at a variety of gapes may show one of two solutions to this. At left, variation in muscle fibre direction, coupled with sophisticated neural wiring, activates only those fibres with optimal orientation. Many mammals show some variegation in the temporalis and masseter muscles like this (McMillan & Hannam, 1992; Anapol & Herring, 2000). At right, instead of this arrangement, the jaw joint no longer just rotates, but slides its centre of rotation forwards towards M. Primates do this to reduce the stretch in M (Carlson, 1977; Hylander, 1978). (d) The jaw of small- and large-gape feeders. At small gapes, M is favoured, while at larger gapes, M is stretched too much and T (temporalis) is more efficient. Thus, small-gape feeders like many ungulates have large M, while carnivores have large T.

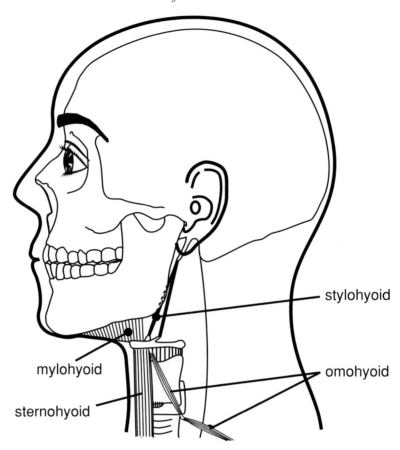

Fig. 2.19 Some of the prominent muscles attached to the hyoid bone.

Tongue muscles

The tongue is very mobile, obtaining this ability from three sources. The first derives from the anchorage of the tongue to the hyoid bone. In reptiles, movements of the hyoid basically determine the movements of the tongue (Smith, 1992). The most prominent tongue movement in lower vertebrates is an anteroposterior motion, which when receiving ingested solids or liquids sometimes requires that the tongue extend out of the mouth for very long distances. In mammals, this anteroposterior movement is also important, being produced largely by muscles that move the hyoid bone and muscles that link this bone to the tongue (Fig. 2.19). They produce anterior (genioglossus muscle) or posterior (styloglossus and hyoglossus muscles) movement, depression (genioglossus and hyoglossus) and elevation (styloglossus and palatoglossus), changing both tongue shape as well as its position. However, the capacity for lateral movement of the tongue, involving unilateral contraction, is extremely important for food manipulation (Fig. 3.7). Other muscles that lie just below the oral surface of the tongue, the intrinsic muscles, have no bony attachments and cannot, therefore, move the tongue around as a body. Nevertheless, these muscles are also extremely important in the manipulation of food particles. There are three sets of intrinsic muscle fibres arranged perpendicular to each other: vertical, transverse and longitudinal (Fig. 2.13).

Facial muscles

The lateral boundary of the vestibule of the mouth is the muscular cheek. The cheek muscle, called the buccinator, is the essential facial muscle for intra-oral control of food particles. It has fibres that run anteroposteriorly in humans, with firm attachments above and below to the maxilla and mandible. Anteriorly, it runs into a fibrous knot of tissue called the modiolus (Fig. 2.20), while posteriorly, it is a continuation of the circular superior constrictor muscle of the pharynx. The interdigitation of the fibres between the two muscles is called the pterygomandibular raphé (Fig. 2.14). When a human subject yawns, this raphé bulges inwards noticeably and can be used as a landmark by dentists seeking to block the inferior alveolar nerve. The buccinator also has strong attachments to the maxilla and mandible. This is because the cheek is an essential device for ensuring that particles thrown laterally by the tongue in the early opening phase of the chewing cycle stay on the working surfaces of the molars and do not fall into the sulci on the lateral side of the teeth. Effective mastication depends on the buccinator contracting during the late opening phase of a chew (Blanton *et al.*, 1970).

The anterior limit of the cheeks is the anterior limit of the postcanines. Anterior to the mouth are the lips, which control the width of the mouthslit and guard the entrance to the mouth. This control is achieved through a large number of muscles on the face that converge onto the corners of the lips. Just lateral to the angle of the mouthslit, these muscles meet at a fibrous knot called the modiolus (Fig. 2.20).[11] The details of the modiolus are complex. Some accounts describe it as a fibrous knot while others have muscle fibres passing through it (Lightoller, 1925). Whatever, there is a circular muscle surrounding the mouth called the orbicularis oris. This muscle, which has no bony attachments, acts as a sphincter. The muscles that converge on the modiolus from above and below are dilators or openers of the mouthslit. The modiolus serves another role. The action of some

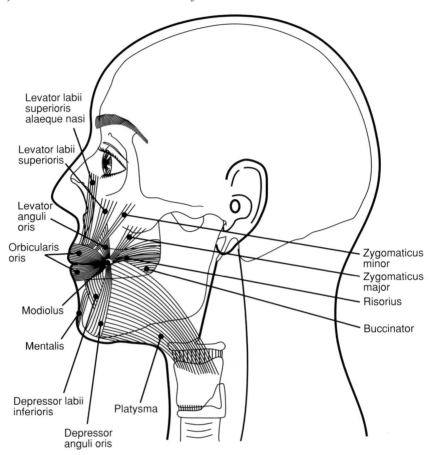

Levator labii
superioris
alaeque nasi

Levator labii
superioris

Levator
anguli
oris

Orbicularis
oris

Modiolus

Mentalis

Depressor labii
inferioris

Depressor
anguli oris

Platysma

Zygomaticus
minor

Zygomaticus
major

Risorius

Buccinator

Fig. 2.20 The facial muscles in the human originally evolved as sphincters and dilators of orifices (mouth, nostrils, etc.). Many muscles converge on the mouthslit, particularly laterally towards its corners. Most are vital for the control of food at the front and sides of the mouth. Many of these muscles double up for the purposes of a facial 'body language'.

upper and lower muscles can fix its position. The buccinator depends for its action on this fixation (Figs. 2.14 and 2.20). Though its fibres run from front to back, its main action is to press against the lateral side of the postcanine teeth. It can only achieve this if it is fixed in some way. By fixing it anteriorly using the modiolus, its contraction tends to move the whole cheek medially, which is the required movement in chewing.

In some rodents (Ryan, 1986), cercopithecine Old World monkeys (Murray, 1975) and even the platypus, the cheek contains pouches that act

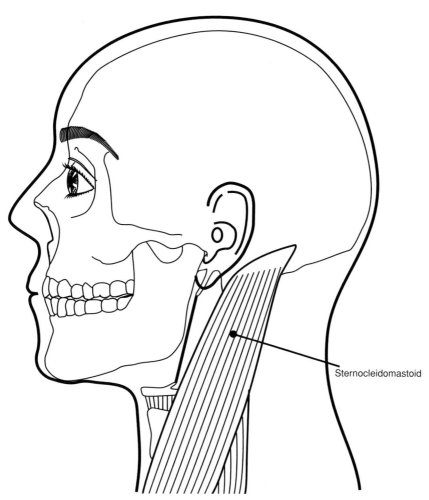

Fig. 2.21 The sternocleidomastoid in humans is a major feature of the neck, having reoriented itself to assist in the head posture of a newly upright mammal.

as food stores, allowing small ingested food particles (e.g. fruits or seeds) to be moved from their source prior to oral processing.

Neck muscles

These are generally set around the neck attaching to the undersurface of the cranium and the uppermost vertebrae. Small delicate deeply set muscles, particularly those located between the upper cervical vertebrae and the back of the head, allow for small movements that align the visual axes. Larger, more superficial muscles, such as the sternocleidomastoid (which is particularly prominent in humans – Fig. 2.21), are responsible for the rapid powerful movements of the head required in feeding. Deeper muscles, such as the longus colli, longissimus cervicis and longissimus capitis, also act to flex the vertebral column in the neck though with a lower mechanical advantage (Basmajian & De Luca, 1985). All these muscles are also responsible for balancing the head and their size reflects the position in which the head is held. In most mammals, support is essentially by a long elastic ligament that runs along the posterior side of the vertebral column called the ligamentum nuchae. However, humans have an habitual upright bipedal stance in which the head is balanced only by continuous active contraction of posterior neck muscles such as the longissimi. Without this action, the head falls forward – such as when dozing over a book like this. That is enough then about structure.

How the mouth operates

This chapter describes how the mouth operates, placing this in the context of a general physiological model. The chapter draws extensively on data from humans because the need for cooperation in many chewing experiments exceeds that which can be obtained from trained animals.

INTRODUCTION

The literature on mastication and swallowing is vast and it is impossible to cite more than a small fraction of the papers that comprise it. Emphasis has been placed here on those papers that describe facts relevant to physiological modelling of the process, however primitive such models might be at this point. Central to unravelling what happens to food particles in the mammalian mouth have been cine- and video-fluoroscopic studies. These date back to Ardran *et al.* (1958) on the rabbit, through Crompton & Hiiemae (1970) on the American opossum, up to the present. These X-ray movies have permitted views of the intra-oral processing of food otherwise obscured by the cheeks. Excellent reviews are available (e.g. Hiiemae & Crompton, 1985; Orchardson & Cadden, 1998; Thexton & Crompton, 1998) as well as studies on particular species. It has required the combination of a wide range of experiments and detailed observation (e.g. Weijs & Dantuma (1981) on the rabbit) to establish the major features of oral processing. However, I have yet to see some synthesis that attempts to weld all this information on mastication and swallowing together. This chapter intends not so much to review what is regularly reviewed in journals and edited volumes, but to show how mastication and swallowing can be related to features of the food input. Many will view the account as speculative in places. However, the aim here is not to write a novel, but to drive towards an understanding of how the description of the process can be tied to its biological role.

Mammalian mastication may have evolved for three reasons. The first and most general reason is that particle size reduction increases the specific surface (the surface area per unit volume) of a solid. The rate of action of enzymes in the gut is proportional to the surface area that they act upon. Mammals consume energy at much higher high rates than reptiles (Schmidt-Neilsen, 1972). Thus, it seems likely that the high metabolic rates of mammals favoured the evolution of mechanical breakdown of solids, prior to any chemical breakdown, because this can expose new surface very quickly. Thus, the evolution of mastication in early mammals may be linked to an increase in metabolic rate.[1]

Some mammals ingest foods that already have a high specific surface. Yet, they still chew them. Familiar examples are ungulates that eat leaves and grasses, which are so thin that they yield little extra surface when they are fractured. The major reason for mastication in these circumstances is likely to be the need to gain enzymatic access to food because it is otherwise indigestible. Seed eating comes into the same bracket because seed casings also obstruct digestion. Mammals that swallow whole mature seeds always defaecate them in an unaltered state – this is how they disperse seeds away from the parent plants. A second reason for mastication in some mammals then is to gain chemical access to foods that are otherwise impervious to the action of their guts. This explanation is not linked to metabolic rate per se and there is good evidence that the pre-Mesozoic reptile *Suminia getmanovi*, dating from 260 million years ago, chewed plant material for just this reason (Rybczynski & Reisz, 2001).

Most mammalian carnivores do very little processing of food in their mouths and swallow foods in very large particle sizes (Savage, 1977). The extra surface area produced by one or two chews is small and vertebrates contain all the enzymes necessary for the digestion of the animal kingdom. However, many mammalian carnivores attack prey that is as big as or bigger than they themselves (Peters, 1983). These predators cannot ingest prey like that in one lump and are therefore forced to reduce the size of their prey simply to accommodate it in their gut. The argument cannot be pressed too far because some vertebrates such as snakes can dislocate their jaw joints to accommodate prey of colossal size. That carnivores do not do this is probably explained by the fact that they cannot afford to store complete carcasses in their guts without compromising their activity levels. This leads to a third reason for evolving food breakdown in the mouth: it could have evolved simply to reduce the particles to a swallowable size. Exactly what is capable of being swallowed by mammals is considered a little later in this chapter.

I start with the first and most general of these reasons – that mastication is essential to the rate of food processing of the mammalian gut – and treat mastication strictly as a comminution process designed to expose new food surface as rapidly as possible.

PARTICLE SIZE DISTRIBUTIONS

Mammals usually ingest food in batches (mouthfuls). When particles of any solid batch are broken down repetitively, they fragment into smaller pieces that range often across several orders of magnitude of size. This comminution process expends about 5% of the planetary energy generated by human industry (Lowrison, 1974). Despite the magnitude of this effort, and the scale of its alleged inefficiencies, the process apparently remains incompletely understood. Yet what has been discovered provides myriad observations of great importance for understanding mastication. By definition, it is mammalian to possess a dentition with a natural comminuting capability, but it is distinctly human to comminute with tools; in fact, this is probably the evolutionary basis of the development of a tool industry (Chapter 7).

The particle size distributions formed during chewing are virtually always skewed, meaning that instead of looking like the symmetrical bell shape of a normal distribution, as fitted by a Gaussian curve, they appear variably distorted. The degree of this skew is initially dependent on what has been ingested, but if the ingested particles are a group of similar-sized particles of a relatively homogeneous food like a bunch of peanuts, then distributions appear to develop as shown in Fig. 3.1. After a few chews, the long tail of the distribution is initially stretched out to the left, but later it moves to the right as particles get smaller with further chewing and as both the feed size and its characteristic influence disappear.

It is difficult to be entirely sure of this pattern because of the limitations imposed by the sizing methods that have, until recently, been available. Most studies on humans over the last century have used sieves, organized so that their size intervals form something like a geometric progression (Fig. 3.2).[2] This is logical because any kind of comminution process, such as the cutting up of vegetables in a kitchen for example, appears to have

PARTICLE SIZE DISTRIBUTIONS

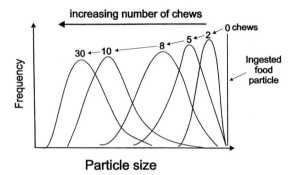

Fig. 3.1 Schematic representation of the development of food particle size distributions produced in chewing. The feed particles are a group of ingested particles with very similar sizes. After just a few chews, several particles are broken, which sends a 'tail' of fragments out to the left-hand side. After several more chews, as more of the feed particles are broken, the distribution begins to look like a normal bell curve, but as comminution proceeds further, the tail drifts to the right-hand side and stays that way. The rate of these events depends on the rate of food breakdown, but this normally takes only about 10 chews in humans (Lucas & Luke, 1983; Olthoff *et al.*, 1984).

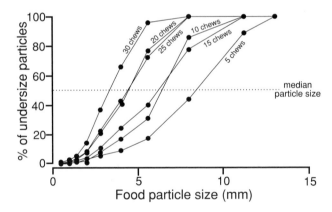

Fig. 3.2 Real particle size distributions resulting from sieving spat samples of raw carrot chewed by humans. The spacing of the mesh sizes is not even, but follows a geometric progression. The plots are cumulative frequency plots in which the *y*-axis reports the percentage of particles in a distribution that are less than a given (sieve) size. Due to the variable skew of the particle size distribution (shown in Fig. 3.1), the median particle size is the best indicator of the degree of particle size breakdown.

some geometric simplicity to it. For example, we could imagine continually halving the sizes of vegetable particles with a knife. Even though we could not cut all of them with each chop, we might expect an overall geometric progression in size to develop.

Epstein (1947) claimed that, if particles were comminuted for long enough, then the eventual result would be a distribution that would be normal if the logarithms of particle sizes were taken. In fact, such 'log-normal' distributions do not materialize. It is more difficult to break smaller particles than larger ones and limits can be reached at which the fracture behaviour of foods changes. Provided that no such fracture limit is reached though (for an explanation of which, see Chapter 4), then a large body of data obtained from humans shows that distributions appear to drift towards 'square-root normality' (Fig. 3.3); (Voon *et al.*, 1986). The same should be expected for any mammal that reduces particle sizes far enough, such as other primates (Walker & Murray, 1975; Sheine & Kay, 1979) or ungulates (Rensberger, 1973).

It is useful to be able to characterize particle size distributions by a mathematical formula. The Rosin–Rammler equation does this, providing a

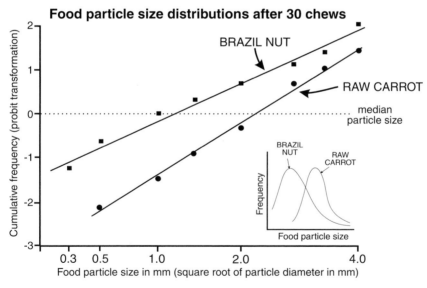

Fig. 3.3 Particle size distributions produced by one subject with raw carrot and brazil nuts. After about 30 chews, a straight line is obtained when the square roots of particle sizes are taken and when cumulative frequencies are converted to 'probit' units. The latter convert the sigmoid shape of normal cumulative frequency plots to straight lines (Finney, 1971).

reliable estimate of both the median particle size, the centre of the distribution, and its spread (Olthoff *et al.*, 1984). Generally, as shown in Fig. 3.3
for brazil nuts and raw carrots, the faster a food breaks down per chew, the
broader its particle size distribution.

ANALYSIS OF PARTICLE SIZE REDUCTION

Epstein (1947) developed a binary method of analysis of comminution
processes which has been of great help in understanding mastication because
it can help factor out the oral variables that influence the rate of food
breakdown. The following definitions stem from Gardner & Austin (1962),
the only amendment being to replace 'per unit time' by the concept of a
breakage step ('per chew'). Once you get the hang of these concepts, they
can be extremely useful. The rate of mastication depends on two factors:
(1) The chance that food particles have of being fractured by the teeth
     during any chew. This is termed the selection function, $S(x)$, where
     $S(x)\,\delta x$, for sufficiently small δx, is that proportion of particles of size
     range x to $x + \delta x$ that are broken per chew.
(2) The size distribution of fragments produced by any particle that frac-
     tures. This is the breakage function, $B(y, x)$, where $B(y, x)\,\delta x$ is that
     proportion of the fragments by volume of size range x to $x + \delta x$ that
     break to below size y per chew (where $y \leq x$).
To illustrate what the selection function means, look at Fig. 3.4 where four
discrete particle sizes are pictured. Particles with crosses have been broken
(i.e. selected) in that chew. The number beside each particle grouping gives
the proportion of particles for each particle size that have been selected.
This is the selection function. Note that the smaller the particles, the lower
the value of this function. This unsurprising result has been found true for
every type of comminuting process including mastication (Lucas & Luke,
1983; van der Glas *et al.*, 1987) where, for any given mouthful of food, the
selection function appears to be related to particle size by a power law:
$S(x) = cx^a$ where a is an exponent of particle size and c is a constant that
depends on the unit of measurement (Lucas & Luke, 1983; van der Bilt
et al., 1987; van der Glas *et al.*, 1987). The best estimate of the exponent
from experiments on humans is $a = 2.0$, i.e. the selection function varies
roughly as the square of particle size. This relationship is indicated in the
graph at the bottom of Fig. 3.4. It should be emphasized though that this
is a statistical average and experiments need to be run several times before
this result is approximated.[3]

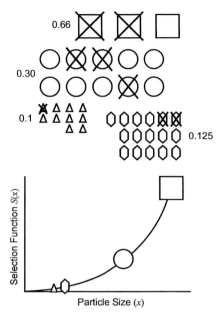

Fig. 3.4 The selection function – a measurement of the probability of particles being fractured. See text for explanation. (Reprinted from *Food Quality and Preference* vol. 13, Lucas, P. W., Prinz, J. F., Agrawal, K. R. & Bruce, I. M., pages 203–213, Copyright (2002), with permission from Elsevier.)

The breakage function is illustrated in Fig. 3.5 where, for simplicity, it is assumed that a mouthful of particles, all of particle size x, has been ingested. Twenty per cent of these particles have been selected in the first chew (i.e. $S(x) = 0.2$). These selected particles and their fragments are identified in the figure by being crossed out. We can grade these fragments by assessing their sizes relative to the size of the parent particles from which they broke. This is commonly done on a cumulative basis. The figure shows a grading level where a particle size y is half the size of x (i.e. $y = 0.5x$). Ten per cent of the fragments that broke from x lie below y, i.e. $B(y, x) = 0.1$ for $y/x = 0.5$. If this procedure is repeated for other values of y, we can build up the distribution of fragment sizes relative to x, the size of their parents.

Measurement of the breakage function means sorting out which fragments break from which parent particle. The only direct method is to take one particle, get a person to chew it just once and spit the fragments out. This has to be repeated many times and the products sized. The results of

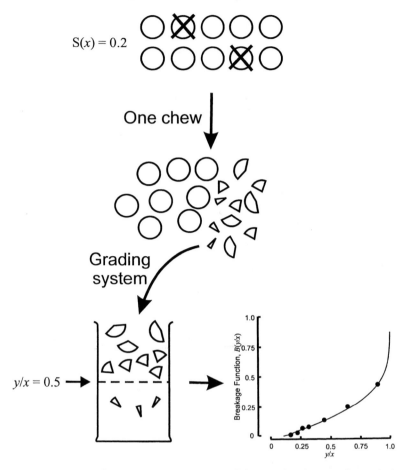

Fig. 3.5 The breakage function – the description of the size distribution of particle fragments produced by a fracture event. The graph shows the fit of one suggestion distribution
(Gaudin–Meloy) to fragmentation data from a human subject. Accurate measurement of
fragmentation requires a lot of repeats in an experiment to get stable results. (Redrawn from
Food Quality and Preference vol. 13, Lucas, P. W., Prinz, J. F., Agrawal, K. R. & Bruce, I. M.,
pages 203–213, Copyright (2002), with permission from Elsevier.)

such tests often resemble Fig. 3.5 and can be characterized by two equations. The first is by Gaudin & Meloy (1962), who claimed a fundamental
derivation relating the size of the parent particle and that of its fragmented
offspring, when the latter is from a single fracture. This relationship has
the form: $1 - B(y, x) \propto (1 - x/y)^m$, where m is an exponent. The other is

an empirical suggestion deriving from Austin (Lowrison, 1974), whereby $B(y, x) \propto x^b$, where b is an exponent.

The selection and breakage functions can be combined. In the example taken above, the fraction of the total volume of fragments of this size range x to $x + \delta x$ that break to less than their half their original size, i.e. to a size less than y per chew, is $S(x)B(y, x)$; in the example, this is $(0.2 \times 0.1) = 0.02$ for $y/x = 0.5$. So 2% of the particles break into fragments that are half the original size per chew. Since the rate of particle size breakdown depends only on selection and breakage functions, then repeating this for all size fractions computes the rate of particle size breakdown without all that chewing, spitting and sieving after different numbers of chews.

Doing this by hand for all size fractions and repeating this for different numbers of chews is not practical, so formal mathematics are necessary. If the percentage of the total volume of particles of size range x to $x + \delta x$ before the nth chew is $P_{n-1}(x)\delta x$, then the percentage of particles below size y after the nth chew is

$$Q_n(y) = \int_y^\infty P_{n-1}(x)B(y, x)S(x)dx + \int_0^y P_{n-1}(x)dx \qquad (3.1)$$

where $\int_0^y P_{n-1}(x)dx$ per cent of particles exist below size y before the nth chew. The percentage of particles of size range x to $x + \delta x$ before the $(n + 1)$th chew, i.e. $P_n(x)\delta x$, can be obtained from

$$P_n(x) = \frac{d Q_n(x)}{dx}. \qquad (3.2)$$

A solution to these equations is possible if the behaviour of the selection and breakage functions with respect to particle size can be specified and if these are not affected by the time that particles have been resident in the mouth. This analysis has been widely adopted in industrial processes, solving for the selection and breakage parameters using computers (Gardner & Austin, 1962). However, this can now end because the problem has been solved analytically by F. A. Baragar (University of Alberta), specifically for understanding mastication (Baragar *et al.*, 1996). The fit to real data looks amazingly good.[4] When the median particle size is taken as the appropriate marker for the degree of size reduction of a batch of food in the mouth, the rate of breakdown in human mastication after about 10 chews follows an exponential curve, such that a log–log plot produces a straight line.

Even if this method of analysis is valid, it still has to be shown that it has some use. This can only be achieved by relating these functions to characteristics of the food intake. In Chapter 1, I linked the surface attributes of food particles with their probability of fracture (the selection function), while the mechanical properties of foods that act to resist the formation of new surface area, were tied to the degree of particle fragmentation (the breakage function). I went further and linked tooth shape to the breakage function and tooth size to the selection function, though I did not use this terminology at that time. The logic of this is not difficult to grasp. The mouth has a large surface area, not much of which is given over to the dentition. Any increase in the size (surface area) of the teeth must enhance the chances of a particle being fractured. This chance can only depend on food particle characteristics that are exposed to the surface of the teeth. Thus, evolutionary changes in tooth size in mammals are linked to change in the surface characteristics of foods in the diet. In contrast, evolutionary changes in tooth shape are likely to be a response to those food characteristics that obstruct the formation of new surface area in food particles by resisting their fracture. These characteristics are the internal mechanical properties discussed in Appendix A.

THE MASTICATORY SEQUENCE

Any masticatory sequence consists of a series of chewing cycles in which the lower jaw moves in orbits. In each cycle, the jaw first opens to allow space for food particles to be collected by the tongue. The jaw then turns and starts to close. Before this happens, however, the tongue throws a set of these food particles towards the teeth, usually (but not always) just to one side of the mouth. The jaw then continues to close in order to fracture these particles. After the jaw opens again, the fractured particles spill out, usually towards the tongue. The postcanine teeth of mammals generally appear designed to encourage this movement. If this spillage did not happen, or if food stuck to the teeth, then because only a fraction of the particles in the mouth get fractured in a chew, bimodal particle size distributions would quickly become evident (one modal size representing unbroken particles, the other the average size of comminuted fragments). None of the extensive data from humans or the limited data for other mammals shows any evidence of such bimodal distributions.[5]

Figure 3.6 shows the orbital path of the lower jaw, viewed from in front, in the human. This movement pattern is not species-specific and could also be found in a very large number of other mammals, including herbivores

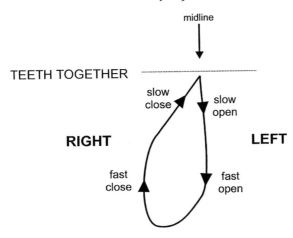

Fig. 3.6 The trajectory of any convenient point of the mandible (like a lower incisor) during a chewing cycle. The path is highly variable, but the essential features are (1) the differing path taken by the jaw in opening and closing and (2) the characteristic changes in jaw velocity at different points in the cycle.

(Hiiemae, 1978) and insectivores (Fish & Mendel, 1982). However, it would be an impossible movement for carnivores, which have a movement path that is almost entirely vertical, and it is not typical of rodents and elephants either, which tend to have a distinct forward component to their jaw trajectory instead.

In most mammals, the opening path is mainly vertical (e.g. in humans: Mongini *et al.*, 1986), but early in closing, the jaw turns laterally towards the side on which food particles are to be fractured. During further closing, the mandible starts to turn again, directing its thrust upwards and medially. The mandible may then pause with the teeth together, or nearly so, or it may slide a small distance first across the upper teeth, before setting out on another opening journey. The cycle is rapid and takes less than a second in the human. However, it can be up to four times faster than this in other mammals (Table 5.2).

The timing of chewing cycles is relatively inflexible, being controlled, just like locomotion is, by neural circuitry called central pattern generators (Lund, 1991). These generators are circuits of neurons within the brain stem that set body rhythms (Luschei & Goldberg, 1981). Everything other than the timing though is very variable. Indeed, Fig. 3.6 does not have to be drawn very exactly, because it is based on an idealized 'teardrop' shape of chewing cycle that is actually rarely seen. This highlights a major difference between mastication and locomotion, the two major motor activities of mammals. Terrestrial locomotion is more stereotyped in form. For example, the gait of a terrestrial tetrapod varies little whether it is walking over hard rock or soft sand. If it did vary greatly between these substrates, then fossil footprints would provide a biased record of the movements that these animals made; instead they supply crucial raw data for analysing their pace of life.[6]

In contrast, movements in every masticatory cycle vary between chews. This is not likely to be due to the mouth having inadequate mechanisms to know where food particles are, or to some jerky motor program driving the muscles, but to the fact that the substrate that the mouth works on is changing as it is (deliberately) being broken down and subject to the statistical patterns described by selection and breakage functions.

For the purposes of description, each chewing cycle is usually broken down into phases. Jaw opening and closing are obvious subdivisions, but each opening and closing phase can be further divided up by differences in jaw velocity. Not every mammal shows this to an equally clear extent, but a slow open – fast open – fast close – slow close pattern of movement in a chewing cycle has been described now in many mammals (Hiiemae & Crompton, 1985).

During early jaw opening ('slow open'), the tongue moves forward rapidly so as to catch food that tends to fall from the lingual sides of the teeth after jaw closure. The tongue may dip its tip into the pool of submandibular saliva that has accumulated behind the lower incisor teeth. Picking this saliva up, somewhat in the manner of an intra-oral lap, the tongue then mixes the food particles with saliva by pressing them against the hard palate. The considerable amount of sensory processing necessary for this probably dictates the slow speed of this part of the cycle. The tongue then releases its pressure on the particles and starts to move backwards. As the food falls from the hard palate, the tongue senses whether particles stay clumped together or else fall apart. If the particles all fall apart, the tongue throws the food laterally into the space between the upper and lower post-canine teeth. The positions of the tongue during jaw opening are depicted in Fig. 3.7.

Foods that fall too far laterally seem to bounce off the cheek as the buccinator contracts (Blanton et al., 1970), so reflecting them, in a damped fashion, back onto the teeth (Prinz & Lucas, 2001).[7] Before the jaw turns and starts to close, of course, the cheeks must start to relax. Then, with both the tongue and cheeks out of the way, the teeth close rapidly towards the food particles. Close to or at tooth–food contact, the mandible slows, probably allowing perception of food fracture properties. As the food

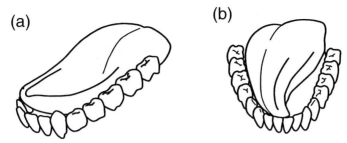

Fig. 3.7 Probable positions of the tongue during early and late jaw opening. In (a), the tongue is collecting food particles in early opening, while in (b), it is throwing them laterally towards the teeth. (After Abd-el-Malek, S. (1955) The part played by the tongue in mastication and deglutition. *Journal of Anatomy* 89: 250–254 (Oxford: Blackwell). With permission.)

fractures, the nearby parotid duct starts to spray saliva over the newly formed fragments, helping to wet their surfaces (Anderson *et al.*, 1985). Then as the whole cycle starts again, these fragments fall towards the tongue.

SENSORY FEEDBACK AND THE CONTROL OF MASTICATION

This complicated sequence of events cannot have an inflexible neural program because the process depends critically on sensing where food particles (and soft tissues like the tongue and cheeks) are (Fig. 3.8). Such knowledge should not be a problem. Animals are exquisitely well organized for the detection of information about their external environment, so much so that some mechanisms, such as those for perceiving light or chemicals, operate at the level of individual photons and molecules respectively (Bialek, 1987). However, current knowledge indicates that mechanisms for detecting the state of the oral environment fall far short of those dictated by the laws of physics. Instead, natural selection seems to have acted to assign the greatest sensitivity to two critical areas: the detection of swallowable particles so as to clear the mouth and to the avoidance of damage to the teeth by wear.

There are many sensory receptors in and around the mouth, as will be evident below, but they can be categorized into those that supply information about jaw displacement (and the rate of its displacement) and those that monitor stress (Fig. 3.8). The overall pattern of control appears to be displacement dominated, with the aim being to keep the jaw moving. However, when there is contact, or anticipated contact (Ottenhoff *et al.*,

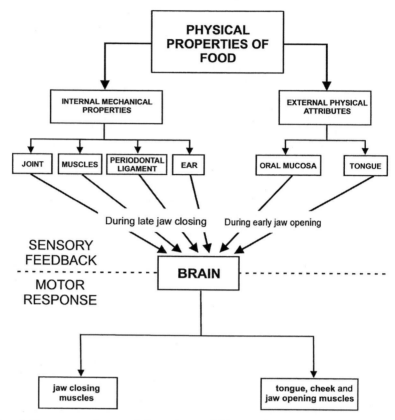

Fig. 3.8 The sources of sensory information available in and around the mouth considered in relation to food properties and the direction of motor responses.

1996), a stress control creeps in. The latter part of a cycle is associated with a slowing of movement, even with pauses (Hiiemae & Crompton, 1985). A stress control seems to pervade movements during the entire chewing cycle when foods of high Young's modulus are chewed (Bourne, 1976).

The events during jaw opening are critical for the success of a chew. Early opening is slow and there is strong evidence for this being mediated by sensory receptors (Ostry *et al.*, 1997). It is not clear where these receptors are – neuromuscular spindles (see below) are almost absent in jaw-opening muscles. Yet this slow speed is a sure sign that food properties are being evaluated. Since there is no fracture involved in opening, it must be surface attributes that are being evaluated on the oral mucosa.

The tongue manipulates food particles in this phase of a chewing cycle and touch sensors on both hard palate (Dixon, 1963) and tongue are probably vital in this evaluation. The two-point discrimination (a test for the 'stereo' sense of touch) of the most sensitive areas of the oral cavity of the human, the anterior hard palate and tongue tip, is about 1–2 mm (Ringel & Ewanowski, 1965; Laine & Siirilä, 1971). In other words, particles smaller than this that rest on the mucosa cannot be discriminated in size. This is very likely related to the general propensity of humans to swallow particles in the low millimetre size range (Lillford, 1991). A swallowing decision that is related to particle size makes sense of the sensitivity range. If 1–2 mm particles are those desired for digestion, then there is no advantage in detecting the exact size of smaller particles since a swallowing decision will already have been made. There is considerable circumstantial evidence for the operation of a low-millimetre oral threshold in a number of mammals, including rodents (Magnusson & Saniotti, 1987; Corlett, 1996), tree shrews (Emmons, 1991; Emmons *et al.*, 1991), cercopithecine Old World monkeys (Corlett & Lucas, 1990; Lucas & Corlett, 1998; Otani & Shibata, 2000; Su & Lee, 2001) a fruit bat (Phua & Corlett, 1989) and many herbivores, such as cattle, sheep, goats and horses (Uden & Van Soest, 1982; Van Soest, 1996). There is also experimental evidence for this in humans (Prinz & Lucas, 1995).[8]

As the jaw turns to close, the displacement sensors that are likely to be most important are those around the capsule of the jaw joint (Klineberg & Wyke, 1973; Klineberg, 1980). At wide gapes, these receptors seem to gauge movement patterns accurately, but at smaller gapes they fail to do so (Öwall, 1978). This is understandable because these sensors are set very close to the pivot. Accordingly, as the jaw closes further, other sensors must monitor its pattern of movement. The neuromuscular spindles in muscles are undoubtedly important and organized in mammals such that they can sense the state of muscular contraction at a wide range of gapes (Matthews, 1972).

Closer to contact with food particles, and to upper and lower tooth contacts, force sensors are also required in order to monitor fracture events. The evidence is that mechanoreceptors in the periodontal ligament (Linden, 1990) initially act in this way. An additional mechanism is undoubtedly provided by dentine sensitivity (Paphangkorakit & Osborn, 2000). As pointed out by these authors, the periodontal ligament is rather pliant, with the result that its sensitivity must be limited with stiffer foods. At higher stresses, these authors provide strong circumstantial evidence for an intra-dental (pulpal) sensor. As there are no specialized receptors

in the pulp (Holland, 1994), it is probable that fluid movement in dentinal tubules under the stressed region stimulates the nerve endings. (This
reasoning requires that the modelling of Spears *et al.* (1993) of force transmission from enamel to dentine be correct; see Chapter 2.) However, wher-
ever the mechanisms are located, the ability to sense a particle between the
teeth is very fine. Depending on method and on the particular teeth being
tested, particles of only 8–15 μm in thickness or diameter can be detected
(Utz, 1986). This distance is only a little above the sensitivity of individual
neurons to displacement (2–3 μm: Anderson *et al.*, 1970). Undoubtedly,
this fine sensitivity is related to the need to avoid dental wear. Most of
the wear is produced at this 'micro level', either by extraneous grit ingested with food (Teaford, 1994) or by the opaline silica contained in many
plant species (Baker *et al.*, 1959; Walker *et al.*, 1978; Lucas & Teaford,
1995).

Slowing of movement during late closing is not likely to be connected
just with perception of food fracture properties, but with avoiding damage
to the teeth. This is generally more difficult when a device, be it artificial or
organismal, is functioning under stress control.[9] Tooth contacts in human
mastication are well known in both chewing and swallowing (Anderson,
1976). In order to avoid tooth wear, electrical activity in the jaw closing
muscles has to be switched off as soon as jaw movement is detected as being
impeded. A reflex (which on the sensory side seems uniquely to contain only
one neuron running all the way back to the midbrain without a synapse)
seems to do this about 12 ms after tooth–tooth contact (or a similar period
after contact with very stiff foods), but due to a force–time lag, the force
invariably builds up for an average of 40 ms beyond this (Gibbs *et al.*, 1981).
So there is at least 50 ms in which tooth wear can take place.

Figure 3.9 shows how the basic events in a chewing cycle are related
to food properties. There is no fracture during early jaw opening and
so palpation of food with the tongue can only detect external physical
attributes. Sensations during opening regulate the vertical amplitude of a
chew. The internal mechanical properties of foods can only be detected
during late jaw closing during fracture. These may affect the degree of
lateral movement during closing.

There is one extra source of feedback (mentioned in Fig. 3.8) that has
been neglected by oral physiologists, though not by food scientists, and that
is sound. Fast fracture in food produces noise (Drake, 1963, 1965; Vickers,
1981) that can be interpreted by subjects in terms of specific food properties,
as shown by Vincent *et al.* (2002). The effect of this on jaw movements
though appears unknown.

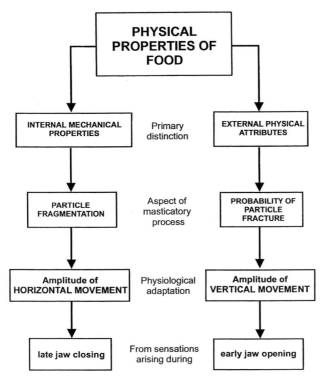

Fig. 3.9 The relationship between food properties and the movement sequence during a chewing cycle. (Derived from Lucas *et al.* (1986e) and Agrawal *et al.* (2000).)

JAW MECHANICS AND THE CONTROL OF JAW MUSCLES

Of all the sensory receptors discussed above, those with the smallest role appear to be the joint receptors. Yet they seem to have substantial innervation. What are they there for? Well, there is another role that could be accorded to these receptors that relates to a general problem called 'jaw mechanics' and which was alluded to in Chapter 2 (Fig. 2.16). The issue can be stated succinctly:

The jaw is a lever, anchored at a joint and moved by muscular efforts acting at fixed locations. Yet the jaw is also loaded by contact with food at various positions along its length and at a range of opening angles. How can the force of loading, the bite force, be organized so as to minimize the effort or without being transmitted through the joint?[10]

The bite force that the food endures has to be distinguished from the resultant of the muscular forces moving the jaw. However, except when chewing very stiff foods (Rensberger, 2000), the distinction between the two is unimportant. From now on, I will mainly refer to bite forces, but the directions of both are almost certainly going to be approximately parallel to the long axis of the teeth, otherwise they would start to tip. Some imbalance is inevitable due to change in the muscle force vector with gape, but any inconsistency needs to be harnessed so as to produce a profitable result, e.g. by keeping teeth in a single functional group together.

A compressive force through the joint can be countered by building up the bones that form the joint so as to prevent total catastrophe: the head of the mandibular condyle breaking through into the cranial cavity. The joints of some mammals, e.g. those of carnivores, are hinges, so limiting any possibility for anterior dislocation, but others are capable of translation and rotation as in humans. Whatever their form, none appears to be built so as to sustain high compressive forces. A greater problem than compression though is the possibility of a force that could separate the bones, a distracting force that moves the mandibular condyle downwards, so rupturing the joint capsule. Such a force could be detected by stretch receptors in the capsule, which may well provide feedback to the muscles affecting their firing patterns. Some of these receptors may be able to sense compression at the joint as well, e.g. those in the lateral ligament of the capsule in humans, a structure that seems to be taut when the teeth are together (Baragar & Osborn, 1984). In this regard, joint receptors could be analogous to mechanoreceptors in the periodontal ligament: they need to sense compression, but actually respond themselves to being stretched (Linden, 1990).

The firing patterns of muscles are not stereotyped, but it is unclear what the efficiency limits for muscles are and what their neural control might be optimizing. The smallest order of physiological organization is the motor unit, which is the number of muscle fibres controlled by a single nerve fibre. However, even if muscle recruitment could be organized that finely, then there would still be light compressive loads at the jaw joint (Osborn & Baragar, 1985).[11] So what is behind the variability of muscular contraction seen physiologically? Osborn & Baragar suggest that the muscles may be acting to produce the maximum bite force at minimal cost in terms of fibre contraction. This approach has produced some surprises: the pull of the lateral pterygoid, probably the superior head of this muscle for the most part, seems important in generating an efficient bite (Osborn, 1995b). Spencer (1998) has recently reviewed the subject as a whole, conducting experiments on humans that test one of the most original models of jaw function, that of Greaves (1978), but there appears to be no current model that can account for all observed patterns of muscle activity.

The internal architecture of the jaw muscles is variable. In some muscles, the fibres lie at an angle to the bite force. This packs more fibres into a given volume than parallel fibre arrangements, but reduces the gape at which the muscle can act. The masseter is a key example of a muscle like this (Anapol & Herring, 2000). Also, the firing pattern of the fibres themselves is variable. They may contract rapidly, but fatigue quickly or vice versa. An important review by Mao *et al.* (1992) has reviewed the distribution of these fibres in different jaw muscles and across mammals in different dietary groups. The muscles of carnivores and rodents tend to be fast acting but quick to fatigue, while those of herbivores are the opposite.

A final aspect of this is that muscle tissue does not generate force equally well at different amounts of stretch (Herring & Herring, 1974). Paphangkorakit & Osborn (1997) show in humans that this has a sizeable effect on the jaw-closing muscle that acts most efficiently to produce a bite. At smaller gapes, it is the masseters, while at wider gapes it is the temporalis muscles. Mammals that use wide gapes, like carnivores, have heavily built temporalis muscles, while herbivores use much smaller gapes and have relatively larger masseters (Fig. 3.10).

SALIVA AND SO MANY SENSES

This section discusses the range of stimuli, including tastes, smells and abrasives that influence the rate of salivation and the properties of saliva itself.

The best time to monitor the quality of the potential food intake is as soon as possible. For this reason, all the special senses of the body operate before, and during the oral processing of food. Smell, taste and texture are all enhanced and/or facilitated by the exposure of new food surface by fracture and fragmentation during mastication. While odours follow the local airflow to the nasal cavities (probably passing backwards and around the soft palate), the most important chemical sense in the mouth is surely that of taste. Taste requires the presence of saliva because only dissolved compounds trigger it. Consequently, it is unsurprising that as food particles are broken down, several sensory mechanisms have developed to pour saliva over them. Saliva is stimulated by the physical presence of foods against the mucosa, by food particles being loaded by the teeth (Anderson *et al.*, 1985), by taste (Spielman, 1990), by odour (Lee & Linden, 1992) and by

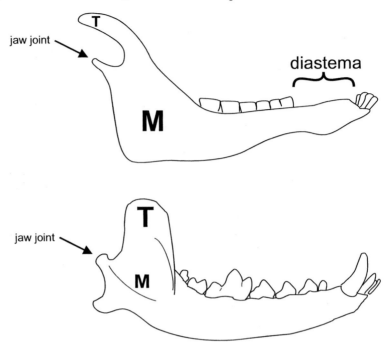

Fig. 3.10 The jaws of a herbivore (above) and a carnivore (below) to show the relative sizes of the masseter (M) and temporalis (T). The sizes are related to the gape angle at which muscles act.

abrasives (Prinz, in press). A good review of much of this is by Hector & Linden (1999).

Tastes have been categorized traditionally as sweet, sour, bitter and salty. However, this view is breaking down due to a revolution in research on this topic. I have no option but to review this here and so need the following short, but necessary, interlude to the text.

The sensation of saltiness is triggered by sodium and potassium ions (Boughter & Gilbertson, 1999). It is difficult to relate to any other aspect of this book, except with respect to geophagy – the consumption of special soils and minerals by mammals – that can cause considerable dental wear. There is no evidence that calcium or phosphate ions in foods can be detected by any specific mechanism, despite the body's requirement for these in mineralized tissue. The sensation of sourness is due to low pH. Food acid erodes enamel. Quite logically, therefore, it seems to be the most potent known stimulant for saliva (Kawamura & Yamamoto, 1978).

Much of the recent work on taste though involves molecular discoveries about 'sweet', 'bitter' and free amino acid taste receptors. Most of it derives from work on rodents, but there is now increasing evidence for it on human taste cells too. Where both human and mouse tissues have been tested, the response is not 1 : 1 because the relevant genes in both species share only about 70% similarity. However, this is a lower figure than for comparisons of most other human/mouse proteins, confirming that taste does adapt to diet (Nelson *et al.*, 2001).

Food sugars are important in the human because of their clinical significance in dental decay. They are detected by a receptor or receptors that give rise to a sensation that we refer to as sweetness. While sweetness is an attractant for mammals with a non-fermenting digestive system, it may repel ruminants for which sugars jeopardize microorganisms in the fermentation chamber of their stomachs. The receptor is useful for both types of mammal, but it probably evokes contrary psychological responses. Sweet taste reception in mice and humans usually seems to involve the conjoint action of two genes (Nelson *et al.*, 2001). A wide range of sweet chemicals can be perceived, including sucrose and fructose, both of which produce a strong response. These sugars are major indicators of the carbohydrate content of plant foods. Sucrose is the ubiquitous transport sugar of plants while fructose is the commonest sugar in fruits. However, not just sugars are sweet – many D-amino acids are as well (Nelson *et al.*, 2002). It is not clear how important this is because the D-isomer is not involved in protein formation. Some proteins themselves are also very sweet, but this is rare.

More importantly, not all sugars excite this receptor. Some sugars, including glucose, galactose and maltose, produce no response from this receptor and yet are perceived as somewhat sweet by humans and other primates (e.g. chimpanzees: Hellekant *et al.*, 1996). Glucose is the transport sugar of animals, but it is also common in fruits because the enzymatic breakdown of one sucrose molecule produces one fructose and one glucose molecule. On current evidence, it appears that glucose must be sensed by some other, as yet unknown, mechanism.[12] Recent evidence suggests something of the nature of this second receptor.[13] Rats can taste oligosaccharides (longer chains of saccharides) that are derived from starch and contained in commercial corn starch syrups (Sclafani, 1991). Laska *et al.* (2001) have recently investigated the attraction of these corn syrups to four primates, one of which, the pigtailed macaque (*Macaca nemestrina*), actually preferred these oligosaccharides and maltose to simple sugars. The most parsimonious suggestion at present is that glucose, galactose, maltose and some oligosaccharides are all detected by a second (unknown) sweetness receptor. Confusingly, both

Sclafani (1991) and Laska *et al.* (2001) employ the term 'starch receptor' but this seems unwarranted. Starch itself is not tasted and there seems little point for most mammals to taste a small proportion of starch breakdown products when their gut will not be able to digest the vast bulk of it. Only ruminants are likely to manage this and rats and macaques are not ruminants. This point is explored further in relation to leaf eating and salivary amylase in Chapter 7.

Free essential amino acids in foods can be assimilated without digestion and one would think that mammals would crave them. Yet only recently has it been discovered that they are sensed on the tongue by a specific mechanism. Proteins are built from L-amino acids and only this optical form triggers the taste receptor (Nelson *et al.*, 2002).[14] Umami, the taste produced by monosodium glutamate, might be transducted by this mechanism, but no one as yet knows. Most people seem to believe that the detection of amino acids in the mouth acts as a proxy for the food's protein content. However, some fruits, such as figs, have large free amino acid concentrations, but no protein (N.J. Dominy *et al.*, unpubl. data). Furthermore, of the hundreds of amino acids known, only a few are 'essential'. Many of the others may actually be toxic, produced as a (probably infrequent) form of defence in young leaves (Coley & Kursar, 1996). Nothing is known about how the amino acid receptor responds to these compounds.

Aside from their normal metabolism, plants produce a wide variety of compounds that appear to have no other specific role than to prevent animals predating them. Some of these compounds are toxic in extremely low doses. They are mostly detected via bitterness, which depends on a large family (between 40 and 80) of receptor proteins, many of which can be found on a single taste cell (Chandrashekar *et al.*, 2000). A protein called gustducin also apparently needs to be present for this taste pathway to function. The range of receptors seems to parallel the range of so-called 'qualitative' plant secondary compounds. 'Qualititative' here refers to the devastatingly powerful effects of some of the compounds, e.g. the alkaloids, some of which (e.g. ricin) can kill a cell with one molecule.

Some fatty acids are essential in the diet and there is now clear-cut evidence in the rat for a fatty acid detector in the tongue. This is not a receptor per se, but makes use of the ability of fatty acids to block an open ion channel (e.g. Gilbertson, 1998). The response is quick and specific and it has been shown using rats of different strains that preference for dietary fat is related to sensitivity to free fatty acids (Gilbertson *et al.*, 1998). It is probable that humans have this detection mechanism. Tittelbach & Mattes (2001) have shown from blood tests that exposure to fat in the mouth boosts its absorption in the blood several hours later to levels far above that of controls. The experiments are complex, but the only sensible interpretation is that fats are detected in the mouth, very probably by Gilbertson's mechanism, leading to an enhanced response later on. Fat substitutes can have the same texture as fats, but without those fatty acids being present there is no response (Mattes, 2001).

Lipids as a whole are the most energy dense of foods and can also be detected texturally in humans as 'creaminess' for example (Lermer & Mattes, 1999). Rolls *et al.* (1999) have shown in an Old World monkey, *Macaca fascicularis*, that a distinct region of the orbitofrontal cortex of the brain responds to the texture of fats in the mouth. This neural response is definitely textural because the cells respond to non-absorbable lubricants in the same way. This is likely to be a tactile response to a lowering of intra-oral friction (the opposite of that induced by tannins, which are discussed next).

Tannins are plant compounds that bind to proteins, generally forming large molecules that may precipitate out of solution, so rendering the protein useless (Waterman & Mole, 1994). If these tannins react with digestive enzymes, then the consequences for digestion can be disastrous because the binding will put those enzymes out of commission. In low doses, tannins inhibit digestion, but in higher doses, they are downright toxic. Usually, tannins are viewed as defences against predation (Dominy *et al.*, 2001), but there is now some evidence that limited doses of them (usually under the banner of 'polyphenols') in the diet improves human health. For example, they can benefit ruminants by speeding their digestion up (Aerts *et al.*, 1999). Yet large doses of this very diverse range of chemicals are definitely toxic (Waterman & Mole, 1994) and levels of ingestion need to be monitored.

About 70% of the proteins in human saliva, largely proteins found in the output of the parotid and the submandibular glands, are rich in proline (Edgar & O'Mullane, 1996). This amino acid is not usually well represented in protein structure (if we discount those proteins involved in enamel formation, that is) and it is not entirely clear why it is there. It might be to protect enamel structure because some of this protein adsorbs onto the enamel surface (Jensen *et al.*, 1992). Enamel has certainly been shown to demineralize very rapidly in deionized water (Habelitz *et al.*, 2002). However, it is (in my opinion) impossible to explain both the quantity and the diversity of these proline-rich compounds simply by this human-based explanation. Alternatively, as is well known, most bind strongly to tannins and some seem to have no other demonstrated role. The result of the binding with tannins is an oral sensation often described as 'astringency' or a feeling of intra-oral dryness. This feeling is probably due to loss of lubrication. While it may be felt as 'dryness', this is more likely to be a consequence of an increase in intra-oral friction (Green, 1993; Prinz & Lucas, 2000). At high tannin concentrations, this increase could simply be due to glycoprotein being lost from the saliva. However, at low concentrations, Prinz & Lucas (2000) and Dominy *et al.* (2001) have suggested that the adsorbed protein coat on the enamel (the 'pellicle' as it is called) could react with them. This forms an ideal way of detecting tannins with the tongue, since the roughness produced on the tooth surface following a tannin reaction is readily identifiable, as when drinking red wine, for example. Why then do some humans appear to like a (very modest) tannin intake? One possibility is that people have always 'known' what scientists have recently begun trying to prove – that polyphenols can be good for you. A more credible explanation though is that of Prinz & Lucas (2000): food particles or lipid films 'attached' to the oral mucosa by saliva might be dislodged by a mouthwash of tannins binding to the salivary glycoproteins. This is much like the use of tea in a Chinese banquet – as a type of cleansing agent.[15]

Increased friction, either tooth–food or tooth–tooth, could increase the rate of tooth wear. The addition of grit to chewing gum results in increased salivation (Prinz, in press). This could be an evolved response to reduce the potential wear threat from the grit, analogous to the way that flooding the mouth with bicarbonate ions prevents loss of mineralized tooth tissues by dissolution (erosion).

THE DECISION TO SWALLOW

At some point during chewing, a batch needs to be swallowed so as to allow a further batch to be ingested. Chewing for too long jeopardizes the rate of energy acquisition: there can be no catching up later if further food input is excessively delayed (Alexander, 1994). Consequently, it is logical to presume that the duration of the 'ingestion–mastication–swallowing' segment of the digestive sequence is optimized for the maximum digestive effect (Alexander, 1991). Accordingly, a physiological trigger should be anticipated that could tip the central nervous system off so as to initiate a swallow at the best moment. There are some problems, however, about generalizing from human swallowing experiments to the physiology of other mammals. When humans swallow, we generally collect food in our mouths until we initiate a swallow voluntarily.[16] Food is then forced rapidly down the pharynx by a piston-like movement of the tongue. Before the food reaches the pharynx though, the raising of the soft palate and the larynx seals off upper and lower respiratory tracts respectively. If the process is mistimed, such that either breathing is in progress or that food particles do not move entirely as a group, then it is possible to send some particles down the lower respiratory tract, so causing choking. In contrast, other mammals have their larynx set much higher than in humans (Thexton & Crompton, 1998). The change in humans appears to have been to facilitate vocal communication. In other mammals, the epiglottis can connect with the soft palate, such that it is possible for them to breathe while swallowing. Their upper and lower respiratory tracts may connect in the midline, allowing food particles to slip down to the oesophagus around this connection without jeopardizing respiration (Hiiemae & Crompton, 1985).

I take the view that, despite differences in gross anatomy, essential features of the swallowing process, in particular the sensory trigger or triggers, would be similar across mammals and that experiments in humans are relevant to these features. Humans initiate swallowing voluntarily. Is there a trigger for this or is it just habit? Studies in humans show that the point at which swallowing starts is not random. For example, if one mouthful of food is given repeatedly, the time to swallow in a given subject varies little between trials. This could simply follow habit – people may chew a certain number of times and then swallow. However, if so, then this is a habit that can be manipulated rather easily because simply changing the size of the mouthful of any food (the food volume) changes the number of chews (or the time taken) before swallowing and by a very substantial margin (Lucas & Luke, 1984a). Rather than habit then, it is more logical to suppose that something about the state of the food is being detected.

THEORIES OF SWALLOWING

Swallowing is a clearing process that is initiated voluntarily, but later becomes reflex once food particles have been cleared from the tongue (Thexton & Crompton, 1998). For optimal efficiency, the process should move chewed food particles as a set from the mouth down to the stomach in one batch. Food clearance lower down the gut is not a problem because there is surplus fluid resulting from glandular secretions and these flush food particles along from one step to the next. In addition, processing in the stomach turns food into slurry, which in itself facilitates transport: peristaltic movements of the gut move pliant material much more easily than stiffer stuff (experiences with diarrhoea versus constipation are informative in this regard). The mouth though is more of an earth closet than a water closet: fluid is in short supply.[17]

Particle size threshold model

A threshold in food particle size during chewing has been the most common suggestion for a trigger. Although there is some evidence that food particles are generally reduced to a low-millimetre size range by the time that swallowing occurs (Lillford, 1991; and see above), experiments and surveys show that the decision to swallow is not triggered by particle size directly. In particular, evidence from Yurkstas (1965) shows that there is an inverse correlation in humans between particle size reduction rate and the size of particles at swallowing. Humans with full dentures, who have very slow particle size reduction rates, often swallow extremely large particles. In experiments on subjects with natural dentitions, changing the mouthful not only changes the number of chews to swallow, it also changes the particle sizes that are swallowed. A very small mouthful is always swallowed at smaller particle sizes than larger ones (Lucas & Luke, 1984a). It appears that the rate of food particle size reduction is probably not being sensed. It does not have any obvious end point anyway and could continue indefinitely (were it not for deformation transitions – Chapter 4).

Dual threshold model

Hutchings & Lillford (1988) made an important conceptual advance by postulating that swallowing might depend on a combination of two thresholds: particle size and particle lubrication. Both of these could involve sensory criteria that have to be satisfied in order for swallowing to take place. The authors offered qualitative descriptions of what they refer to as the 'breakdown paths' of various foods. For example, an ingested raw oyster satisfies both thresholds on entry to the mouth and could be swallowed immediately. In contrast, a slice of a mandarin orange, while not being reduced in size by chewing, needs sufficient juice expressed from it in order to be swallowed. The slice satisfies the particle size criterion immediately after ingestion, but not that of lubrication, so it is chewed until its surface is lubricated by saliva and juice. A peanut appears to have to be chewed both to reduce its size and to lubricate it.

These examples indicate that, if there is a particle size threshold, then this depends on the food in some way: oysters and mandarin slices are much larger than peanuts. Although Hutchings & Lillford's study was based on many years of practical experience, it was qualitative in nature and presents considerable problems about how to detect these thresholds. Nevertheless, Prinz & Lucas (1995) devised an experiment involving a varying liquid : solid mix with just that intention: to try to make a putative food particle size threshold 'move' independently of a lubrication threshold. Subjects were offered a spoonful of brazil nut particles suspended in yogurt. In these solid–liquid mixtures, both the percentage of brazil nuts by volume and the size of the brazil nut particles were varied independently. The overall volume of the mouthful was kept constant. The experiment also included yogurt without any nuts in it as a control and the whole experiment was performed blind both to the subjects and experimenter, with the mixtures being coded and given in random order. The number of chews, if any, taken by the subjects before swallowing and the time taken to make those chews were both recorded.

The results of this experiment can be described by reference to the schematic diagram in Fig. 3.11. When concentrations were high and particles were large, the standardized chewing rhythm kicked in almost immediately.

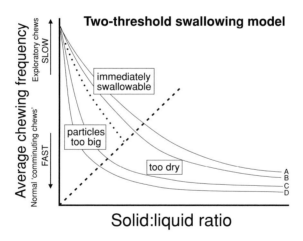

Fig. 3.11 Average chewing rates on nut–yogurt mixtures. Variation in the speed of chewing movements is related to difficulty in detecting small particles. The solid lines describe experimental data from human subjects, each referring to a single particle size where A is smallest and D largest. When the ratio of nuts to yoghourt was high, subjects chewed at average speeds, but when it was low, subjects began to 'explore' the mix for particles, so slowing their average chew speed. The dotted and dashed lines describe the suggested particle size and particle lubrication thresholds. See text. (Reprinted from *Archives of Oral Biology* vol. 40, Prinz, J. F. & Lucas, P. W., pages 401–403, Copyright (1995), with permission from Elsevier.)

However, when concentrations were low, and particularly when particles were also very small, chewing motions were very slow. These are called 'exploratory chews' in Fig. 3.11. Ordinarily, chewing proceeds with a un-changing rhythm and it is only when exploratory manoeuvres of some sort are required that this rate decreases. Three domains are indicated, con-trolled by two thresholds, one for particle size and the other for particle lubrication. The solid lines indicate results for similar particle sizes in vary-ing concentrations, where A is the smallest size and D, the largest. The vertical axis describes the rate of chewing, where faster chews are closest to the origin and slower chews higher up.

Superimposed on the results in Fig. 3.11 are boundaries that could de-marcate Hutchings–Lillford thresholds. The experiment could be viewed as support for these thresholds. That for particle size comes out very close to the expected range ($\sim$1.4 mm). However, there seems to be far too much fluid in the mix (4 parts liquid : 1 part solid) to explain how a ball of peanut fragments, for example, can be swallowed safely.

Bolus model

The search for another model for swallowing is also due to Hutchings & Lillford and concerns the nature of the food bolus and the forces that could stick food fragments together. Lillford (1991) illustrates bolus formation by showing a sequence of cooked meat samples that had been chewed for different numbers of chews by one human subject and then spat out onto a dish. Early in the sequence, the single beef particle starts to be broken up into fragments. Later, these fragments begin to stick together to form a bolus, even though the fragmentation process is continuing via further chews. At the point at which the person wanted to swallow (but was instead instructed to spit), all fragments were essentially stuck into a single ball. When the subject was asked to chew on further, rather than swallow (in this example, for 50 chews, when wanting to swallow after 30), the spat sample then consisted of myriad separate fragments floating away in saliva. The study, which was not quantified, seems to show that there might be a unique point at which a bolus will form: if a swallow is too early or too late, particles might be left behind.

Prinz & Lucas (1997) attempted to produce a quantified model of the forces involved in bolus formation. As food particles are reduced in size and simultaneously wetted, the forces in the mouth tend to make food particles either aggregate or to stick to the oral cavity. Surface tension and viscosity (Cottrell, 1964) were considered to provide the relevant forces and, in order to keep the theory simple, the model was constructed in two dimensions. It was also limited to food particles with spherical shapes, so that what applied to one two-dimensional plane through a ball of particles would be presumed to apply to any other. Also, the foods were assumed to release negligible fluid when they were fractured and, thus, that the only wetting agent to consider would be the saliva.

The mucosal lining of the oral cavity is always lubricated by a mucus-rich, salivary film. This film is ever-present and does not reduce in thickness after swallowing (Edgar & O'Mullane, 1996). A spherical food particle entering the mouth could easily be attracted to the lining by a surface tensional force given by:

$$F_A = 4\pi r \lambda \qquad (3.3)$$

where r is the radius of this food particle and λ is the surface tension of this salivary film (Fig. 3.12a). As can be seen immediately from this equation, the larger the food particle is, the larger that the adhesive force will be. After food particles are broken, a spray of saliva from the parotid gland of the same side as that on which they are chewed wets them. The particles then fall back onto the tongue, which has just collected submandibular secretions. The tongue then presses the particles against the hard palate during the early part of the opening phase of a chewing cycle. Imagine that these spherical particles are pressed together into a ball. If the particles stick together at this point, it is likely to be because of a viscous

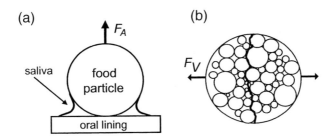

Fig. 3.12 (a). A spherical food particle sits on the oral mucosa, adhering to it initially by virtue of a surface tensional force created by a mucus-rich salivary film. (b) A section through a spherical ball of food particle fragments, packed by the tongue, but separated by saliva that infiltrates the inter-particular space. This ball will be held together by a force deriving from viscous adhesion. (Reprinted from *Food Quality and Preference* vol. 13, Lucas, P. W., Prinz, J. F., Agrawal, K. R. & Bruce, I. M., pages 203–213, Copyright (2002), with permission from Elsevier.)

force that acts at very close range. The force required to separate these discs is

$$F_V = 3\pi\eta D^4/64d^2t \qquad (3.4)$$

where η is the viscosity of the oral fluid filling the spaces between the food particles, D is the radius of the ball, t is the time-span over which the separation takes place and d is the average distance between particles (Fig. 3.12b).

The prediction of the model is very simple. Bolus formation should begin when the net cohesive force, $F_V - F_A$, is greater than zero and the optimal period to swallow is when $F_V - F_A$ is maximized because this would be the safest time, with the least chance of leaving particles behind.

Testing the model requires particle size reduction rates for foods. Prinz & Lucas chose brazil nut kernels and raw carrot because they appear to represent the extremes of breakdown rates in the human mouth. Turgid carrot particles break down slowly, while brazil nut breaks down quickly (Lucas & Luke, 1983; Lucas et al., 1986d). Selection and breakage functions for these foods were available and inserted in the comminution equations given earlier to produce particle size distributions. The number of chews taken to swallow and the particle sizes at swallowing were known for a large number of subjects too (Lucas & Luke, 1986), allowing a full test of the model. The last component of the model was the most difficult: how to simulate the packing of particles by the tongue. A computer model was built specially to do this (detailed in Prinz & Lucas, 1997).

The results are shown in Fig. 3.13 where the net 'bolus-creating' force, $F_V - F_A$, is plotted against the number of chews. Note how just low these forces are. Initially, the force is negative, which means that food particles are more likely to stick to the oral cavity than to each other. However, after just a few chews, the force rapidly becomes positive and peaks between about 20–30 chews. The peak is shallow with raw carrot particles because these break down so slowly, but for both carrot and brazil nuts, the cohesive force gradually declines from the peak as too much saliva is introduced, letting particles 'float' away.

Predictions of the model

The model predicts several features of oral processing behaviour, some of which have been observed in humans and some of which remain to be tested (Prinz & Lucas, 1997; Lucas et al., 2002). The model predicts Lillford's (1991) result with beef. 'Forced' chewing beyond the normal point of swallowing will start to swamp the particles causing them to float apart.

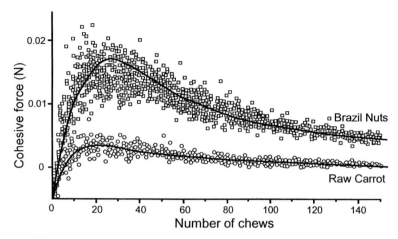

Fig. 3.13 The bolus-making force, plotted on the vertical axis, as it varies during a masticatory sequence. Early on, the forces that attract food particles to the walls of the oral cavity are greater than that which would stick them together, so the overall bolus-making force is negative. However, the viscous cohesive force predominates after just a few chews. It declines only when excess saliva weakens it by separating food particles and allowing them to float away. (Reprinted from *Food Quality and Preference* vol. 13, Lucas, P. W., Prinz, J. F., Agrawal, K. R. & Bruce, I. M., pages 203–213, Copyright (2002), with permission from Elsevier.)

The model also suggests why humans swallow fruit pieces very quickly: the time to swallow should be promoted by the expression of fluid from within food. If fluid expressed from the fruit were sugar-rich, then the adhesive force would increase and swallowing would be promoted even further. Food acidity should also promote early swallowing by stimulating extra saliva.

While fruit flesh often forms an almost instant bolus, leaf fragments do not. Their outer covering, the cuticle, is so hydrophobic as to be effectively unwettable (Barthlott & Neinhuis, 1997; Neinhuis & Barthlott, 1997). This is not just a function of the waxiness of the cuticle, but also of its micro-roughness, which makes the contact areas of the leaf surface with water, grit particles, fungi and other microorganisms so small that these will fall off the leaf.[18] This keeps the lamina clean and free from infection for as long as possible. The cuticles of leaves interfere in bolus formation because leaves have such a high specific surface that the cuticle dominates exposed surfaces of leaf fragments throughout mastication.

The general value of the model for explaining swallowing patterns in other mammals can be doubted and not just because of anatomical differences. It cannot connect the efficiency of oral processing up to later steps in the digestive process in any convenient way, something that other models can do (Alexander, 1991, 1998). However, the existence of a food bolus itself though cannot be denied – Lillford's (1991) pictures show it and 'artificial bolus makers' can apparently demonstrate it *in vitro*. Unless there is some neural evidence for a 'swallow trigger' though, the model really remains just a possibility.

There is a flicker of hope for such a swallow sensor though in the textural fat sensor of Rolls *et al.* (1999) that resides in reinterpretation of their data. These authors actually show cortical neurons in an Old World monkey that respond purely to lubricants such as oils (both vegetable and mineral) and the like, i.e. to a stimulus that probably involves a drop in intra-oral friction or, at minimum anyway, to a surface sensation in the mouth. Unfortunately, the authors do not give a location for this stimulus in the mouth, their concerns being to ascertain that the neurons in the cerebral cortex did not fire in response to either taste or odour. The most likely location is the tongue where sensory nerves have been demonstrated that are exquisitely sensitive to forces in the low millinewton range (Trulsson & Essick, 1997). As stated previously, Rolls *et al.* have established that there is a group of neurons with a specific frictional role, but it does not seem logical that these cells exist to detect fat texture if there is already a chemosense for these compounds. I suggest a possible alternative: could these neurons relay information about the developing state of the food bolus and provide the information critical for swallowing decisions?

4

Tooth shape

OVERVIEW

Virtually the entire dental literature deals with the function of teeth from a geometrical perspective. In this chapter, the first of three forming the core of this book, I advance the alternative hypothesis that the mechanical behaviour of foods is the major influence on tooth shape. The principal mechanical properties of foods involved are the Young's modulus and toughness. However, I will also be referring to more specific influences such as the level of stresses within foods, the yield strength, strain and Poisson's ratio. To understand these concepts clearly, the reader is advised to work through Appendix A before reading further. While some of the influences on tooth shape can be quantified fairly precisely, others are too complex for this. The analysis then has to borrow from simpler circumstances in which the basic principles have been established. The chapter starts by assuming that foods are homogeneous solids, but the value of understanding features of food structure such as their internal connectivity should quickly become apparent. To keep things simple, all through the chapter, the basic analysis assumes linear elastic behaviour in foods. Only general principles of dental–dietary adaptation are emphasized because Chapter 7 deals with real dentitions.

INTRODUCTION

The toughness of foods plays a crucial role in shaping teeth. Unfortunately, toughness is an energetic concept that still seems alien to most biologists. When most people think of fracture, they think in terms of fracture strength, whereas, in fact, there is no such thing as a unique fracture stress for either a solid material or a biological tissue. I am not referring here to the fracture stress in tension say, as compared to that in compression or shear, or to the effects of varying the rate of loading and temperature, but to

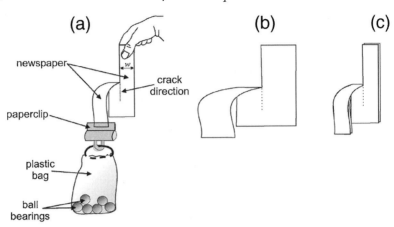

Fig. 4.1 A simple experiment to ascertain that cracking is controlled not by stress but by energy. (a) A sheet of newspaper has been cut into the shape of a pair of trousers, one trouser held by hand and the other attached to a bulldog clip from which hangs a plastic bag. Ball bearings can be placed in the bag until the paper starts to fracture in the direction indicated by the dotted line. (b) Changing the width, w, of the legs lowers the stress, but will not vary the number of ball bearings needed to make the crack start, whereas doubling the thickness by tearing two sheets at once, as in (c), *will* require a doubling of the number of bearings.

the fact that there is no such thing as a unique strength for a material under any given loading circumstances. Griffith (1920) established this about 80 years ago. However, with the exception of two popular books by Gordon (1978, 1991), this fact has not been well marketed to the general public. Even some engineers seem to feel that fracture mechanics is an unnecessary burden and an impractical complication. The net result is the retention of the Galilean concept of fracture strength as gospel truth by those in other sciences (see Chapter 5).

It is now known that energy, not stress, controls fracture. Simple examples illustrate this. The trouser-tearing test can be performed without special equipment by tearing up a newspaper (Fig. 4.1). Hold one leg of a sheet of the newspaper and attach a bulldog clip on the free-hanging lower leg. Add a plastic bag that can be filled with weights such as ball bearings (Fig. 4.1a). Once a sufficient amount of bearings has been added, the paper will tear. The stress on the leg imposed by the mass of ball bearings, M, will equal Mg/wt where g is the gravitational constant, w is the leg width and t, the paper thickness. By experiment, it will be found that varying w, which changes the leg stress, will make no difference to the force at which fracture starts (Fig. 4.1b). On the other hand, increasing the thickness of the paper by tearing two to three sheets simultaneously instead of just one will require double or triple the number of ball bearings respectively (Fig. 4.1c). The onset of fracture is proportional to Mg/t, but independent of w and, therefore, of stress. The meaning of Mg/t is not 'force per unit thickness', as might appear at first glance, but 'energy per unit area' and the understanding of it is in terms of conservation of energy. If the work done in extending the crack in the paper a distance l is Mgl and all this work is converted into a crack of area lt, then, with the toughness of the paper defined as R,

$$Rlt = Mgl.$$

This reduces to

$$R = \frac{Mg}{t}.$$

In a universal testing machine, we do not usually measure l directly, but use the displacement of the crosshead of the machine instead. This travels twice as far as l because both legs are stretched. In addition, the force, F, is given directly, so that

$$R = \frac{2F}{t}.$$

The same lack of dependence on stress is true for crack growth in any situation. For example, there is no dependence on the thickness of the peel in the peeling test (Appendix A) (Kendall, 2001). Increasing peel thickness decreases the stress, but does not alter conditions for its removal. The phenomenon of notch sensitivity provides another example (see Appendix A) and there are also many indirect methods that demonstrate it. One is to show that the fracture stress for a given material varies with particle dimensions. Kendall (1978a,b) showed this very clearly for 'compressive strength': smaller particles are stronger than larger ones. This has been corroborated in many other studies (e.g. Darvell, 1990) and is a well-known effect in foods.[1]

So forces, always acting as stresses, are just the means to provide energy to a particle: it is what happens to this energy that influences whether this particle fractures. Ordinarily, the energy is stored as elastic strain energy within the particle and is available for release into a crack under the right conditions. Alternatively, loading may produce permanent (plastic) deformation of the particle. From a fracture point of view, this is energy lost and an efficient comminuting mechanism needs to avoid this if at all possible because a large amount of energy can be involved. Energy may be lost in other ways as well, e.g. as heat or kinetic energy (as when glass shatters and small pieces fly away), but both these effects can be an unavoidable consequence of needing to produce fracture very quickly. The last possibility is that the work done on the particle can somehow be channelled directly into fracture, so avoiding energy transfer to the bulk of the particle. This is the only one to offer the (illusory) target of 100% efficiency in loading.

LOADING GEOMETRY

While energy has no direction, the loading that imparts this most definitely has. There are three basic types of 'pure' loading that could lead to the fracture and fragmentation of solids: tension, compression and shear (Fig. 4.2). A tensile loading involves a force acting perpendicular to the surface of a solid and directly away from it (Fig. 4.2a). The opposite of a tensile load is a compressive load. This also involves a force acting perpendicular to the surface of a solid, but straight towards it, rather than directly away (Fig. 4.2b). The third possibility is that when the solid is loaded in a direction parallel to its surface. This is loading in shear and is an angular action (Fig. 4.2c). The movement can be linear, as in Fig. 4.2c, or involve rotation, as in torsion (twisting). Pure torsion is a form of shear, but any twisting in practice nearly always involves tension too.

These loading patterns are not the equivalent of internal stresses: all these loadings can lead to combination of compressive, tensile and shear strains inside particles as a reaction to them. Provided that Poisson's ratio is not zero, compression will lead to tension at right angles to the load and shear at 45° to it. The Poisson's ratio of solids is usually considerably less than 0.5, meaning that particles will densify (reduce their volume) under a

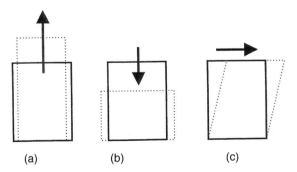

Fig. 4.2 The three basic loading geometries of loading: (a), tension, (b), compression and (c), shear.

compressive load. In such solids, the efficiency of inducing tensile stresses indirectly via such a load is generally low. Pure shear is defined as a state in which the sum of tensile and compressive stresses in the specimen is zero, i.e. the hydrostatic stress is zero. (Remember in such sums that tensile stresses are given the opposite (positive) sign to (negative) compressive stresses; see Appendix A.) However, all these 'pure' loads are rare outside of mechanical testing. Pure shear is extraordinarily difficult to produce anywhere, even in a laboratory test. Pure tension and compression are extremely inefficient methods of generating critical stresses. Why load all of a particle evenly if the effect is aimed at a particular location? Bending, for example, results in tension just on one side of a particle. The tensile stress is not even along the bent length but peaks on the outer surface midway between its supports. Bending is an efficient way to produce tension because it acts locally. In constrast, indentation produces localized compression.

There are many ways to load a particle, such as by pulling, pushing, bending or twisting, and all of these may be seen during ingestion. However, once inside the mouth, the options reduce. Only pushing (compression) is readily available unless there is some external restraint for the teeth to act against. A tensile load inside the mouth would require moving parts in both upper and lower jaws in order to get a grip. That is very complex to arrange and has not evolved in mammals.

FRACTURE GEOMETRY

Most, if not all, of the literature on tooth function suggests that tooth shape can be understood by examining the details of loading, or the distribution of stresses produced in the food particle. This is incorrect. The factors that promote the production and growth of cracks within that particle are the crux of the matter. In support, the next section shows how the geometry of fracture is controlled by food particles, not by teeth.

Elastic fracture

Many solids maintain an elastic response to a loading right up until close to fracture. If a crack forms and runs right through such a particle, dividing it into two, then the fragments can be fitted back together to form the size and shape of the original particle. Compression, e.g. by the molar teeth, cannot lead to crack growth directly because cracks or flaws in particles are closed by compression rather than opened. Compression tends to heal cracks and its comminuting effect has to act indirectly via shear or tension. However,

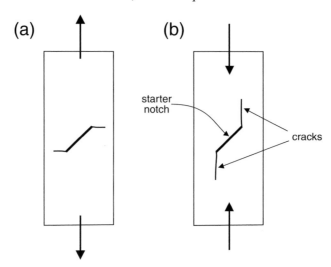

Fig. 4.3 A plate made of a homogeneous material has had a sharp-ended starter notch cut into it at an angle that could favour crack extension in shear or tension in equal measure. When loaded either in tension (left) or compression (right), crack growth from the ends of the slot rapidly aligns along a direction in which tensile stresses are maximized: at 90° to the tensile loading (left) or parallel to the compressive loading (right).

is it shear or tension that is responsible for crack growth? The experimental and theoretical answer is unequivocal. Figure 4.3 shows the behaviour of a homogeneous laboratory specimen in tension (Fig. 4.3a) and compression (Fig. 4.3b), when a crack extends from the ends of a notch. The notch has been inclined at 45° to the load to give the crack every opportunity to grow in shear. It is found in experiments that, once a crack appears at the notch tip on loading, the crack follows a short curved path and lines up perpendicular to the direction of the maximum tensile stress. This path is perpendicular to the load when that is tensile or parallel to the load when it is compressive. Theoretical analysis supports these results by showing that cracks grow essentially in the direction in which the elastic strain energy stored within them is released most rapidly, where this rapidity is assessed per increment of crack area growth, not per unit of time (Lawn, 1993). This direction is always tensile in homogeneous elastic solids. If the applied load is more complex, with a mixture of tension and shear for example, then it is possible for the crack to travel for a short distance with a shear component (as shown by the first crack extension in Fig. 4.3). However, this is fleeting (Mai & Cotterell, 1989) and continued growth of the crack tends to follow a direction perpendicular to the main tensile stress (Lawn, 1993). The theory

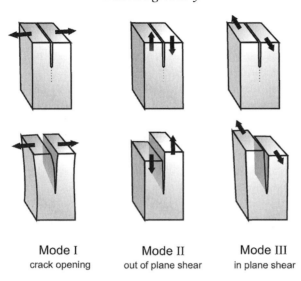

Mode I
crack opening

Mode II
out of plane shear

Mode III
in plane shear

Fig. 4.4 The three modes of fracture. Only mode I fracture, where crack opening is associated with tensile stresses ahead of the crack, is common in homogeneous materials. It is very difficult to get fracture in either of the shear modes even in 'pure' laboratory tests because it is not easy to avoid an opening component at the crack tip.

is difficult (if not impossible) to solve for all except the first crack extension, but experiments seem conclusive: cracks tend to grow by opening up.

Generalizing from the above, there are three possible directions in which a crack could grow. These are termed modes of fracture, each defined to correspond with the three components of stress in front of the tip of a crack that could result in its further growth (Fig. 4.4). Mode I fracture entails tensile stresses in front of the crack tip that encourage the crack surfaces to pull apart. This 'crack-opening' mode is usually what is evaluated in toughness tests. An example is when a large carnivore pulls meat off a carcass. Mode II involves shear stresses in the plane of the section. An example is provided, in theory anyway, by the action of a paper punch or an incisal bite. Mode III involves out of plane shear stresses – such as those apparently produced by tearing a piece of paper (trouser-legging) or by a pair of scissors (see Appendix A). In practice, fracture in neither modes II or III is easily tested in isolation because it is extremely difficult to prevent crack opening (mode I) involvement (because, as explained above, this favours faster energy release). In practice, there is almost certain to be a strong opening element. In fact, there is little reason to invoke modes II or III as having any relevance to the action of the teeth. Even a guillotine or a pair of scissors produces mode I cracking in an elastic solid (Atkins & Mai, 1985).

Thus, I cannot see any general grounds whatsoever for supposing that the design of mammalian teeth is to 'maximize shear' (Shaw, 1917). Sharp-bladed teeth will not produce shear cracks in either mode II or III. Yet it can hardly escape anyone's attention in reading the dental literature that shear lies at the magical heart of dental function for most researchers. This is an error. The direction of maximum shear stress in a solid is not going to be the direction in which cracks in food particles generally grow. Even if it were, then such a theory could not explain dental diversity: all teeth would look the same.

Some may feel that all this is restricted to homogeneous foods. Could there be something special about the structural make-up of some food particles that makes a heightening of shear stress within them advantageous for cracking? Well, heterogeneous food particles will, by definition, contain interfaces between their structural components. These interfaces might have a very low toughness. If so, then cracks will tend to deflect along them. If this deflection lies at an angle to the main direction of the bite force, then this will involve shear and some mode II or mode III component (as indicated, for example, in Fig. 4.18). However, the point is that the direction of deflection is the natural path of the crack. No aspect of dental design, such as the presence of a sharp blade, is required to produce this, and the original path of the crack will be recovered if only a slight weakness appears in a favourable direction. Otherwise, a crack oriented at an angle to the main direction of bite force will tend to arrest (stop).

It is generally unlikely that any tooth would possess features designed to promote such natural crack directions. If it did, then the most logical dental strategy would be to attempt to open up the interface rather than induce slip. There are some instances where natural fracture paths in potential foods are designed to facilitate fracture. Very often, seeds have these in order to open up and germinate. However, even seeds have design features to avoid being opened. To paraphrase: they need 'to open, but not be opened' (Lucas *et al.*, 1991b). Fleshy fruits are also exceptional: they *are* designed to be eaten, but only the flesh and only by target consumers. In fact, these fruits appear to be craftily designed in order to keep competitors (i.e. fruit thieves) out (Cipollini & Levey, 1997). Otherwise, any potential food particle is part, or all, of an organism that is designed to resist fracture. The threat may not come from a predator (it could be wind, for example) but any natural path of fracture in that particle is actually likely to form part of its defences. If this logic is accepted, then teeth that let cracks follow that path will fail to fragment the tissue. An example is to try to bring down a tree by following the grain of its wood, rather than aiming across it with a bladed axe or saw. The latter is a much tougher path, but it is the only way to fragment it. This is the way that beavers (*Castor fiber* – a rodent) do it. Leaves show this protective design too: cracks deflect along their veins as a way of stopping the loss of a whole leaf.

To sum up, we can assume that elastic foods fracture under the influence of tensile stresses. When these cracks are straight, then it makes sense also to refer to this as mode I fracture. However, most commonly, a crack zigzags through a food structure. There is no point at all considering modes of fracture when this happens, which is to say: *the mechanism of fracture prevention in foods is what matters for understanding tooth form, not fracture geometry.*

Plastic fracture

When tissue volume is conserved during deformation (i.e. Poisson's ratio is 0.5), then the structural units in that tissue must be sliding over each other: in other words, this is shear. Plastic deformation is exactly like this. Tissues that show large amounts of plastic deformation do not really crack even though there are changes in their external surface area as they distort. However, very few plant or animal tissues display plastic deformation to this extent because it would be incompatible with their structural role in the organism. Limb bone provides an example. It is known that it could have a far higher crack resistance than it does (a comparison between the antlers of deer and their limb bones shows this – Currey & Brear, 1992), but achieving it would either reduce Young's modulus considerably, so lowering the rigidity of the locomotory system, or cause yielding that would interfere much more by producing uncontrolled change in the shape of bones. (Shape control is less crucial for antlers and these do show substantial plastic deformation – J. D. Currey, pers. comm. in Purslow, 1991a.) Plastic deformation is possible though in any solid when loading is confined to a sufficiently confined space such as during indentation (Atkins & Mai, 1985). Even so, any crack(s) that start at or near the plastic–elastic boundary of this indentation grow along lines of tensile stress (Lawn, 1993).

So I conclude this section, as the last, by predicting that the geometry of fracture will not explain anything at all about mammalian teeth. The word 'shear' does not explain tooth shape any more than do 'crushing' and 'grinding'. Semantically, words like these are simply shorthand descriptions of successful fracture events, often connoting a whole range of conditions surrounding such events, but denoting only one thing: fracture. They do not explain the factors required for successful fracture – they cannot if their meaning assumes it – and so they have no place in an analysis at all. There is a very large set of 'breakage' synonyms, many more than 100 in the English language, which can be dragged into functional descriptions of dental actions. None has any analytical value, being used as a kind of descriptive shorthand. The fault in seeing otherwise, and the reason why food properties are rarely invoked as causes of variation in tooth shape, is very clear from the following quote:

The fact is that mammals use their teeth to acquire and prepare their food and that their molars are in this sense adapted to certain kinds of activity and beyond this to certain types of diet. (Simpson, 1936)

The activities to which Simpson refers are 'shearing' and 'grinding'. He gives complex definitions for these in terms of tooth shape and movement, but which do not include foods at all. The interpolation of 'activities' such as these between teeth and diet is the central reason why studies of dental–dietary adaptation have remained stagnant. Without foods, all that can be done is to eulogize teeth for being so good at what they do.

All further sections of this chapter assume that the shape of teeth is an evolved response for overcoming the toughening mechanisms inside foods that frustrate their fracture and that these mechanisms lie at the heart of the diversity of dental form.

FRACTURE LIMITS

Ashby (1989, 1999) showed how to analyse the possible limits to the success or failure to fracture particles in a systematic manner. When a force produces a displacement, work is done. Forces always have to act as stresses, but displacements are not always equivalent to strains because they could easily be off the axis of loading. Immediately, these simple considerations generate two possible limits to successful fracture: a food particle could fail to be fractured either because there is insufficient stress or because there is not enough displacement. An example, modified from Lucas *et al.* (2000), illustrates the difference.

Suppose an elephant were attempting to remove a branch from a tree by bending that branch downwards. The animal might be able to bend it all the way down to the tree trunk, but not succeed in breaking it. This is displacement limitation: there was sufficient stress, but the elephant ran out of displacement. The alternative is that the branch barely bends because the elephant could not generate sufficient stress. There was a lot of displacement still available but not enough stress. This is stress limitation.

Mastication appears most often to be a displacement-limited activity because that is what restricts the efficacy of loadings that push tooth surfaces together. This may apply to success/failure to fragment a particle once, or to a limitation on the number of times that a particle can break into separable pieces, within any particular chew. Whichever, this is a displacement limit. In contrast, ingestion is never going to be displacement limited. An elephant is not going to stop attempting to break the branch just because bending it vertically downwards fails to detach it. It could simply pull, or pull and twist the branch simultaneously, persisting until successful. In the same way, a primate that is ingesting a food particle by incision will only attempt to load a particle between upper and lower incisors if there is sufficient displacement to cause fracture. Otherwise, unless the particle is very small, it would select another method. I will assume then for the theoretical analysis, that mastication is displacement limited while ingestion is stress limited.

BASIC ROLES OF THE DENTITION

Teeth at the front of the mouth have different shapes to those at the back. The reason for this could be due partly to these different limits, but also due to the differing roles that anterior and posterior teeth play in food intake. Figure 4.5 suggests a hierarchy of activities necessary for the management of solid particles. A primary requirement is *grip* for without this there can be no management at all. The secret of grip is friction. Grip allows the tongue to direct and *transport* food particles. The teeth of most non-mammalian vertebrates are designed only for this function. Grip for transport may require *crack initiation* within a food particle. Most mammalian teeth are designed for *fracture*. If this need goes further to produce food *particle size reduction* (i.e. fragmentation), then there are two alternative strategies. Both a final particle size and shape might be required, a strategy that we could call *sculpture*. If no particular food particle shape is required other than that some regularities may emerge from repeated size reduction, then the process can be termed *comminution*. To emphasize the difference, sculpture is fracture designed to produce a fragment(s) of a particular shape while comminution is multiple fracture producing random fragmentation (Lowrison, 1974).

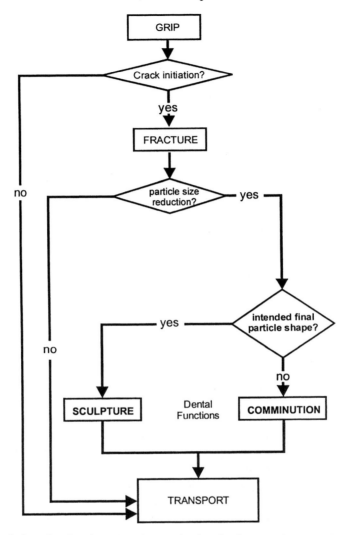

Fig. 4.5 A chart showing the various basic tasks that the dentition has to perform in the mouth. See text for explanation.

Putting the last two sections together, the use of the incisors and canines in a first bite, with or without subsequent sculpture (fracture designed to produce a fragment(s) of a particular shape), involves different demands to comminution (multiple fracture producing random fragmentation) by the postcanine teeth. Mastication will be discussed first because conditions there are better defined than during ingestion.

The easiest way to set up the principles of construction of the basic tooth features is to use a specific example. However, the analysis is quite general, depending only on the relative ease of initiating and propagating cracks within the food particles and the need to produce separate fragments rather than just cracks. I need to assume that the loading is close to optimal, so that some combination of force, displacement and/or work is minimized. Loading an isotropic homogeneous food with a minimal force requires that the contact area between tooth and food particle be minimized because this keeps contact stress high. However, the shape of a food particle cannot be precisely controlled during ingestion and it then changes continually as it is fragmented. The exact loading conditions inside the mouth will change from chew to chew. Thus, control of contact areas relies completely on tooth shape.

Large particles

Imagine a 'large' food particle – one that requires reducing in size in order to be swallowed – lying on a flat supporting surface. The cheapest method of starting a crack in this particle, regardless of its properties or structure, is to push a point into it (Fig. 4.6a), a point being a surface, both dimensions of which are very small. A point contact produces a very large stress in the food particle for minimal load. In dental studies, of course, pointed features of tooth crowns are called cusps. What influences the design of these cusps? If cracks start at cusp tips, then it may be thought that cusps should be sharp. However, somewhat ironically as far as some of the dental literature is concerned, a sharp cusp is likely to suppress cracking. In even the least tough materials, if the stress can be confined to a small enough volume, then the solid will deform plastically rather than crack. This is exactly what happens in indentation. Imagine that the cusp tip behaves just like a spherical indenter. The fracture load that produces cracking is proportional to indenter radius (Frank & Lawn, 1967). Since the area of indentation is proportional to the square of the radius, this means that sharper indenters elevate stresses, and thus tend to cause plastic indentation rather than cracking. Assuming this to apply to teeth,

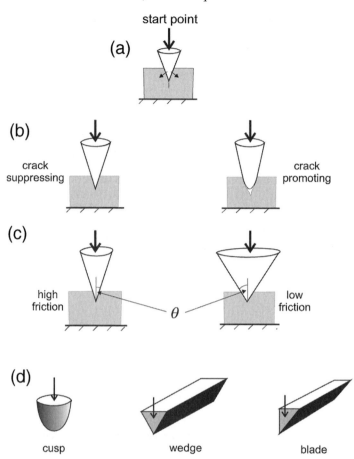

Fig. 4.6 Considerations for understanding the design of dental features and basic features. The main arrow gives the direction of the force. (a) The apparently logical starting point – a sharp narrow-angled cusp indenting a food particle with minimum stress. (b) A sharp cusp may suppress cracking in the food particle by raising the local stress; in contrast, a blunter cusp is more likely to promote cracking. (c) A narrow-angled cusp will encounter higher friction than one with a wide angle. (d) The basic elements of a dentition: a cusp, wedge (symmetrical to the direction of the force) and blade (asymmetrical). None of the features can be derived without considering the mechanisms by which cracks grow in foods.

then blunt cusps are more efficient at producing fractures than sharp ones (Fig. 4.6b).

It may also appear logical to suppose that the cusp be narrow-angled, again so as to reduce the force. Though this may be so, it results in higher friction. If the included angle of the cusp is 2θ (Fig. 4.6c), then the increase in the force due to friction as the cusp indents the particle is given by

$$F_{\text{fric}} = F_{\text{nofric}}(1 + \mu \cot \theta) \qquad (4.1)$$

where F_{nofric} is the force in the absence of friction, μ is the coefficient of friction and the term 'cotθ' refers to the cotangent of the half-angle, i.e. $1/\tan\theta$ (Hankins, 1925). Widening the cusp angle reduces friction and, while the force itself will increase due to the larger area indented by the cusp, it is more likely to promote fracture rather than plastic deformation. Also, both a wide angle and a blunt cusp tip are also features that help protect the cusp from fracture.

The other tooth surface, the lower one in Fig. 4.6, is just a flat surface. However, there is no reason why it should not also act to promote fracture and can, therefore, also be cusped. Clearly, the direct opposition of two cusps, one upper and one lower, is an unstable arrangement because, after fracturing the food, the cusp tips will contact and damage each other. An array of cusps, alternating as in Fig. 4.9, seems logical. The spaces between the cusps can be curved so as to retain food fragments generated by propagating cracks. This allows repeated fracture within one compressive movement.

Without knowing more about the food, prediction about cusp form is stuck here because some foods could crack easily, while others not crack at all. What food properties influence this? To derive these requires some assumptions. A crack in a food particle that is laid across these cusps may start: (a) remote from cusps, e.g. by bending against a three- (or more) point cuspal support or (b) adjacent to a cusp tip as the particle is indented (Fig. 4.7).

I make two basic assumptions about fragmentation:

(1) *Food particles are loaded relatively late in the closing phase of jaw movement.* It follows that the displacement during which a particle can break up is small and will limit fragmentation. Displacement limitation will apply. (I ignore the possibility of fragmentation during jaw opening when food stuck to the upper and lower teeth could fracture as the teeth separate).

(2) *Fragmentation of particles on loading follows the storage of elastic strain energy in these particles, the release of which later pays for growing cracks.* This assumption might seem unnecessary, but it excludes the local loading of thin rod- and sheet-like food particles by tooth cusps (considered later).

Figure 4.8a shows the geometry for the three-point bending of a simple beam-like food particle. The example is drawn by Ashby & Jones (1996).

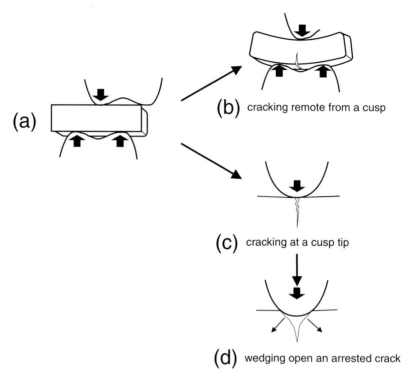

(a)

(b) cracking remote from a cusp

(c) cracking at a cusp tip

(d) wedging open an arrested crack

Fig. 4.7 A schematic diagram to illustrate cracking. Three cusps are shown in (a), and cracking can occur remote from a cusp (b) or at a cusp tip (c); in the latter case, the crack may arrest and can then be wedged open by the cusp's further movement (d). (Reprinted from *Food Quality and Preference* vol. 13, Lucas, P. W., Prinz, J. F., Agrawal, K. R. & Bruce, I. M., pages 203–213, Copyright (2002), with permission from Elsevier.)

If E is the elastic (Young's) modulus of the food, then the deflection shown in Fig. 4.8b is

$$\delta = \frac{3Fl^3}{4Ebt^3} \tag{4.2}$$

while the maximum stress (sometimes referred to as the modulus of rupture), halfway between the supports on the tension side, is

$$\sigma_F = \frac{3Fl}{2Ebt^2}. \tag{4.3}$$

The force, F, can be removed by combining Eqns 4.2 and 4.3. Moving all the food properties to the left-hand side, then the criterion for the initiation

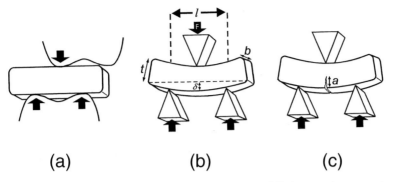

(a)   (b)   (c)

Fig. 4.8 Cracking remote from cusp tips. (a) Cusps modelled as producing a three-point loading of a food particle. (b) The relevant particle dimensions as the particle bends through a deformation δ. (c) The crack, of length a, develops remote from the cusps. (Reprinted from *Food Quality and Preference* vol. 13, Lucas, P. W., Prinz, J. F., Agrawal, K. R. & Bruce, I. M., pages 203–213, Copyright (2002), with permission from Elsevier.)

of cracking at less than an arbitrary displacement δ (i.e. with displacement limitation) is

$$\left[\frac{\sigma_F}{E}\right] > \frac{2\delta t}{l^2}. \tag{4.4}$$

For failure to the point that an engineer would be interested in, which is usually the point of crack initiation, a combination of food properties that determines the response of the food is (σ_F/E).

However, the interest is not just in crack initiation, but in the resistance to crack propagation because this influences the formation of separated fragments. The first possibility is that the food particle cracks distant from a cusp – at the point of the maximum tensile stress. Once cracked, the stress field in Fig. 4.8b is given by

$$K_{\mathrm{IC}} = c\sigma_F(\pi a)^{0.5} \tag{4.5}$$

where c is a dimensionless constant that depends on the geometry of the loading. Then, dividing through by E and noting $K_{\mathrm{IC}} \approx (ER)^{0.5}$ (see Appendix A), where R is the food toughness, gives

$$\left[\frac{R}{E}\right]^{0.5} \approx \left[\frac{\sigma_F}{E}\right] c(\pi a)^{0.5}. \tag{4.6}$$

If the number and size of fragments is limited by the displacement available, then $(R/E)^{0.5}$ is the combination of food properties that influence crack propagation, and thus the possible fragmentation of the food. It forms a displacement-limited fragmentation index that corresponds to (σ_F/E) at the onset of cracking.

In the above scenario, the cusps are really acting only as supports. There is no need for them to be sharp and it would be better if they were blunt in fact because this would increase their longevity. Their height needs to be sufficient so that the food particle does not 'bottom out' before failing. This could be achieved by making cusp height proportional to typical values of (σ_F/E) in the diet. For example, materials with a high Young's modulus generally have a low value of (σ_F/E). This suggests that extensive consumption of such foods would result in selection for low molar cusps because the deflection, δ, would be small at fracture.

Fractures may also start at, or very near, the cusp tip (Fig. 4.8c). Then

$$(ER)^{0.5} = c\,\sigma_F(\pi a)^{0.5} \tag{4.7}$$

where c is probably close to 1.0 (Marshall & Lawn (1986) give $c = 1.33$ for indentation). If these cracks run straight through the material, then fragmentation is limited only by stress, not displacement. However, these indentation cracks may arrest before fragmenting the particle. An arrested crack can only be continued by displacement of the cusp into the particle so as to wedge the particle open (Fig. 4.7d). Unsurprisingly, this follows a displacement criterion with

$$\left[\frac{R}{E}\right]^{0.5} = \frac{0.866uw^{1.5}}{a^2(1 + 0.64w/a^2)} \tag{4.8}$$

where $w = 0.5b$ and u is the width of contact of the cusp at the point at which the crack of length a starts to grow (Hoaglund et al., 1972).

To summarize, if stress is limiting, then $(ER)^{0.5}$ is the probable criterion for the fragmentation response of a food solid while if displacement is limiting, it is $(R/E)^{0.5}$. Note that both of these indexes have dimensions. For reasons given above, mastication is likely to be dominated by displacement limits, but evidence for this will be offered a little later.

Cusped postcanine teeth will only succeed in fragmenting food particles if the value of $(R/E)^{0.5}$ is low. If instead the index is high, then cracks in a food particle will quickly arrest. Even if a cusp penetrates a particle of such a food so as to reinitiate a stalled crack, that crack is then unlikely to spread laterally. For such foods, tooth form should adapt such that the pointed tip of a cusp is extended in one dimension so as to form an edge, a surface with one dimension that is very small. A wedge is an example of an 'edged' design. It is oriented symmetrically with respect to the force (Fig. 4.6d), so that stalled cracks can be reinitiated by the edge. Provided that the long dimension of the wedge exceeds any of the particle dimensions, this ensures particle subdivision.

If cracks arrest quickly, then the sharpness of the wedge becomes very critical in re-establishing the crack. In this way, the edge of the wedge will remain close to or against the crack tip all the way through the food particle. However, while the food particle will be subdivided into two, the end result will be that the wedge contacts the opposing tooth surface, so damaging it. Wedges are not durable and need to be redesigned as blades. The difference between a blade and a wedge is that, in the latter, the force is aligned parallel to one surface (Fig. 4.6d). If the opposing tooth surface is similarly designed, then the surfaces parallel to the force can pass each other. These surfaces need to contact each other as they pass or the particle will not separate into two (Fig. 4.9). Wedges appear unlikely as components in a dental battery, with cusps and blades being more frequent, but a possible role will be assigned to them a little later. So the two basic components of the working surfaces of teeth then are likely to be cusps and blades. Blades would be expected in matched sets, one on the upper dentition and one on the lower, but cusps would not. Being adapted to foods that crack, cusps do not have to contact after food fracture.[2] They will produce fragments ahead of their path of travel in a chew and rather than have a flat resting surface for particles that lie under cusps, it seems more logical to design this lower surface as a retainer allowing fragments to be rebroken in the same chew. A basin-like structure, i.e. a fossa, serves this purpose. The postcanine dentition is thus expected to consist of patterned arrays of cusps and blades. The analysis so far is summarized in Fig. 4.9.

The effect of particle shape and size

When a particle of any solid gets sufficiently small, then it ceases to crack and yields instead. The size at which it does this is called the 'brittle–ductile' transition. The food property group that controls this deformation transition was worked out by Kendall (1978a) and is $(ER)/\sigma_y^2$, where σ_y is the yield stress. The transition particle size at which cracking behaviour in a solid starts to be frustrated by plastic deformation, is given by

$$d = c\frac{ER}{\sigma_y^2} \qquad (4.9)$$

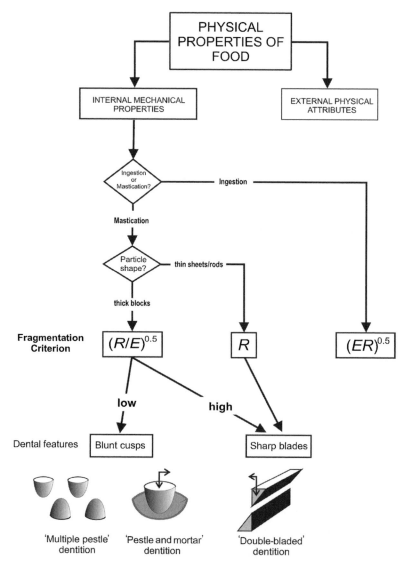

Fig. 4.9 Chart showing the basic relationships between food properties and dental features. Blunt cusps are associated with foods of low $(R/E)^{0.5}$ that crack relatively easily. Cusps can be aligned as upper and lower alternates (to avoid mutual damage) or into fossa as 'pestles and mortars'. Blades have to contact the supporting surface on the other side of particle because they represent optimal designs to break foods with high $(R/E)^{0.5}$ that obstruct cracks. To prevent excessive damage, the supporting surface has to have a reciprocal matching shape, forming a 'pestle and mortar' type of arrangement. The upper and lower dentition would, therefore, be expected to possess pairs of edges. If these were like wedges,

where d is a linear dimension and c is again a constant. For compression, Kendall (1978b) showed that $c = 32/3$. However, its value depends greatly on the general conditions of loading. Atkins & Mai (1985) have generalized the arguments and show that for indentation, the constant is likely to be >2000.

Below the transition size, energy stored in food particles cannot be transferred into cracks. Thus, a particle that cracks above this size threshold, and which might be comminuted with a cusp, will not crack below this size limit and might require a blade. This transition might set a limit to food particle comminution (e.g. the low millimetre range identified by Lillford (1991)), but it also suggests another simple way of looking at the range of food properties. If any food particle were sufficiently large, then it should crack due to work done by the teeth on the particle being stored as elastic strain energy, which then 'pays' for any crack that might initiate. Below the threshold size, energy storage is not sufficient to do this and so the work done by the teeth must pass directly into crack growth. The food property group that controls resistance to mastication is no longer $(R/E)^{0.5}$, as in the analysis above, because that depended on elastic energy storage, but R alone. In those circumstances, chewing would become like a kind of (always relatively inefficient) toughness test.

Experimental validation

Corroboration of the simple theoretical approach advocated above is currently being sought by experiments on human subjects. The fragmentation of particles that results from one chew is a direct measure of the breakage function. If this can be expressed in a form dimensionally consistent with the food property grouping $(R/E)^{0.5}$, then it can provide a test of the above theory.

Fracture changes the surface area of particles. However, this is not itself a measure of the breakage function because the surface area of a particle of any given shape depends on its volume. The surface area per unit volume of a particle is often called its *specific surface* in comminution studies. Taking

---

Caption for Fig. 4.9 (*cont.*) symmetrical about the direction of the force, then there would be considerable tooth–tooth wear. Blades are asymmetrical allowing their faces to brush past each other. The arrows on the 'pestle and mortar' and 'double-bladed' units indicate that some lateral movement will be required – in the former because the loose-fitting arrangement will reduce friction, the latter to make sure that blades contact as they pass.

two food particles, a large one labelled x and a smaller one called y, with the ratio of any of their linear dimensions being λ, then the ratio of their specific surfaces is

$$\frac{\lambda^2}{\lambda^3} = \lambda^{-1}, \qquad (4.10)$$

which means simply that the smaller a particle is, the greater is its surface area to volume ratio. This is trivial, but suppose now that both particles x and y are subjected to displacement-limited fracture. If the displacement needed to start a crack in x is u_x, while that to crack y is u_y, then linear elastic fracture scaling (Chapter 5) shows that the ratio of these displacements

$$\frac{u_y}{u_x} = \lambda^{-0.5}. \qquad (4.11)$$

There is, in fact, a long tradition in the mining and powder science industries of using an index with these units (i.e. the reciprocal of the square root of particle size) called Bond's work index (Bond, 1952, 1962). A considerable body of careful work supports it as an indication of the work required in industrial comminution processes (e.g. Rose & Sullivan, 1961).[3] Thus, the square root of the specific surface (or, alternatively, the reciprocal of the square root of particle size) will be used to define the relative success of fracture under these conditions. Is the fragmentation of particles in chewing, described in this way, related to $(R/E)^{0.5}$?

Agrawal *et al.* (1997, 1998, 2000) tested this on human subjects. These studies have been summarized by Lucas *et al.* (2002) and are only described briefly here. The methodology was basic. The toughness and Young's modulus of 38 relatively homogeneous foods (foods whose heterogeneity did not appear to affect fracture growth) were measured using mechanical tests described in Appendix A. The stress–strain curves of these foods were relatively linear. The measurement of toughness included energy expended both in elastic (reversible) fracture and small irreversible (plastic) deformation even though the theory above was constructed only including the former. Standard bite-sized pieces of these foods were then produced which five subjects each made one bite on, using the distal premolar–first molar teeth. This biting procedure was repeated once for each food. Saliva dissolves some of these foods very rapidly and so each particle was shielded from the oral environment during biting by placing it inside a sealed latex bag (as described by Mowlana & Heath, 1993). The actual bite was thus on the latex, which transferred the stress through to the food particle. It is possible that this covering disturbed the results, but the latex appeared to grip the food particles effectively. On opening the bags and tipping their contents onto a Petri dish, this loading had obviously damaged all foods, with most lying in fragments. These fragments were not rearranged other than to make sure that fragments did not cover each other. Images of these fragments were taken by a video camera held directly above the centre of the dish and fed into a computer that analysed the apparent surface area of these particles. The initial surface area was taken to be the area of the face of an unbroken particle; volume was assumed not to change after a chew.

The relationship of $(R/E)^{0.5}$ to change in the square of the specific surface in these foods is shown in Fig. 4.10. The line represents inverse proportionality (because $(R/E)^{0.5}$ is resisting crack growth) that appears to provide strong support for the theory. Foods with an index >25 mm$^{\frac{1}{2}}$

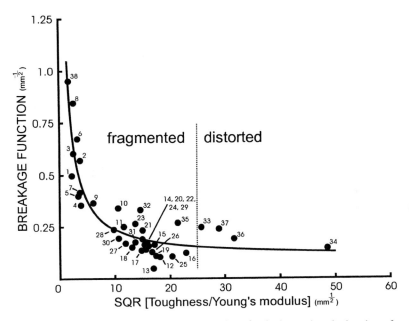

Fig. 4.10 Experiments on six human subjects with 38 foods shows that the breakage function, the estimate of fragmentation, and the food property index $(R/E)^{0.5}$ are inversely proportional to each other. This food index appears to describe resistance to fragmentation during human mastication, although above an index of 25 mm$^{\frac{1}{2}}$, particles were distorted rather fractured. Data points 1–9 are nuts; 10–27, cheeses; 28–32, fruits and vegetables; 33–36, breads; 37 – a type of soyabean curd; and 38, monocrystal sugar. (Redrawn from *Food Quality and Preference* vol. 13, Lucas, P. W., Prinz, J. F., Agrawal, K. R. & Bruce, I. M., pages 203–213, Copyright (2002), with permission from Elsevier.)

were plastically distorted and not broken into discrete fragments. This may indicate the limit of effectiveness of a cusp-based dentition like that of the human.

Further papers by Agrawal *et al.* (1998, 2000) on a subset of these foods relate $(R/E)^{0.5}$ positively to the electrical activity of the anterior part of the temporalis muscle, a portion of the jaw-closing musculature that lies close to the skin, and also to the degree of lateral movement by the mandible during the closing phase of chewing. Although there is, as yet, no published information on the limiting particle size, i.e. on the deformation transition controlled by $(ER)\sigma_y^2$, preliminary evidence obtained by K. R. Agrawal (pers. comm.) on food fracture behaviour during incisal bites in humans suggests that such a transition in some foods is often in the low millimetre range. This gives a value for the constant in equation 4.8 that suggests loading with a strong compressive element (Kendall, 1978b; Atkins & Mai, 1985).

SPECIFIC INFLUENCES OF FOODS ON DENTAL FORM

Tooth sharpness

The sharpness of teeth has been one of the overriding obsessions of studies of the mammalian dentition, being measured as the radius of curvature, ρ, of the tip of a structure (Lucas, 1982a). However, it is not the only relevant measure of the shape of a tip because this takes no account of how rapidly this tapers from the bulk of its supporting structure (Freeman, 1992; Sheikh-Ahmad & McKenzie, 1997; Evans & Sanson, 2003). A survey of the sharpness of mammalian teeth found that it tended to be affected by tooth size and that the bladed molar teeth of carnivores and insectivores were usually noticeably sharper than those of herbivores (Popovics & Fortelius, 1997). Experiments by Freeman & Weins (1997) on models of bat canines proved that sharpness was a strong influence on the forces needed to penetrate apple flesh. How can this be understood? The principal influences on tooth sharpness are probably the extensibility (i.e. the strain to fracture) of foods and their Poisson's ratios because both of these can influence the sharpness of crack tips.

Imagine a crack, oriented at right angles to the load, at the centre of a large rectangular strip of a homogeneous elastic material subjected to uniform tension (Fig. 4.11a). Modelling this crack as a very narrow ellipse, Inglis (1913) showed the effect of this crack was to elevate the primary tensile stress (the stress in the direction of the load, perpendicular to crack length)

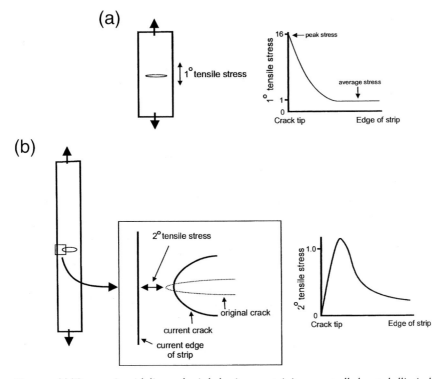

Fig. 4.11 (a) Tissue strip with linear elastic behaviour containing a centrally located elliptical crack (crack length being 7.7 times larger than crack width) being pulled uniformly. The graph on the right shows that the effect of a sharp crack is to elevate the primary tensile stress at the crack tip far above the average in the specimen (Inglis, 1913). (b) The elongation is now 100% and the Poisson's ratio is 0.3. The crack tip has rounded considerably due to a secondary tensile stress that develops in the direction of the crack. A higher Poisson's ratio would magnify the blunting.

at the crack tip far above the average prevailing in the tissue as a whole. This is shown in Fig. 4.11a, where if the crack length is $2a$ (the '2' implying that the crack is double-ended) and its width is w, then the stress at the crack tip is raised to $[1+2a/w]$ times the average stress in the tissue. The radius of curvature at the crack tip being ρ, then the stress multiple can also be expressed as $[1 + 2(a/\rho)^{0.5}]$. Thus, the blunter the crack tip, the lower the stress and the less chance there is of crack propagation.[4]

Now if this tissue is very extensible (i.e. with a large strain at fracture), the crack tip will blunt as the strain increases. The cause of the blunting is another tensile stress – I will call it the secondary tensile stress – that

Table 4.1 *Poisson's ratios reported for various plant and animal tissues*

Tissue	Highest Poisson's ratio (usually at low strain)	Source
Animal		
Cow teat skin	1.4–1.6	Lees *et al.* (1991)
Aquatic salamander skin	2.5	Frolich *et al.* (1994)
Wall of dog carotid artery	0.3–0.4	Dobrin & Doyle (1970)
Intersegmental membrane of female locust	~1.0	Vincent (1981)
Plant		
Cork	~0.0	Gibson *et al.* (1981)
Calophyllum inophyllum leaves (parallel to veins)	0.3	Lucas *et al.* (1991a)
C. inophyllum leaves (perpendicular to veins)	~0.0	Lucas *et al.* (1991a)
Watermelon	0.5	Miyazato *et al.* (1994)
Standard cellular solid	0.33	Gibson & Ashby (1999)
Human tooth tissues		
Enamel	0.3	Reich *et al.* (1967) cited in Waters (1980)
Dentine	0.23	Renson & Braden (1975)
Engineering materials		
Metals	0.3	Ashby & Jones (1996)
Unfilled rubber	0.5	Vincent (1981)

develops parallel to the long axis of the crack. Right against the crack tip, the secondary tensile stress in the material is zero (because there is empty space within the crack side and there is nothing there to be pulled on). However, a very short distance in front of the tip, this stress rises above the average primary stress, subsequently declining rapidly (Fig. 4.11b).[5] It is this secondary stress that blunts the crack.

The propensity to blunt cracks depends not just on extensibility, but also on Poisson's ratio. The higher Poisson's ratio is, the more a crack tip can blunt for any given stress. Normally, Poisson's ratios in engineering solids are around 0.33, but they can differ very markedly from this in biological tissues. In animal soft tissue, the ratio can be >1.0 (Table 4.1), the major cause being the arrangement of collagen fibres in the connective tissue. These are often loosely connected in the form of networks and may realign in relation to the stress (Purslow *et al.*, 1984). Network geometries can produce very high Poisson's ratios (Fig. 4.12a) so that when the longitudinal strain at fracture is high, there is considerable elastic blunting of natural cracks. Figure 4.12c shows the effect of this on an unnotched and notched tensile specimen.

All extensible materials can get some protection from fracture by this elastic blunting effect, which can elevate their toughness considerably. Its efficacy can be demonstrated experimentally by constraining the curvature of the crack tip so as to prevent this blunting. Lake & Yeoh (1978) show, for example, that the apparent toughness of rubber is reduced in proportion to the sharpness of a razor blade pressed against the crack tip. A free-running crack in rat skin has a toughness of 14–20 kJ m^{-2} (Purslow, 1983), while that cut with scissors (blade sharpness ~1.6 μm) is only ~0.59 kJ m^{-2} (Pereira *et al.*, 1997). This is about a 20-fold reduction, converting a tough material into nothing like one.[6]

Finally, Purslow (1991a) has presented a quantitative argument suggesting that a J-shaped stress–strain curve (Fig. 4.13), such as exhibited by most vertebrate soft tissues, adds to this crack-blunting effect. This point will be returned to in Chapter 7.

Generalizing all this to dental–dietary adaptation, the postcanine teeth of carnivorous mammals need to have sharp-bladed features to suppress the toughness of vertebrate soft tissues. To practical limits, the sharper the blades, the lower the cost, but whatever the sharpness, they need to travel all through the tissue to stay in contact with the crack tip (to prevent it blunting) and will end up contacting the opposing tooth in the other jaw. Tooth–tooth wear is, therefore, inevitable.

In Fig. 4.14, soft tissue is going to be subdivided between two flat blades. However, as the blades close, the food quickly extends beyond the ends of the blades, a function both of great extensibility and a high Poisson's ratio, producing a slit in the tissue but failing to subdivide it. One solution to this is to make these bladed teeth much longer than the material that they are going to comminute. However, the need for them to be 'outsize' can be reduced greatly by a concavity of the blades so that their ends contact well before their centres. The carnassials of carnivores are like this (Fig. 6.2). The same principle applies to mammals with an invertebrate diet. The cuticles of insect larvae are sometimes very extensible and Vincent's (1981) study confirms that very high Poisson's ratios are possible in some circumstances. The sides of tribosphenic molars can look very similar to those of carnassials with inflexed blades (Fig. 7.3). However, the toughness for insect cuticles appears to be much lower than the soft tissues of vertebrates (Appendix B). Many have commented on the inflexion of blades on the molars of

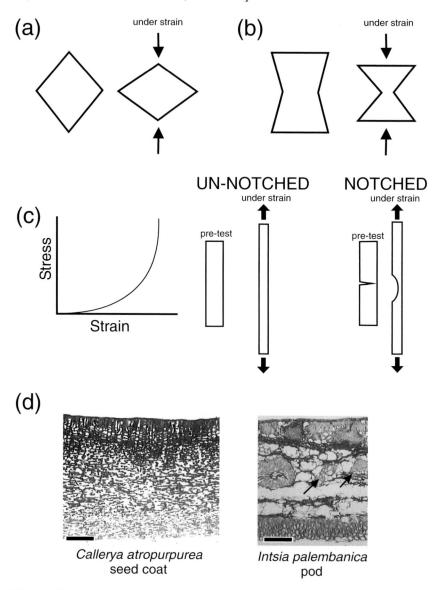

Fig. 4.12 Tissue geometry, Poisson's ratio and non-linear elasticity. Open cellular structures have a massive effect on all major mechanical properties (Cox, 1952; Gibson & Ashby, 1999), including Poisson's ratio. (a) An open 'diamond' framework that produces a high Poisson's ratio, the value depending on the strain (Frolich *et al.*, 1994); (b) an alternative arrangement whereby the walls can tuck into the structure under load so increasing density with zero Poisson's ratio; (c) simulates the effect of both non-linear elasticity and Poisson's ratio of

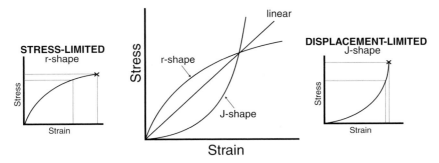

Fig. 4.13 The stress–strain curves of materials can be characterized as either r-shaped, J-shaped or linear. The cross indicates the fracture point. Simply due to curve shape, as higher loads, rubber maintains a relatively constant stress at a wide range of strains and the opposite for skin (Purslow, 1991a). It is suggested in Chapter 7 that r-shaped curves correspond to stress-limited defences, while J-shaped curves (much more common in biological tissues) represent displacement-limited defences.

carnivores (Fig. 6.2) (Savage, 1977) and insectivores (Fig. 7.3) (Crompton & Kielan-Jaworowska, 1978) and their role in trapping food, but the true diet-based explanation of this need seems to reside in the above.

Plant tissues do not have any ability to blunt cracks elastically in this manner. Their Poisson's ratios tend to be low because of their cellular construction (Wilsea *et al.*, 1975), with values averaging 0.3 (Table 4.1). Also, they take much smaller strains to fracture and most tissues have almost linear stress–strain curves. In consequence, the molars of herbivores do not have to be as sharp as those of carnivores and insectivores. There may be no general need to have blades at all on herbivore teeth and certainly no need

---

Caption for Fig. 4.12 (*cont.*) ~1.0 on a strip of vertebrate tissue under tensile loading. The strip extends greatly at low stress. This blunts any notch e.g. produced by an indenting tooth. The effect is seen in rubber (Lake & Yeoh, 1978) and fresh spring roll skin (Sim *et al.*, 1993), but is greater in vertebrate soft tissues because Poisson's ratios are higher. Sharp-bladed teeth frustrate this mechanism. (d) The soft seed coat of *Callerya atropurpurea* (Leguminosae) has a thin outer layer (at top) that is fully connected, but deeper in the coat, connections formed during development have been lost, so producing an open framework. Pressure on the outside of the coat with a blunt object like a cusp does not induce much tension in this framework. Poisson's ratio will be very low and a bladed tooth is required to open the coat by indentation. However, if the coat could be subjected to e.g. trouser-legging, as in ingestion using the forelimbs, then cracks would grow very cheaply – more cheaply than with a blade (Appendix B). (e) The pod of *Intsia palembanica* (Leguminosae) shows a similar pattern of defence to the seed coat: groups of fibres (arrowed) course internally but are only loosely connected to other structures. Scale bars, 100 μm.

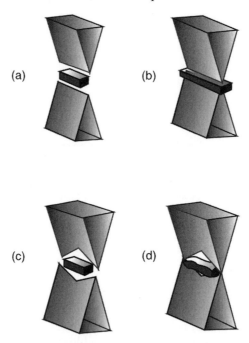

Fig. 4.14 (a) A piece of animal soft tissue lies between two flat blades. Loaded as in (b), it deforms beyond the ends of the blades due to high Poisson's ratio and high strain. In (c), an inflection or concave curvature allows the ends of the blade to contact prior to much strain on most of the particle, so trapping it while the strain is low as in (d).

for inflexions or concavities. Experiments on plant tissues with scissors with blades of varying sharpness suggest that this has only a relatively minor effect on toughness values, e.g. an 800% reduction in blade sharpness appears to raise apparent toughness values in across-grain tests on woods by only 37% (Darvell *et al.*, 1996; Lucas *et al.*, 1997). Occasionally, a free-running crack in plant tissues can cost less than one driven by a blade, implying that natural cracks can be sharper than those induced by scissors blades. This is particularly true for cracks that propagate between cells (see next section).

Very low Poisson's ratios in plant foods can also influence tooth form. In a food that behaves like cork (Gibson *et al.*, 1981), with a zero ratio, it will be impossible to generate tension indirectly via the compressive effect of an indenting cusp or blade: cracks will not 'open up' and there is likely to be considerable friction against the tooth as the plant material densifies under load. The optimal solution is again a sharp bladed tooth. However, the included angle of the blade might be much smaller (limited by the possibility of its own fracture) because most of the energy loss is likely to be sub-surface and not affected much by lubrication: in cork, the coefficient of friction increases with load for just this reason (Gibson *et al.*, 1981).

Structural connectivity

It might be thought logical that a food particle would evade fracture best by having a tightly connected structure that provides maximum resistance to indentation by teeth. In fact, slight disconnection can elevate toughness provided that the scale of this detachment is kept within critical limits. The limits are due to disconnected regions acting like flaws that must be kept small to avoid great reduction in the stress at failure. The strategy works well in artificial composites (Atkins, 1974). Gordon (1980) discusses it with respect to 'tensile' structures, showing how the separation of load-bearing elements by a very low-modulus matrix can prevent crack propagation and produce notch insensitivity (Fig. A.9). In the limit, fracture mechanics is defeated like this and Galileo rules (see Chapter 5).

Structural disconnection is common in vertebrate soft tissues (Purslow, 1983, 1991a), but not in plant tissues.[7] Some leaves show notch insensitivity if they are gripped in their entirety (veins and all) and then pulled (Lucas *et al.*, 1991a; Vincent, 1992), but otherwise, continuum mechanics seems to predict the fracture behaviour of fully connected plant tissues very effectively (Gibson & Ashby, 1999). There are exceptions though (Fig. 4.12d). The soft seed coat of *Callerya atropurpurea* (Leguminosae) has a fully connected, but relatively thin, outer layer. However, most of the inner tissue has lost connections that were formed during its development. Accordingly, pressure on the outside of the coat with a blunt object does not produce much tension in this inner tissue – it is too disconnected for that – and a blade is required to open it. The pod of *Intsia palembanica* (Leguminosae) shows a similar pattern of defence, with internal fibres only loosely connected to other structures.

Patterns of fracture in relation to structure in plant tissues

Without the need for a musculoskeletal system, plants protect every cell by developing a semi-crystalline wall inside the cell membrane. This cell wall is a mechanical composite, consisting of cellulose, hemicelluloses and, sometimes, lignin. Individual cells are glued together loosely with pectins or firmly by lignin. The major component taking the load is the stiffest element and that is cellulose, consisting of nanometre-sized crystalline fibrils combined into microfibrillar bundles. Hemicelluloses and lignin act as a binding matrix to glue these together. While hemicelluloses are hydrophilic, lignin is a heavily cross-linked hydrophobic polymer that waterproofs tissues and connects up all the wall elements. This interferes with transport processes across the wall: if lignification is extensive, the cell dies. Lignification starts in the middle lamella and proceeds to the primary, then the secondary, walls.

The purpose of the cell wall is to provide each cell with the potential to support its shape. However, many cells have only a thin primary wall. Cell stiffness in this case is achieved, not by the wall alone, but by maintaining cell contents under high pressure (between 1 and 5 MPa; Tomos & Leigh, 1999). The critical factor that prevents bursting is the toughness of the wall. To this end, the cellulose microfibrils in it are laid down in lamellae, with the orientation of one layer unrelated to that in the next (McCann & Roberts, 1991; Carpita & Gibeaut, 1993). The primary cell wall is usually thin, but can be locally thickened, as in collenchyma, a tissue that supports growing tissues. More generally though, cells with thick walls also contain a secondary cell wall. This wall usually consists of three distinct parts, S1, S2 and S3, laid down in that order. The bulk of the secondary cell wall ($\sim$50% or more of its thickness) is usually the S2 layer (Fengel & Wegener, 1989). The cellulose microfibrils throughout this layer are parallel to each other and, in elongated cells specialized for mechanical support such as sclerenchyma fibres, they tend to wind at a small angle (say, 10°–30°) to the cellular axis (Preston, 1974). When there is secondary cell wall present, lignin is nearly always added to the hemicellulose glue.

Extensively lignified cells are just dead mechanical supports.[8] The mechanical advantage of forming such complex structures does not lie in their exemplary stiffness (i.e. Young's modulus) because this is bound by what is called a 'rule of mixtures': any composite cannot be stiffer than the stiffest component or less stiff than the least stiff component. There is also no overall gain in strength (however that is defined) from composite structure either because it is bounded in a similar way. The single major property to escape these limitations is toughness, which is set by structure much more than by material content. The toughness of an artificial composite, for example, is typically 1000–10 000 times tougher than any of its separated components (Atkins, 1974; Harris, 1980).

In general, the major nutrients of plants are inside the cells, so a mammal must push a crack through the cell wall to release these. The ease of doing so depends greatly on wall toughness and the amount of cell wall in the tissue. Nutritionists know the cell wall content of tissues as 'fibre'

(Van Soest, 1994, 1996) and measure it as a proportion of their dry weight. The term 'fibre content' will be avoided here because of a potential confusion with a plant cell type that is called a 'fibre'. Instead, I will define the amount of cell wall in a plant tissue, V_c, as a proportion of its fresh volume.[9]

Lucas *et al.* (1995, 1997, 2000) reported toughness data on 82 plant tissue and plant-derived materials (Appendix B), varying from watermelon flesh ($V_c = 0.0031$; its thin-walled cells being >0.5 mm in diameter) to seed shells, the walls of which are so thick that they can occupy 95% of the tissue (i.e. $V_c = 0.95$). The study used scissors cutting tests (Appendix A) because this technique can be used on a wide variety of tissues, regardless of their shape and size, and can help to tease two major causes of toughening in plant tissues apart.

When a crack runs through an object, it disrupts a small volume of material around the crack tip, permanently disturbing it. This plastic zone is one of several zones of disruption that can be defined, depending on the material (Lawn, 1993). Here I will distinguish small-scale fracture events within the plant cell wall (which is what cracks), expressed whatever the thickness of the tissue, from disruption in the surrounding cellular framework. The latter must be progressively diminished whenever the thickness of a specimen is reduced below the normal size of this plastic zone. Thus, when a specimen is thick enough, toughness is independent of thickness (inset in Fig. 4.15a), but as it is progressively thinned, toughness will drop in proportion to thickness because the volume of tissue being affected also diminishes. There will be a threshold thickness, which is usually around 0.5–1.0 mm in plant tissue, where a plot of toughness versus specimen thickness seems to plateau. The exact threshold is difficult to define with precision, but plots like the inset in Fig. 4.15a can yield two values fairly accurately: an 'intercept' toughness, derived by extrapolating a regression line to zero thickness, and the slope of this line. The former is sometimes called the 'essential' work of fracture (Atkins & Mai, 1985) and probably represents elastic fracture mechanisms that act within the cell wall. The slope of the regression line on the other hand provides a total estimate of the extent of the plastic behaviour that plant tissues can display.

The intercept toughness for all 82 tissues is plotted in Fig. 4.15a, where it is seen to be proportional to V_c. By extrapolating from the densest seed shells, which have $V_c = 0.95$, to $V_c = 1.0$, an indirect estimate of the toughness of the cell wall itself can be obtained. This is 3.45 kJ m^{-2}, a value that can be interpreted as the intrinsic toughness of the cell wall.[10] This figure is a surprise because it is exceptionally low for a composite.

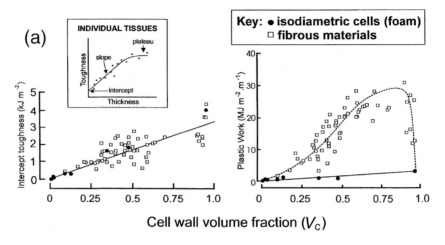

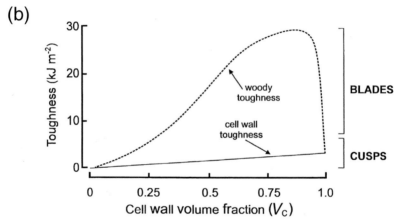

Fig. 4.15 Two separable components of toughness in plant tissues. (a) The toughness of plant tissues in relation to the amount of cell wall they contain (defined as V_c, the fraction of fresh tissue volume that the cell wall occupies) and also tissue geometry. The inset shows schematically how cutting tests are conducted on individual tissues, so potentially defining an intercept, slope and plateau. The intercept toughness describes elastic fracture in cell wall, while the slope describes the ability of cells to deform (buckle) plastically. The main graphs show results for a wide range of plant materials. At left, intercept toughness is proportional to V_c, irrespective of tissue geometry. At right, plastic buckling in woody fibrous tissues absorbs considerable amounts of energy. Tissues with isodiametric 'foamy' cells cannot do this. (b) A attempt to pull these two effects, cell wall and cellularity, together by assuming that plasticity is confined to a spherical zone, of diameter 1 mm, around the crack tip. The overall toughness of woody tissues, calculated as the sum of 'woody toughness' and 'cell wall toughness', is shown to be up to 10 times that of foamy tissues of the same V_c, but peaks at $V_c = 0.8$–0.9, declining thereafter.

Copying natural principles, artificial composites can be made that are 100 times as tough as this (Gordon & Jeronimidis, 1980). Even wood tested by standard methods (Jeronimidis, 1980) is 10 times tougher than this estimate for the cell wall. How can this be?

The answer, according to Jeronimidis (1980), is that a truly cellular mechanism provides most of the toughness of woody tissues. It is known that, at critical stresses, structures like the helically wound S2 wall collapse irregularly and irreversibly into the cellular lumen (Page *et al.*, 1971), so absorbing large amounts of energy (Gordon & Jeronimidis, 1980; Jeronimidis, 1980). It appears that this plastic buckling mechanism, operating probably just in front of the crack tip, is the factor that elevates the toughness of woody tissues well above cell wall toughness levels.

The effectiveness of the events in the cellular framework in setting toughness levels is clear in Fig. 4.15b where the toughness–thickness slopes for 82 tissues are plotted against V_c. The plastic toughening mechanisms appear to behave very non-linearly. Tissues with very low V_c have no secondary cell wall and, therefore, no potential to buckle plastically at all. From about $V_c = 0.2$–0.8, the mechanism absorbs energy roughly linearly with V_c. At the peak toughness, at $V_c \approx 0.8$, buckling provides about 10 times the toughness that cell wall alone can. Above the peak, toughness starts to plateau and falls precipitously at $V_c > 0.9$ or so. This is almost certainly because the cell wall is then so thick that the buckling mechanism is frustrated by there being insufficient space (almost no cell lumen) for the wall to buckle into. Thus, woody tissue such as many seed shells with $V_c = 0.90$–0.95 do not really behave mechanically like woods. Despite these shells having an almost identical chemical structure to woods (Preston & Sayer, 1992), the 'woody' toughening mechanism hardly operates at all. They are exceptionally stiff and with a high yield strength, but they are not tough at all. In fact, the graph in Fig. 4.15b, which attempts to put all the toughening mechanisms back together, suggests the toughness of the densest seed shells reverts to levels due to the cell wall.

All this has important repercussions for tooth shape. Foods with very low V_c, like fruits, will not have much crack resistance, so teeth designed to break them can be cusped. Woody tissues with $V_c = 0.2$–0.8 are progressively more resistant, so blades will be needed (Fig. 4.15b). When $V_c > 0.8$, crack resistance is lost again because woody toughness does not operate, and so cusps may suffice once more. This could explain very well why the dentitions of both frugivores and seed-eaters very often have rounded cusps, while those of leaf-eaters have sharp blades (Kay, 1975b, 1978; Kay & Hylander, 1978). If so, this mechanical analysis shows plainly

why there can be no recourse to the chemical 'fibre content' of plant foods
to explain the dentitions of herbivores: this will not describe the difficulty
of fracture. Lignin on its own is often touted as a tough material by nutritionists, but it cannot be so. Toughness is a consequence of the composite
cell wall and tissue structure, not of individual chemical components (Lucas
et al., 2000).[11]

The best evidence for the predominance of plant tissue structure over
cell wall content in setting toughness is from the seed shell of *Mezzettia
parviflora* (Lucas *et al.*, 1991b, 1995). The shell has three distinct zones, an
outer one composed entirely of parallel fibres (zone I in Fig. 4.16), an inner

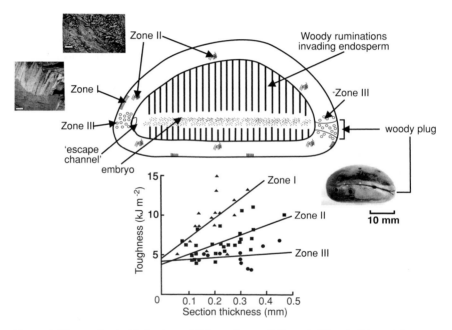

Fig. 4.16 The mechanical behaviour of *Mezzettia parvifolia* seeds. The shell is mainly composed of bundles of fibres that course randomly (zone II, scale bar, 100 μm), but those
placed most superficially are normal to the surface (zone I, scale bar, 100 μm). There is
also a narrow band of stubby cells (zone III – appearance shown in Fig. 4.17c) that runs all
round the shell and connects to a woody plug. This band connects to a woody plug that
facilitates germination while being so narrow that a mammalian predator cannot utilize it to
enter the seed. Each zone has a similar intercept toughness and V_c, but very different capacity for plastic buckling. Invertebrates operate at a smaller scale than the cellular aspect of
toughness – the circular holes seen were produced by scolytid beetles that enter randomly
anywhere around the seed. (Redrawn from Lucas *et al.* (1995) The toughness of plant cell
walls. *Philosophical Transactions of the Royal Society London, B* **348**: 363–372.)

one in which bundles of fibres appear to course randomly (zone II) and lastly, a band of short stubby cells (called brachysclereids) encircling the seed (zone III). All these zones have roughly the same cell wall volume fraction ($V_c \approx 0.94–0.95$). Figure 4.16 shows tests on each zone with specimens of different thickness. The intercept toughness for the three zones is about the same, being very close to that of cell wall itself as would be anticipated. However, the influence of cellular organization is very clearly shown in the toughness–thickness slopes. Fracture of the fibres in zone I at right angles to their long axis gives the highest slope; zone III the lowest slope (not significantly different from zero, in fact) with zone II intermediate. The cause of these differences is structural: long fibres in zone I can buckle whereas short stubby cells in zone III cannot. The randomly directed fibres in zone II are often not oriented such that they can buckle and so lie between these extremes. Nothing about cell wall chemistry explains this – it is purely a function of cellular organization, thus demonstrating how important tissue structure is.

The relationship of plant tissue structure to defence against predation in plants in discussed in Chapter 7, but it is perhaps pertinent here to note that the anisotropy of woody tissues does not allow plants any easy compromise between protection from attack from all directions (as often implied by seed shell structure) and from one direction only (as in attack on wood in trees). The former is the probable reason for fibre arrangements such as zone II in seed coverings. Yet random fibre directions are not the midway compromise that they might appear. Their Young's moduli provide one reason for this.

The Young's moduli of woods, loaded 'along the grain', are proportional to V_c. However, when the load is applied across the grain, it is proportional instead to V_c^2 (Gibson & Ashby, 1999). Given that $V_c < 1.0$, this means that a mammal that attacks wood by indenting or bending it will tend to encounter a lot less stiff a structure than if it just pulls on it. The toughness though is the same. Williamson & Lucas (1995) add the netting analysis of Cox (1952) to this so as to explain why the Young's moduli of the random fibre networks of seed shells are only 2–6 GPa, nothing like the 30 GPa if the cells could all be loaded along their long axis.

The above discussion of plant tissues all assumes that cracks travel through cells. Yet a free-running crack will select that path in a material in which energy is released most rapidly (Lawn, 1993). If Young's modulus is constant, then this path can be predicted by comparing the relative toughness of intracellular and intercellular paths. However, if the modulus varies too, as it does in woody tissues because of their considerable anisotropy, then the strain energy release rate in different directions becomes extremely complicated to calculate (Atkins & Mai, 1985).

Experimental observations on woods show that cracks pass directly across cells without deviating when $V_c < 0.2$ (Ashby *et al.*, 1985). However, when $V_c > 0.2$, cracks zigzag, alternating between being 'stopped' by an interface and finding a sufficiently cheap path across cells (Ashby *et al.*, 1985). The toughness of the middle lamella, the intercellular 'glue', is not known in thin-walled tissue like parenchyma because no one has succeeded in driving cracks between cells, but in woods and seed shells, the cost of a direct path is low, $\sim$0.1–0.2 kJ m^{-2}, and independent of V_c (Jeronimidis, 1980; Ashby *et al.*, 1985; Lucas *et al.*, 1991b).

If an intercellular path provides an avenue for particle fragmentation, as it does in many seed coats, then it makes sense for a seed predator to try to direct cracks that way. The point about tooth design here is that such a path will be precluded by sharp bladed teeth because these will drive cracks straight across cells at far greater cost. So blunt cusped teeth do the job much better if this path leads to fragmentation, as shown by the experiments shown in Fig. 4.17. However, in any tissue where fibres are aligned in one direction, blades are required to cross them. These blades need to be sharp, not because that is critical for suppressing any particular toughening mechanism, but for retaining the desired crack path, which will deviate from it very easily (as part of the plant's defences).

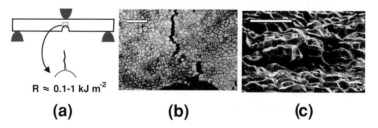

(a)   **(b)**   **(c)**

Fig. 4.17 Intercellular fractures in plant tissues from blunt starter notches. (a) Shows a three-point bending set-up with a blunt notch. In (b), a sharp crack propagates from this notch, running between cells. *Schinziophyton rautanenii* seed shell (courtesy of C. R. Peters). Scale bar, 100 μm. (c) Shows a typical 'builder's rubble' surface of between-cell fracture in zone III of *Mezzettia parviflora* seed shell (see Fig. 4.16 for explanation). Scale bar, 100 μm. The cost of these cracks (i.e. their toughness) depends on how circuitous the crack is, but 0.1–1.0 kJ m^{-2} compares to 20 kJ m^{-2} if sharp bladed teeth push cracks through these cells. For woods, the difference in cost can be >100. The endosperm of *Mezzettia* seeds is traversed by woody struts called 'ruminations'. These are reminiscent of the plicidentine found in some non-mammalian vertebrate teeth (Shellis, 1981).

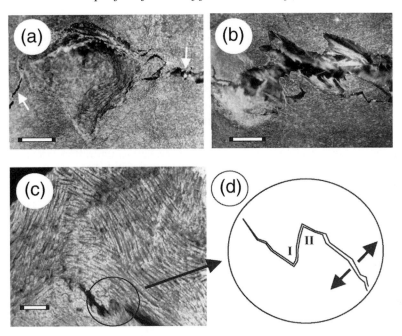

Fig. 4.18 Interlocking fibres in a seed shell. (a) An indentation in a cusp may propagate an intercellular crack in a seed shell with low $(R/E)^{0.5}$. Scale bar, 1 mm. In (b), the crack has deflected (producing a strong mode II component from an original mode I crack) and then arrested because of interlocking fibres. Scale bar, 1 mm. The probable crack arrest mechanism is shown in (c) (scale bar, 100 μm) and (d) in which fibre bundles interlock. The suggested evolutionary response would be the development of cusps with pronounced marginal ridges that first penetrate the seed and then wedge it open against this resistance. These ridges are seen in the cusp design of (a), where they send off cracks (arrowed) that will grow further as the cusp penetrates the shell. In so doing, the cusp and ridges act rather like a convex wedge.

The overall cost of intercellular fracture is affected by crack deflections. A circuitous intercellular crack path (Fig. 4.18) may cost 10 times that of a straight fracture between cells. At maximum, the cost is ~1 kJ m^{-2}, which is still a lot cheaper than an intracellular path (compare this to *Mezzettia* seed shell zone I toughness values for thick sections in Fig. 4.16). More important than this though is what the interlocking nature of woody fibres can do to obstruct crack growth in what is otherwise a very low-toughness tissue. Figure 4.18a shows an indentation by a cusp that has ruptured a seed shell, but crack growth has been obstructed sufficiently for the shell not to break cleanly (Fig. 4.18b). Figure 4.18c shows how this is mostly because interlocking fibres 'catch' each other across the crack, so impeding its growth. The logical response to this in a seed predator is to convert a circular cusp form in one that has ridges (as shown in Fig. 4.18a), which act to wedge open the crack after the initial fracture. Such an explanation makes considerable sense of the presence of ridges on the sides of cusps in many mammals. It is possible to go further and suggest that such a cuspal design is actually like a wedge, where the wedge is somewhat convex in outline (rather than straight), the opposite of a concave (inflexed) blade.

Living non-woody plant tissue has a very low toughness as simple calculations show. Take parenchyma, the tissue containing most of the food value of plants. This thin-walled and homogeneous tissue always cracks through cells. Often $V_c \approx 0.05$, giving a tissue toughness of $(0.05 \times 3450) \approx 172.5$ J m^{-2}, conforming to results from experiment (Vincent, 1990). The lower toughness limit for parenchyma is probably about that of ripe watermelon fruit flesh, being 10–30 J m^{-2} (Appendix B). The Young's modulus of parenchyma depends on its turgidity. Turgidity involves the active transport of water into a plant cell to pressurize it and maintains the shape of a cell as though it was a water-filled balloon with the help of only a very thin primary cell wall. However, most turgid parenchyma has a Young's modulus of ∼1–10 MPa (Mohsenin, 1977) and, as shown in Fig. 4.10, tissue like this (e.g. many fruits and plant storage organs) can be broken down rapidly by cusps.[12]

Fracture can be associated with extensive bursting of cells and fluid release (Peleg _et al._, 1976). The energies involved in fracture can be so low that teeth may not even be needed because the flat surfaces of soft tissues, like tongue and palate, can fracture cells. Fruit bats can do this by compressing fruit flesh between the tongue and hard palate and two ape species, the chimpanzee and the orang-utan, sometimes compress similar tissue between the labial surface of the incisors and lips (Walker, 1979). If the soft tissues of the mouth can fracture these foods, then it is also probable that the soft tissues of the gut can do so too, so much fragmentation of such fruit in the mouth is immaterial.[13] However, many types of fruit flesh do not release their contents so easily. There can be a variety of reasons for this. If fruit is unripe, it may crack only in one defined plane (Peleg & Gómez-Brito, 1977). At ripeness, the middle lamellae may break down, allowing cells to slide past each other. The mode of failure then changes from intracellular fracture to plastic shear (e.g. Heyes & Sealey, 1996).

If the major role of the posterior teeth is not to reduce particle size, but burst as many cells as possible, then the optimal arrangement is to have blunt tooth surfaces because they contact more cells than sharp ones.

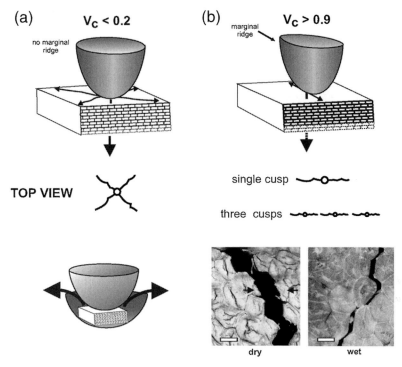

Fig. 4.19 The left column (a) shows rounded cusps acting on thick turgid parenchyma, while at right (column b), cusps are acting on thick shell. (a) Many cells fracture under a symmetrical blunt cusp. Fracture lines need have no direction because the object is not tissue fragmentation. A cusp fitting into a loosely fitting fossa is the design that expresses most juice. (b) Cracks in seed shells are affected by their moisture content. In 'wet' shell, cracks pass neatly between cells. In drier tissue, however, they pass into the cell wall, even though the major part of the path remains intercellular. Scale bars, 20 μm.

However, if cells are not tightly bound, as in some ripe fruit flesh, and slide across each other easily, then the tissue needs to be enclosed or it will simply flow out of the way. Reciprocal curvatures of tooth surfaces, such as a loosely fitting cusp-in-fossa or pestle-and-mortar arrangement, provide the necessary compression but with channels for fluid escape that would otherwise impede tooth movement. The cusp would have no need for any marginal ridges. Figure 4.19 contrasts features of tooth cusps designed for low V_c and high V_c plant tissues. For the latter, moisture content is a minor variable, although it does affect the 'cleanness' of intercellular fractures in woody seed shells: cracks run into the cell wall in dry tissue but pass cleanly between cells when the tissue is saturated (Williamson & Lucas, 1995).

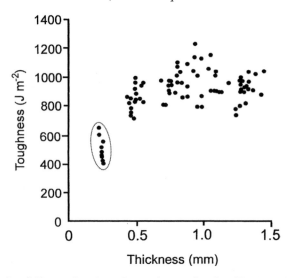

Fig. 4.20 Results of Choong (1997) on the toughness of stacks of laminae of mature leaves of *Castanopsis fissa* with a scissors test. Individual laminae (points surrounded by an ellipse) have variable toughness, but this is always much lower than the toughness of two to four leaf layers. Above ~0.8 mm, toughness appears to plateau. The midrib and secondary veins were excluded from these tests.

Thin sheets and rods

When particles become very thin, these tissues will no longer store enough energy to make cracks propagate. The transition thickness for plant tissue appears to be between 0.5 and 1.0 mm (Fig. 4.20). This is important. The lamina thickness of most mature leaves ranges between 0.15 and 0.45 mm, almost certainly below the critical deformation transition thickness, so cracks do not spread.[14] Just for this reason, leaves have to be fragmented with blades rather than cusps. Structural heterogeneity adds another: the mesophyll of mature leaves is where the nutrition is, but a thicker-walled epidermis surrounds it and the leaf usually contains thick woody veins that prevent effective pressure on thinner-walled tissues (Choong, 1996).

INGESTION

There is much less to say about the shape of teeth designed for ingestion than mastication because these shapes are fairly simple, linking back to reptiles because of their usual single cusp form. Most mammals have small peg-like incisor teeth that simply grip particles in order to ingest them.

Their working surfaces can sometimes be broad and rough, as in ungulates and some primates, to grip leaves. Among living mammals though, broad 'spatulate' incisor teeth (Fig. 2.3) in both upper and lower jaws are unique to the Anthropoidea, a taxonomic group containing the apes and the Old World and New World monkeys.

Canines are extra large pegs, very prominent in carnivores, for controlling prey items, even for killing them, prior to ingestion. Whether for killing or controlling, carnivores need to avoid fragmenting food particles with their canines because they would lose their prey in so doing. Blades are needed to fragment animal soft tissues, so the incisors and canines of carnivores are not like this otherwise their prey would escape (albeit with elongated wounds).[15]

Canines are generally conical, but with a recurved (actually spiral) form. The upper and lower canines of carnivores are generally long and projecting teeth, acting at larger gapes than the other teeth (Chapter 5; Fig. 6.2). The degree of projection and the relative sizes of upper and lower canines vary greatly, but the uppers are always the longest. The essential load on the tooth is not just due to indentation of their tips, but to an antero-posterior force (Simpson, 1941; Smith & Savage, 1959). The cross-sectional shape of the canines tends to reflect this, typically having oval cross-sections with the long dimension being antero-posterior. Carnivores that break bones must deal with the vagaries of the direction of the bite force depending on the orientation of the bone vis-à-vis the tooth. These tend to have circular cross-sections (Van Valkenburgh & Ruff, 1987).

A classification of ingestion

The major problem with investigating ingestion is the variety of ways in which it can be achieved. The only classification that I know is that of Osborn *et al.* (1987) and Ungar (1992), which has been employed successfully in field studies of primates (Ungar, 1994; Yamashita, 2003). Figure 4.21 is a modification of the Osborn–Ungar scheme showing various ingestion methods, the actual details of which vary with food type.

There are two major possibilities:
(1) A food particle is *fractured* between the anterior teeth without any other applied force.
(2) The particle is *gripped* by the anterior teeth while the animal uses tension (and a variable amount of torsion perhaps) against an external restraint to fracture the particle. This external restraint could include pulling a

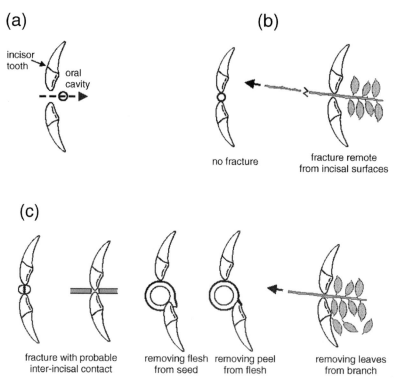

Fig. 4.21 Categorization of incisal use by primates. (a) Ingestion without incision. (b) Using the incisors for grip with any fracture being remote to the incisal surfaces. (c) Fracture close to or at the incisal surfaces. From left to right: fracture of a small seed, fracture of a stalk or shoot, peeling a fruit, removing fruit flesh from a seed, peeling a fruit and taking leaves off a branch as the latter is drawn through the teeth. This is not a comprehensive list of ingestion methods; primates, for example, often bite through leaves detaching their distal parts while leaving the leaf bases on the plant (Dominy, 2001). (Adapted from Osborn *et al.* (1987) and Ungar (1992) with permission.)

fruit, leaf or branch off an attached plant or pulling against an item held by a forelimb.

As stated early in this chapter, the first possibility is displacement limited and would presumably not be employed unless the particle would fracture under such a loading. Otherwise, the second possibility would be chosen. Thus, it is probable overall that ingestion is subject only to stress limitation, not displacement. Agrawal & Lucas (2003) have offered a critical test of the stress-limited hypothesis, involving fracture of food particles between the incisors in humans. They compared *in vivo* conditions as a crack propagates from an indentation produced by the anterior teeth (the stress at fracture multiplied by the square root of the depth of incisal penetration) with the *in vitro* measurements of $(ER)^{0.5}$. There was a strong linear relationship, supporting a recent view expressed in food science that this is probably close to what humans sense as food resistance during ingestion (Vincent *et al.*, 2002).[16]

Factors affecting ingestion

Successful ingestion depends critically on grip and therefore on high friction. This is generally opposite to the postcanines where friction adds to the load burden. To this end, the front teeth are less likely to be lubricated as well by saliva than posterior teeth and they may stick out of the mouth to avoid being wetted altogether. Grip on plant leaves (by anterior and posterior teeth) tends to be complicated by features of the cuticle (Chapter 3). Enamel decussation may improve grip because of the raised ridges that it generates through wear (Fortelius, 1985; Boyde & Fortelius, 1986; Rensberger, 2000). The exposure of dentine, as in ungulates, is important because dentine alone has a much rougher surface than enamel, the coefficient of friction being three times higher against a standard (Table 6.2). A mixed dentine–enamel surface provides much greater grip than enamel alone. Some herbivores, both living and fossil, have replaced their anterior teeth by surface-dry keratinized lips. J. F. Prinz (pers. comm.) has found that the coefficient of friction between opposing dry mucosal surfaces may be as high as 1.5. A saliva coating brings this down to very low values, however, except perhaps when food tannins interact with it. Some ungulates have lower teeth acting against an upper lip, while others have lost their front teeth altogether, probably to enhance grip.

The enamel of some canine teeth, e.g. extinct sabre-toothed carnivores, can have rough serrations that probably help grip (Abler, 1992), particularly on skin. Mammalian hair is itself very tough, as evidence from horse hoof (Bertram & Gosline, 1986) and human fingernails (Pereira *et al.*, 1997) shows, and it is possible that tooth serrations are related to consumption of very hairy prey.[17]

The ease of ingestion of foods with high or low Poisson's ratios might be very different. Vertebrate soft tissues, for example, would exhibit maximum toughness under pulling or twisting because they would blunt cracks (Fig. 4.12), while something like a cork would fracture much more easily than under an indenting tooth. Animals that remove bark, like some rodents and primates (Vinyard *et al.*, 2003), have to indent it with their anterior teeth. The properties of barks should be important for understanding their anterior tooth form.

CONCLUDING REMARKS

This chapter has only touched on a few of the possibilities that relate the mechanical behaviour of foods to tooth shape, but the only major feature of food to have been excluded is that of tissue age. Toughness depends on structure, but structure takes time to develop. So young tissues, whether plant or animal, always make for foods that are much easier to process in the mouth than older ones. Dental features predicted for the consumption of some foods, such as leaves, may not apply if those leaves are very young.

5

Tooth size

The molar teeth of early mammals were less than a millimetre in length while those of a living African elephant are several hundred times larger. What is the reason for this? Well, it is probably true to say that, despite being a prime focus of investigation for a century or more, the adaptive significance of such tooth size differences is not understood. The initial approach here is to consider constraints not just on tooth size, but on the size of all mouthparts, that are imposed on the digestive system by the energy requirements of the mammalian body. The argument leads quickly to what engineers call 'scaling arguments'. Biologists call them this too, but they also use the term 'allometric analyses'. Both terms describe changes in the shape of animals required by change in their size. The predictions made here appear to be novel because they scale the mouth directly to the food particles that it ingests and not to whole body size (which is the norm in such studies). However, the arguments are couched in standard allometric form because this is the nature of the available data. Postcanines are treated separately from incisors and canines and both general arguments and specific roles are covered. The overriding philosophy is that physical properties of mammalian diets explain not only tooth size, but also the size of most orofacial structures.

INTRODUCTION

Tooth size was postulated in Chapter 2 to be an evolved response to variation in the external physical (or surface) attributes of foods. Paradoxically, only the last sections of this chapter follow this logic, the early part examining instead whether tooth size adjusts to changes in the mechanical properties of foods. This apparent illogicality is explained by the difficulty in separating size from shape in solid objects. Although the distinction may seem very clear, with the former being dimensional measurements and the latter angles and ratios, the distinction is actually blurred. For example, a critical parameter of tooth shape is sharpness even though this is just a radius of curvature, a linear measurement (Chapter 4). As soon as the size of a tooth is related to the size of anything else, like a neighbour or the mouth and body that contain it, we can begin to think of this as a description of shape, not just size. Rather than get stuck on this, I will let arguments lead the way.

THEORETICAL RELATIONS BETWEEN BODY MASS, FOOD INTAKE AND METABOLISM

The limits to the general size of both the oral cavity and its critical contents, the tongue and teeth, stem from the energetic requirements of the mammalian body and the arrangement of its digestive system so as to supply these. Previous arguments (given in Chapter 3) have impinged on this issue briefly, but they now need to be discussed in detail. I have remarked that the mammalian digestive system is composed of a sequence of steps and that the rate of the slowest step will determine the rate at which the whole sequence runs. It follows from a *reductio ad absurdum* that the rate at which mastication and swallowing must run is that which will satisfy the body's metabolic needs. This must constrain the size of all the anatomical components taking part in these activities, including the oral cavity and the teeth. Firstly though, how can the energy requirements of mammals be calculated?

Large-scale studies of the metabolism of mammals were initiated during the era following the First World War (Kleiber, 1932) and have continued since. These studies have established that the basal metabolic rate of animals, i.e. the rate of energy consumption at rest, is generally related to the body mass (M) of a mammal raised to the power 0.75, i.e. to $M^{0.75}$ (Kleiber, 1961). This relationship is not easy to understand and criticisms of it as a purported fact have been frequent. For example, heat loss is the principal problem that mammals have to contend with because they maintain a body temperature that is usually far above their surroundings. Heat loss is via surfaces. If we assume an invariant density to the mammalian body, such that its mass is proportional to its volume, and also that the body shape of all mammals is similar, then the surface areas of different sized bodies will be proportional to $M^{0.67}$. This prediction follows from simple geometry. If we measure any dimensions of a large and a small geometrically similar mammal and compare them, these will obviously be in a fixed ratio. Suppose that the larger mammal has linear dimensions λ times that of the small mammal. The surface areas of the large mammal will be λ^2 bigger than that of the smaller mammal and its overall volume, λ^3 bigger. Accordingly, surface areas rise (λ^2/λ^3) as fast as volumes. Given that their masses are proportional to their volumes, this is equivalent to writing that surface areas scale as $M^{2/3} = M^{0.67}$.

The exponent for metabolic rate appears to exceed this exponent, thus appearing to destroy a trivial explanation. However, this did not satisfy Benedict (1938), who felt that the data were sufficiently inaccurate that a 0.67 exponent was just as reasonable as 0.75. Unfortunately, this particular challenge has been left behind as the data have been extended and line-fitting techniques improved (LaBarbera, 1989). The higher exponent has now been supported very strongly and a dependence of energy requirements on geometry disproved.[1]

Most recently, some proponents of a method generally called 'phylogenetically independent contrasts' have doubted that any fixed body-mass exponent exists. Essentially, their contention is that there is a significant degree of inertia to evolutionary change in organisms that results from common ancestry. Thus, after any given time period, closely related animals would resemble each other much more strongly than more distantly related ones. An elaborate method of compensating for this presumed inertial bias has been worked out and the adoption of this approach to energetics can indeed destroy the simplicity of Kleiber's result (e.g. Lovegrove, 2000). There is no space to discuss this here, but I will assume that this nihilism is wrong. Readers who are on the bandwagon are of course free to disagree with me and can doubt many of the relationships printed in this chapter. Of course, it would help in the defence of the 0.75 exponent if it had received some fundamental explanation. Many have tried, but nothing yet seems sufficient.[2] Nevertheless, I will base the early part of this chapter on existing measurements rather than theory and assume that the rate at which mammals need energy scales roughly as $M^{0.75}$.

Early attempts to relate masticatory function to body metabolism focussed only on the area of the working surface of postcanine teeth in mammals, searching for evidence that postcanine tooth area also scaled as $M^{0.75}$. There is no good reason for this argument, but it was proposed (Pilbeam & Gould, 1974; Gould, 1975), criticized immediately (Kay, 1975a), persisted with (Pilbeam & Gould, 1975; Martin, 1979) and then dropped for good after an onslaught of data showed that it was impossible to sustain (Fortelius, 1985 and many others). Kay (1975a, 1978) and, later, Calder (1984) pointed out that other aspects of the process have be included to make a coherent theory. These additional factors include the rate of chewing as a primary factor. Larger mammals chew more slowly than smaller ones, which may put a premium on making every chew more effective in these mammals as compensation. It is clear from this early work that pure empiricism, even if it hits the front pages, is not an approach that will last. More appropriate are fundamental arguments, such as that advanced by Fortelius (1985, 1990), which are centred, as is this book, on the food intake as the stimulus for adaptive change.

Relations between body mass, tooth size and energy requirements

Fortelius (1985) argued as follows. Assuming all foods to have a standardized energy density, Kleiber's work shows that the masticatory process needs to provide the gut with volumes of food per unit of time proportional to $M^{0.75}$. Larger masticatory systems have jaws that traverse greater distances than smaller ones and so they will tend to have slower chewing cycles. In fact, many body rhythms, such as the rate of respiration and the heartbeat, decline with body mass by an exponent of -0.25, i.e. $\propto M^{-0.25}$ (Peters, 1983). Fortelius assumed this exponent might also apply to chewing rates and argued from it that mammals should ingest a mass (volume) of food proportional to $M^{1.0}$ each time the masticatory process operates. In other words, if the need is proportional to $M^{0.75}$, but the rate of processing scales as $M^{-0.25}$, then mammals should ingest an amount of food per unit of time proportional to $M^{0.75}/M^{-0.25} = M^{1.0}$. (An implicit assumption here is that all mammals feed for a similar proportion of every 24-hour period.)

As Fortelius remarks, if mammals need to ingest food volumes proportional to their own bulk, then every body structure directly involved in acquiring food should exhibit geometric similarity to accommodate this food bulk. Thus, the volume of the oral cavity (strictly, the intra-oral volume) has to accommodate this food batch and would, therefore, be expected to scale as $M^{1.0}$. The opening of the mouthslit has to be large enough to admit this food into the mouth, but it is an area, not a volume, and would thus be proportional to $M^{0.67}$, not $M^{1.0}$. The maximum gape angle of the jaw, a dimensionless measure, would be unaffected by size (scale as $M^{0.0}$) provided that jaw lengths scale as $M^{0.33}$, which they should. In order to achieve the necessary gape, the lengths of jaw muscles, both those that actively open the jaw and those permitting the jaw to open by stretching, should also scale as $M^{0.33}$. Those mammals that regulate the ingested mouthful of food using their incisors, like primates, should have their relevant lengths (their mesiodistal dimensions) proportional to $M^{0.33}$. The height of a carnivore's canines should scale with the same exponent in order that all carnivores can inflict the same amount of damage on their prey. And so on.

Once food is in the mouth, it needs to be processed by mastication. Fortelius continues by predicting that the size of the postcanine teeth of mammals should also show geometrical similarity, i.e. that the areas of the working surface of the postcanines should be proportional to $M^{0.67}$, since this surface then matches that of the food. However, this is a premature conclusion because it omits an examination of the central feature of the process: the reaction of the food batch to mechanical loading.

HOW THE EFFORT REQUIRED TO FRACTURE FOOD VARIES WITH ANIMAL SIZE

Elastic fracture

To add mechanics to the analysis, I need to make four initial assumptions:
(1) Foods are ingested as single particles, sized in proportion to a mammal that feeds on them.
(2) Food toughness is constant.
(3) Foods have linear stress–strain curves up to fracture, with negligible yielding.
(4) Foods are homogeneous.

Thus, the mouthful that any mammal ingests, whether this mammal is large or small, is considered to be composed of a single, geometrically similar, linear elastic, homogeneous food particle with invariant material properties. The adaptation of the size of the jaws and teeth is with respect to this particle.

Figure 5.1 shows the loading of a large food particle termed x, and a smaller one, labelled y. These particles are configured as cylindrical columns loaded in compression, although any geometry or loading regime could be considered. The results of loading the particles are shown on a force (F)–displacement (u) graph where the loading slope is called the *stiffness* of the particle and given by F/u. The work done can be calculated from the triangular area under the graph in Fig. 5.1, which is $\frac{1}{2}Fu$. This work is transferred entirely to the particle and stored as elastic strain energy prior to food fracture.

The compressive displacements that x and y endure act in proportion to their linear dimensions while the forces produced act over their surface areas (proportional to the squares of their linear dimensions). The work done on the particles is given by the product of the force and the

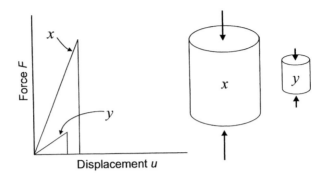

Fig. 5.1 The compressive force–displacement relationship for two intact solid cylindrical specimens of the same shape and material, but differing in size. Each specimen is being compressed within a linear elastic range.

compressive displacement and is proportional, therefore, to the cube of linear dimensions. Thus, defining the ratio of any linear dimension of x and y is λ, then the force, compression and the work done on x is (Atkins & Mai, 1985):

$$\lambda^2(\text{force on } y) \times \lambda(\text{elastic compression of } y)$$
$$= \lambda^3(\text{elastic strain energy absorbed by } y). \qquad (5.1)$$

This relationship says that forces act over areas to produce stresses. The production of identical stresses within the two particles requires that the linear dimensions of x must be λ^2 times those of y. Since compressions depend on linear dimensions, the work that is done (force multiplied by compressive displacement) is λ^3 larger in x. This work, the product on the left-hand side of Eqn 5.1, is absorbed by the particles as elastic strain energy and carried by their volume, which is also λ^3 times bigger in x. This has broad support from experimental evidence on industrial materials (Atkins, 1999).

Galileo (1638) spotted the importance of size to mechanical problems in Renaissance times. D'Arcy Thompson (1961) discussed Galileo's 'principle of similitude' extensively in a now-classic account of scaling. It is based totally on similarity of stress levels. The most frequently given example in the literature is that of weight-bearing columns: the larger the column, the fatter it will have to be to do the same job. The explanation is that the weight of the column, proportional to its volume, grows with size increase as λ^3, while the cross-sectional area of the column, which takes the load, is only raised as λ^2. Thus, in order to take the weight, the cross-section must grow disproportionately large in bigger columns. This analysis has been applied to the shapes of mammalian limb bones (D'Arcy Thompson, 1961) and to the size of muscles and their ability to produce forces so as to move these bones (Hill, 1950). It remains one of the central examples of shape change with size in animals and the mechanical arguments stemming from it pervade biology. According to Gould (1966), Galileo effectively founded the study of allometry by making such observations.

Studies of how to keep deformation within limits are perfectly appropriate whenever similarity of stress or displacement, keeping strains within elastic limits, is adopted for the design of a structure. For example, artificial structures such as skyscrapers can be engineered safely by keeping these principles in mind. As long as the loads on any part of the structure stay well below that which would produce buckling, or do not take the material beyond its yield stress (a stress unaffected by size), the structure should retain its shape indefinitely. Engineers may, in some circumstances, be less conservative and design against the onset of fracture (such as the 'leak before burst' criterion for pressure vessels – Atkins & Mai, 1985), but they do not generally consider designs that prevent fast crack growth (propagation) because this is far too risky. Buildings, bridges, aeroplanes or oil refineries are considered to have failed long before this point is reached.[3]

This 'construction mechanics', then, is based largely on the control of deformation. It contrasts greatly with the 'destruction mechanics' of the masticatory apparatus of mammals. Changing the shape of a food particle by chewing is pointless unless the particle fractures and then fragments. It is fracture/fragmentation, and not elastic response, that forms the design criterion for mastication because the oral cavity is engineered for the total destruction of foods, not something half-hearted like shape change. The mechanical relationships that prevail during the deformational domain do not apply after fracture and I will show this now by extending the mechanical analysis to the scaling of slow-growing elastic fractures. Gurney & Hunt (1967) solved this and their reasoning is followed here.

Figure 5.2 shows a cracked particle loaded in three-point bending. This loading pattern is broadly relevant to how the alternating cusps of the upper and lower postcanine teeth in humans could load a food particle during the closing phase of a chewing cycle. I have used it as an example in Chapter 4 and will continue with it here. Again, it is arbitrary and the geometry does not really matter because the argument is akin to a dimensional analysis.

Note that the particle contains a central crack caused by this loading. Suppose a small displacement du on this particle, produced by a bite force F, now extends this crack to produce a new increment of crack area, dA.

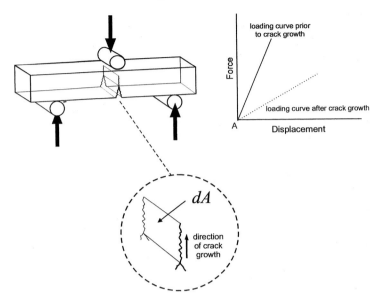

Fig. 5.2 A cracked particle, loaded in three-point bending, in which an existing crack will propagate under a small displacement, *du* (not shown), so as to extend the area of the crack from a sharp notch by an increment *dA*.

As before, the work done on the particle is $\frac{1}{2}Fdu$, but part of this work has gone now into developing the crack. If the toughness (the work required to grow the crack by a unit area) is R, then the work done in growing the crack is RdA. As a result of crack growth, the energy stored in the particle will drop. However, this contained energy is not entirely depleted and has not fallen to zero. The dotted line in Fig. 5.2 shows that the particle is now more compliant, i.e. its force–displacement slope being shallower. The drop in strain energy within the specimen after cracking is $\frac{1}{2}d(Fu)$.

The sum of the work done by all forces during such static conditions should be zero. Thus,

$$Fu - RdA - \tfrac{1}{2}d(Fu) = 0.$$

By putting $\frac{1}{2}d(Fu) = \frac{1}{2}(Fdu + udF)$, dividing both sides by F^2 and then rearranging, we get

$$F^2 = 2R / \frac{d}{dA}\left[\frac{u}{F}\right]. \tag{5.2}$$

We now consider two particles again, a large one (x) and a small one (y). However, this time scaled cracks are introduced into them such that the

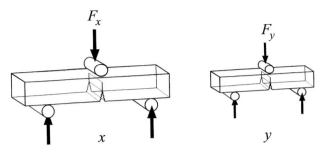

Fig. 5.3 Two geometrically similar cracked particles of the same material loaded by forces F_x and F_y in three-point bending.

ratio (large/small) of any linear dimension of the outer surface of these particles and of the dimensions of their cracked surfaces is also λ. The crack area is A_x in the large particle and A_y in the small one (Fig. 5.3). From the above elastic situation, the compliance of x and y, the inverse of their stiffness, must be related by

$$\frac{u_x}{F_x} = \lambda^{-1}\left[\frac{u_y}{F_y}\right]. \tag{5.3}$$

These compliances change with respect to growth in crack area by

$$\frac{d}{dA_x}\left[\frac{u_x}{F_x}\right] = \lambda^{-2}\frac{d}{dA_y}\left[\frac{u_x}{F_x}\right] = \lambda^{-3}\frac{d}{dA_y}\left[\frac{u_y}{F_y}\right]. \tag{5.4}$$

Note that crack growth rates, i.e. growth with respect to crack area, in the two particles are related by λ^{-3}. We can now use this result in combination with Eqn 5.2 to find the relationship between the forces on the two particles producing crack growth, which is

$$F_x^2 = \frac{2R_x}{\dfrac{d}{dA_x}\left[\dfrac{u_x}{F_x}\right]} = \frac{2R_y}{\dfrac{1}{\lambda^3}\dfrac{d}{dA_y}\left[\dfrac{u_y}{F_y}\right]}. \tag{5.5}$$

The forces are, therefore, related to the toughness of the particles by

$$F_x = \lambda^{1.5}\left[\frac{R_x}{R_y}\right]^{0.5} F_y. \tag{5.6}$$

If both large and small particles have the same toughness, i.e. $R_x = R_y$, then

$$\frac{F_x}{F_y} = \lambda^{1.5}. \tag{5.7}$$

Note straightaway that the ratio of these forces is now *not* proportional to the ratio of the areas over which they act. Nevertheless, to obtain the stresses, the force must be divided through by area. For y, this area is still λ^2 greater than that for x, so the ratio of the stresses is

$$\frac{\sigma_x}{\sigma_y} = \lambda^{-0.5}, \tag{5.8}$$

and, going back to the compliances, the scaling of the displacements is

$$\frac{u_x}{u_y} = \lambda^{0.5}. \tag{5.9}$$

The reason for the importance of fracture should now be clear. The ratio of the forces on large and small particles is not λ^2, as in Eqn 5.1, but $\lambda^{1.5}$ (Eqn 5.7) because larger particles fail at smaller stresses (Eqn 5.8).

The force, displacement and energy absorbed in crack formation in particle x (Atkins & Mai, 1985) are

$$\lambda^{1.5}(\text{force on cracked } y) \times \lambda^{0.5}(\text{displacement of cracked } y)$$
$$= \lambda^2(\text{energy absorbed by cracked } y). \tag{5.10}$$

The right-hand side of Eqn 5.10 says simply that the energy absorbed by growing cracks depends on crack area, which is to say a ratio of λ^2. In contrast, prior to fracture (Eqn 5.1), energy absorption depended on λ^3. It is because of this dimensional discrepancy that cracks are so dangerous: they can feed on an abundance of stored energy (Atkins, 1999).

Now take a large mammal, denoted by X, and a small mammal, denoted by Y, ingesting single food particles termed x and y respectively. Going back to Fortelius' (1985) analysis and the first assumption, if these particles are typical food items, then they should be in direct proportion to the sizes of these mammals, i.e. $x/X = y/Y$. These mammals need to grow crack areas in these particles at the same rates per chew. Both the body masses of the mammals and those of the food particles that they are dealing with differ by λ^3, where λ is now the ratio of any of the linear dimensions of either the mammals or their food particles. We can deduce from Eqn 5.7 that the larger mammal X need not produce bite forces that are λ^2 greater than mammal Y, but instead only produce forces that are $\lambda^{1.5}$ as big. Its jaw adductor muscles are generating this bite force and their ability to do this is proportional to the physiological cross-sections of these muscles. Thus, these muscle areas will also be related by $\lambda^{1.5}$.

We can now convert any of these ratios to classic biological scaling by remembering, from above, that the volume of mammal X is λ^3 that of mammal *Y*. Provided that these mammals have the same density, then their body masses will be related by the same ratio as their volumes. Thus, masticatory adductor muscle cross-sectional areas are predicted to be proportional to $M^{0.5}$ (i.e. $\lambda^{1.5}/\lambda^3$). This contrasts with geometric similarity, which would have muscular cross-sections proportional to $M^{0.67}$. Fracture scaling (if I may start to call it that) does not require muscle areas this large because larger mammals are eating weaker ($\equiv$ larger) food particles. I predict, therefore, that any large mammal will have proportionately smaller muscle cross-sectional areas than its absolutely smaller counterpart.

Jaw-closing muscles act to move the lower jaw such that the teeth press on the food. The working surfaces of the teeth make an area of contact with the food that dissipates the load. Nothing very definite needs to be specified about the actual area of contact. The important point is that the area of tooth–food particle contacts, whatever they may be, cannot increase at a faster rate than the scaling of bite forces produced by the jaw-closing muscles or this would set off 'force inflation'. Thus, I predict that the working surface areas of teeth that fracture and fragment food particles will be proportional to $M^{0.5}$.

I now need to introduce a bit of bio-jargon. When part of an organism is predicted to scale by a factor less than that which preserves geometric similarity, the result is called 'negative allometry'. Such a body part would be proportionately smaller in an absolutely larger animal. The converse, where the larger animal is predicted to have a proportionately larger body part, is termed 'positive allometry'. Lastly, when there is no change of shape with size, it is called 'isometry'. I will adopt this jargon from now on because it can ease description once these simple definitions are grasped. It should be clear now that the fracture scaling of body parts results in negative allometry. For example, while geometrically similar muscle areas scale isometrically, i.e. $\propto M^{0.67}$ (Hill, 1950), fracture theory predicts that they should be $\propto M^{0.5}$.

Virtually all allometric analyses predict either positive allometry, isometry or negative allometry, but never a mix. The interesting feature of a fracture analysis is that it *does* predict a mix: some aspects of the mouth are geometrically scaled so as to accommodate food particles ($\propto M^{1.0}$), while others are negatively allometric because they are determined by the fracture scaling of these same particles. The arguments thus become complex. However, this complexity is also to the benefit of this analysis because it predicts a genuine change in the shape of mammalian body parts with size that other theories really do not. Mixed rules are, as far as I can see, the

Table 5.1 *Selected body-mass exponents of masticatory muscles and oral structures predicted from an elastic fracture scaling analysis*

Variable	Predicted exponent of body mass
Angle of jaw opening (gape)	0.0
Area of mouthslit	0.67
Area of the hard palate	0.67
Cross-sectional area of mandible	0.5
Diastema	0.08
Height of carnivore canines	0.33
Intra-oral space	1.0
Jaw length	0.33
Mesiodistal length of herbivore incisors	0.33
Muscle cross-sections	0.5
Muscle lengths	0.33
Muscle mass	0.83
Postcanine tooth area – herbivores	0.5
Surface area of the dorsum of tongue	0.67
Tongue volume	1.0

only way to rescue allometric analyses from the substitutes' bench in the study of shape in biology.

What are the predictions? I give one example of how to calculate these and then place the rest in a table, discussing them with the evidence. The lengths of jaw-closing muscles need to be geometrically similar ($\propto M^{0.33}$) to stretch sufficiently for the equally wide gapes that food particles ($\propto M^{0.33}$) enter the mouth. However, the (physiological) cross-sectional areas of these muscles that generate the bite force need to scale only as $M^{0.5}$. Thus, adductor muscle mass is suggested to scale as $M^{0.5} \times M^{0.33} = M^{0.83}$. The result is that a muscle like the masseter should be short and fat in a small mammal, but longer and skinnier in a larger one.[4] Table 5.1 summarizes a number of the predictions like this from the theory.

A caveat

Alexander (1985) touched on a big problem with scaling theories that suggest departures from geometric similarity. Over the range of body sizes seen in mammals, body designs at the size extremes become absurd. Elastic similarity, a theory associated with McMahon (1973, 1975), is somewhat like this. Alexander (1985) shows that a size jump from a shrew to an elastically similar elephant results in the latter having such fat legs that these would not have space to fit onto its body. This appears (appeared to me at any rate) to 'kill' the theory, but this may not be so. Allometric arguments are often given an implicit upward direction, from small to large, because that is the general direction of evolutionary change (Alroy, 1998). However, from a non-biological perspective, it is also reasonable to argue this the other way around from big to small. Starting with the elephant, what would the legs of an elastically similar shrew look like? The answer is 'remarkably thin', much thinner than they actually are, but its legs *would* be able to fit onto its body. The point is not that I wish to resurrect elastic similarity, but that allometric analyses do not predict the absolute size of any organ in any one organism, merely their size relative to that in other organisms. The logic behind Alexander's caveat seems in fact to explain a lot about dental evolution when there is reduction in body size.

The scaling theory based on food fracture given here is even more complex than envisaged in Alexander's objection to non-geometric scaling because I am suggesting a mixed isometric and negative allometric scaling pattern and just for one small part of the body, the oral apparatus. It could not be applied to the locomotor apparatus or musculoskeletal elements of the trunk because they are not designed according to fracture similarity. Food fracture is a uniquely complex and essential aspect of mammalian life.

EVIDENCE FOR FRACTURE SCALING

The data in the literature use varying estimates of body size. The usual measure is body weight, but head–body length is often employed as well. Damuth (1990: his Fig. 12.1) studied a large sample of ungulates and showed that head–body length is isometric to body weight and could be used to represent whole body size for dental studies. Less satisfactory is the use of whole skull length though, simply because this is dominated by the mesiodistal dimensions of the teeth. Skull length is negatively allometric to whole body size in mammals (Martin, 1990). Basicranial head length (the length of the brain case) is too: larger mammals have relatively smaller brain volumes, so their brain cases are also relatively smaller (Fortelius, 1985). There is also a considerable headache introduced by variation in the line-fitting techniques. These are beyond the scope of this book, but have been reviewed by LaBarbera (1989). He advocates the reduced major axis (Kermack & Haldane, 1950) because this method incorporates variation in both the body part under consideration and body mass when constructing trend lines. A least-squares regression, the conventional line fitted by a statistical package and reported by many researchers, can be converted to a reduced major axis value very easily (the slope itself, but not its confidence limits). However, another technique, the major axis, is also found in the literature and this is difficult to relate to the other techniques. Of course, all matters little if the data points cling tightly to the line and the larger the range of body sizes incorporated in the data, the more obvious the trend is going to be.

There are more problems than this. Valid tests of allometric predictions about body functions that are directly involved in feeding are complicated by the fact that diets are not independent of animal size. There have been several debates about how general or particular the dataset ought to be (e.g. Kay, 1975a versus Pilbeam & Gould, 1975). Animal and plant feeders do not predate similar mechanical targets, so it is reasonable to separate them. The need to distinguish between types of plant-eaters is more complex. Very few small vertebrates, including primates (Kay & Covert, 1984), are folivorous (leaf-eating) because it is difficult to fit a ruminant digestive system into a small animal. Despite this, a very small bird called the hoatzin has it (Grajal *et al.*, 1989) and microtine rodents (lemmings and voles), which all weigh <0.3 kg, possess something very similar (Hume, 1994). Nevertheless, very small mammals may well be excluded from certain diets. Are any diets excluded for large mammals? The answer is yes. Few large mammals are insectivorous, unless they feed on social insects, because it is difficult to catch them fast enough to fulfil energy requirements (Kay & Covert, 1984). The largest terrestrial mammals generally consume structural parts of plants, although even these (e.g. lowland gorillas: Rogers *et al.*, 1990) devour fleshy fruits whenever they find them. Bearing in mind that it is difficult to know exactly how to homologise diets, I will simply keep to a traditional herbivore–carnivore split in making a general review of the evidence.

ASSUMPTIONS ABOUT THE FOOD INTAKE

As with the rest of this book, I will consider the food intake first, and follow with the required changes in tooth size.

The rate of ingestion and the size of the oral cavity

There is ample evidence that the rate of ingestion of food volume is proportional to $M^{0.75}$ (Peters, 1983; Shipley *et al.*, 1994). However, the exponent for the scaling of chewing rates (the time needed to complete an average chewing cycle) is not well established. Large mammals chew more slowly than smaller species, but taken together, the data of Hendrichs (1965) and Fortelius (1985) suggest an exponent lying between $M^{-0.19}$ and $M^{-0.23}$ (Fortelius, 1985). Kay (1978) makes a similar statement, but does not present the raw data. In contrast, Druzinsky (1993) suggests a much lower exponent of $\sim M^{-0.13}$. The stride frequencies of limbs scale very similarly to this in mammals, which led Druzinsky to suggest that jaw movements are less like body rhythms such as heart rates and more like locomotor patterns. However, the reduced major axis slope for all the data that he tabulates, his own and that of Fortelius combined, provides 35 species and a body-mass exponent of $\sim M^{0.17}$ (Table 5.2). This does not support either Fortelius' nor Druzinsky's argument very well and there is considerable variation not explained by the fit.

I do not know either if the internal volume of the oral cavity is proportional to body size (i.e. scaling as $M^{1.0}$). However, the weight of the gut in mammals, which would be proportional to gut volume, has a body-mass exponent of 0.94 (Peters, 1983) and the volume of the rumen and the capacity of the whole gut in ruminant animals is also isometric to body size, i.e. $\propto M^{1.0}$ (Demment & Van Soest, 1985; Owen-Smith, 1988), which provides massive support for Fortelius' viewpoint. The (nominal) surface area of many other compartments of the gut in mammals appears to scale as $M^{0.75}$ (Peters, 1983; Martin *et al.*, 1985), suggesting that the food passage time down the gut scales very similarly to other body rates (i.e. $\propto M^{-0.25}$). While some evidence pertains to this (Robbins *et al.*, 1995; Van Soest, 1996), there is an urgent need for more. Fermentation can add enormously to the time of food retention, but models of mammalian gut function otherwise seem to assume a constant food passage velocity (Alexander, 1991).

The analysis of feeding can become far more complex than this though, needing to take into account differences between mammals in the time devoted to feeding every day (Mysterud, 1998). However, the assumptions about gut sizes and food passage times are not critical to the arguments advanced here. There seems no point to stop and scratch my forehead over this, so I go back a step to the mouth.

Indirect evidence about the volume of the oral cavity comes from gape. Many mammals can open their mouths to very similar gapes, about 60–70° according to Herring & Herring (1974), which is in accord with predictions.

Table 5.2 *Chewing frequencies of mammals in relation to their body size*

Species	Body mass (g)	Chewing frequency (s^{-1})
Myotis lucifugus (little brown bat)	7	4
Suncus murinus (Asian house musk shrew)	41	5.46
Tupaia glis (common tree shrew)	150	4.2
Rattus norvegicus (Norway rat)	200	5.21
Pteropus giganteus (Indian flying fox)	480	1.72
Saimiri sciureus (squirrel monkey)	550	2.8
Aplodontia rufa (mountain beaver)	583	4.83
Tenrec ecaudatus (common tenrec)	700	2.62
Cavia porcellus (guinea pig)	900	6.13
Mustela putorius (domesticated ferret)	1 350	4.59
Otolemur crassicaudatus (thick-tailed galago)	1 500	3.18
Pedetes capensis (springhare)	2 300	3.15
Oryctolagus cuniculus (domestic rabbit)	2 500	3.61
Didelphis marsupialis (common opossum)	2 500	2.56
Felis domesticus (domestic cat)	2 500	3.25
Macaca mulatta (rhesus monkey)	3 500	2.99
Marmota monax (woodchuck)	3 790	1.72
Ateles sp. (spider monkey)	6 000	3.07
Capra hircus (domestic goat)	20 000	2.4
Sus scrofa (miniature pig)	22 300	3.03
Canis familiaris (domestic dog)	36 287	3.16
Capra ibex (ibex)	40 000	1.68
Hemitragus jemlahicus (thar)	50 000	2.12
Homo sapiens	60 000	1.3
Oreamnos americanus (mountain goat)	60 000	1.28
Capra falconeri (markhor)	64 000	1.72
Ammotragus lervia (Barbary sheep)	66 000	1.71
Equus hemionus (Asiatic wild ass)	210 000	1.25
Bos grunniens (domesticated yak)	250 000	1.15
Tapirus terrestris (Amazonian tapir)	272 155	1.45
Bison bonasus (European bison)	300 000	1.26
Bos taurus (domestic cattle)	476 272	1.45
Camelus bactrianus (Bactrian camel)	500 000	1.1
Equus caballus (domestic horse)	650 905	1.27
Loxodonta africana (African elephant)	2 812 273	0.97

Source: Data from Hendrichs (1965), Fortelius (1985) and Druzinsky (1993).

The hard palate traverses the length of the oral cavity. Ravosa (1991) measured its anteroposterior dimension in 108 species of primates, finding it proportional to $M^{0.316}$, close to isometry. Druzinsky (1993) also conducted a mini-survey of palatal length in the mammals in his chewing frequency study with an identical result.

Ingested particle size

Is it correct to assume that larger mammals feed on larger single food objects? Among carnivores, larger predators seem to feed on equally large prey (Peters, 1983). Frugivores may show a similar pattern. Fleshy fruits are designed for feeding on by vertebrates because these animals can disperse their seeds effectively. Janson (1983) shows that fruit size is an important component of dispersal syndromes and Gautier-Hion *et al.* (1985) show that this characteristic affects fruit choice by mammals. Thus, for example, small bats eat smaller figs than larger bats (Wendeln *et al.*, 2000). However, the net needs to be cast wide geographically to establish this because it is not always true at a local level (e.g. for primates: Ungar, 1996). On a broad evolutionary scale, the dispersion of tree species in tropical forests can best be explained by larger frugivores eating larger fruits, very often containing larger seeds, which they transport shorter distances than small frugivores like bats and birds achieve with small, smaller-seeded, fruits (Howe, 1989). Eriksson *et al.* (2000) document a substantial evolutionary increase in fruit size during parts of the Paleocene–Eocene, which could be related to the evolution of larger mammals capable of dispersing the seeds of these fruits.[5]

A major stumbling block though is leaf-eating, where the relationship between food item size and mammalian body size breaks down completely. Plants do not want their leaves to be eaten and it would be absurd to expect them to scale these to the size of the likely herbivorous threat because this strategy would probably encourage rather than deter the predator. There is no general tendency for bigger plants to have bigger leaves, flowers/inflorescences or fruits/infructescences containing bigger seeds than smaller ones (Cornelissen, 1999). This is not to argue the impossible. A tiny plant (size in plants often being assessed, somewhat confusingly, by height) obviously cannot have gigantic leaves. However, the biggest flower (*Rafflesia*) sits on the floor of the rainforest, while the biggest leaves in that environment are on (often young) plants that are of about human height (Turner, 2001).

The term 'browser' means 'consumer of leaves other than grasses'. The major browsing pressure for the last 65 million years is thought to have come from tiny invertebrates, not mighty mammals (Coley, 1983). Even so, the ecological factors that determine tree leaf size are probably more to do with light reception and climate than with defence against predation (Turner, 2001). In contrast, the major grazing (grass-eating) pressure has probably been mammalian and the average size of grass leaves is probably influenced by this, becoming tiny (in volume) compared to their consumers.

It would appear certain that the first assumption of the theory, that foods are ingested as single particles, sized in proportion to mammal that feed on them, is well and truly breached at this point. Nevertheless, plant-eating mammals still need to find food quantities proportionate to their size and probably forage on the basis of food patches instead of individual leaf size. If the tongue stacks these items during chewing such that stresses could be transferred across them when loaded by the teeth, a batch of leaves could perhaps be treated as a unit. Even so, comparisons would have to be limited to other leaf-eaters.

Food toughness

It seems long to have been assumed that larger mammals take in food of lower quality than smaller ones, especially herbivores (Bell, 1971; Jarman, 1974). Gould (1975) gave this as one possible reason for predicting a positive allometry of tooth size in mammals. Quality is an imprecise term, but one sense of the word connotes 'energy content'. Foods of lower energy density would definitely have to be eaten in larger quantities than richer food. However, another use of the term 'low quality' is to connote something about the mechanical properties of foods, as in 'tough' meat or 'fibrous' plant food. These uses of the word are relevant here and provide an alert about the validity of assuming that toughness is constant. A tougher food intake means higher bite forces (Eqn 5.6).

What about the converse: a reduction in food toughness? A reduction in bite forces, i.e. of the jaw musculature, is a logical evolutionary response, but this would amount to little if tooth size did not reduce faster. Why do frugivores (meaning 'fruit flesh' feeders) often have such small cheek teeth (Freeman, 1988)? The answer could be that their foods are not tough (Appendix B). This reasoning could apply equally well to piscivores (fish-feeders). The toughness of muscle tissue in mammals and fish can be compared very crudely by considering their collagen contents. Sato et al. (1986) show in 22 species of fish that collagen constitutes 0.34–2.19% of fresh muscle weight, while in Atlantic salmon, it is 2.9% (Aidos et al., 1999). In mammals, figures of 1.9–9.6% are given for the muscles of cows (Nguyen & Zarkadas, 1989) and 2.84–5.89% for those of domesticated pigs (Zarkadas et al., 1988).[6] These figures do not determine relative toughness, but they do indicate that the fish muscle will be much less tough than mammalian muscle, even if their collagen is chemically identical (which it is not: fish collagen is generally much more soluble). Accordingly, tooth size might be reduced in piscivores for this reason. The argument might be extended much further to nectar feeders that probably hardly need cheek teeth for most of their diet. Freeman (2000) shows that nectarivorous bats have greatly reduced cheek teeth.

So, variation in food toughness affects predictions about tooth size. In particular, a reduction in food toughness probably means a reduction in tooth size. The application of this to cooking and human evolution in spelt out in Chapter 7.

Food homogeneity

This is an obviously false assumption – all foods are heterogeneous. The issue though is anisotropy, i.e. food structure at a scale relevant to fracture processes. The relevant scale for fracture-resistance mechanisms in plant tissues is very small, at the cellular level. Full toughness in a plant tissue is already expressed in tissues that are >1 mm in thickness (Chapter 4). This is the approximate size range of swallowed particles and so does not affect the validity of the assumption. However, there are fracture protection mechanisms in animal soft tissues that act at a much larger scale and which do reflect directly on this homogeneity assumption. Accordingly, I will consider predictions of fracture scaling given in Table 5.1 for herbivores separate to carnivores.

HERBIVORY

Jaw muscle size

Unlike for limb muscles (Alexander, 1985), there are no systematic data on the sizes and shapes of jaw muscles. An ideal group for study would be the bovids (the cattle family), but I do not know of enough data to test the theory, either on the lengths of opening and closing muscles or their physiological cross-sectional areas. However, there are some data on muscle masses. Cachel (1984) presented body-mass exponents for masticatory muscle masses in 31 adult primates. These exponents vary from 0.755 (standard error ±0.12) for the anterior digastric, an opening muscle for which my whole mass predictions are weakest, to 0.959 (±0.102) for the anterior part of the temporalis. The rest of the temporalis, the posterior part, has a body-mass exponent of 0.903 (±0.115) while that for the masseter is 0.869 (±0.122). The exponent predicted by fracture theory is 0.83 (Table 5.1), which is clearly lower than most of these values, although not really precluded by them. In contrast, limb muscles are at least isometric or show positive allometry (Alexander, 1985), which helps perhaps to indicate that different constraints must be operating on the jaw muscles than on those in the limbs.

However, fracture scaling theory is much more specific than this. It predicts that the masseter, for example, should change from a short fat block in a small mammal to a longer narrower band in a very large one. Fitting the latter onto the skull of a large mammal might involve raising the jaw joint higher above the plane of the working surface of the teeth to accommodate muscle length. This is a curious prediction because much effort has been spent in trying to explain why the height of the jaw joint varies such a lot in mammals. Fracture scaling suggests that it might be size-dependent. Some data pertain to this in primates. Although there is no clear evidence that larger primates have higher jaw joints (Lucas, 1981; Smith, 1984), the highest joints do tend to be in the largest species, e.g. all the living great apes including humans, the extinct apes *Gigantopithecus* and *Sivapithecus* and the large extinct gelada baboon, *Theropithecus oswaldi*. Grazers (grass-eaters) have higher jaw joints than browsers too, which may reflect body-size differences (Mendoza *et al.*, 2002).

Bite force

There are very few mammals for which maximal bite forces are known accurately. Much of the literature seems more concerned with trying to find record forces than with systematic investigation. The data in Table 5.3, obtained by such diverse methods as contraction induced by electric shock, voluntary bites on transducers and the results of bite mark analysis (described in Lucas *et al.*, 1994), give no great confidence in any exponent.

Demes & Creel (1988) took another approach, estimating the physiological cross-sectional areas of the masseter and temporalis muscles and the lengths of their lever arms from landmarks on skulls. From this, they calculated 'bite force equivalents' for species of hominoid primates. These values were then checked against real bite force data on the long-tailed macaque and the human (Table 5.3) with good agreement. Hill *et al.* (1995) confirmed the validity of this for Japanese macaques (*Macaca fuscata*) using a 'bite mark' analysis. Individual monkeys were seen trying to open a woody gall with their teeth, sometimes succeeding, otherwise failing. In doing so, they left many bite marks on gall fragments that could be used to calculate the bite forces that they were developing with their anterior teeth. These calculations agree well with the bite force equivalents as estimated by Demes & Creel's method. Anyway, the important point here is

Table 5.3 *The maximum bite force reported for a variety of mammals*

Species	Location	Maximum bite force (N)[a]	Body weight (kg)	Source
Norway rat (*Rattus norvegicus*)	Incisors	50.2	0.56	Robbins (1977)
Cat (*Felis domesticus*)	Canines Carnassials	227.9 274.4	3.8	Buckland-Wright (1975)
Long-tailed macaque (*Macaca fascicularis*)	Molars	333 (females)	3.6	Hylander (1979a)
Japanese macaque (*M. fuscata*)	Incisors	487	11.0–8.0	Hill *et al.* (1995)
Rhesus macaque (*M. mulatta*)	Premolars	242 (males) 185 (females)	7.7 5.4	Dechow & Carlson (1990)
Human (*Homo sapiens*)	Molars	400	65	Carlsson (1974)
Orang-utan (*Pongo pygmaeus*)	Molars	2000 (females)	36	Lucas *et al.* (1994)

[a] The data include muscle-twitching experiments, but are not limited to them.

that Demes & Creel (1988) found that tooth areas are isometric to these 'bite force equivalents', which supports a critical assumption made in the fracture scaling theory – that bite forces match tooth sizes in mammals with similar diets.

Incisor tooth size

Those anterior teeth involved in sculpting the bite size, the incisors of herbivores, appear to be isometric to body mass – e.g. incisal widths of bovids ($M^{0.33}$: Illius & Gordon, 1987) and of cercopithecoid primates ($M^{0.312}$: Hylander, 1975; although this slope, from least-squares regression, is certainly an underestimate). These results are in accord with expectation (Table 5.1).

Postcanine tooth size

The largest dataset is on small mammals, mostly rodents, with the area of the first lower molars of 288 species plotted against head–body length (Creighton, 1980). The slope is equivalent to $M^{0.61}$, considerably below isometry but well above an exponent of 0.5. There are many other studies, but Fortelius (1985) summarizes them well and adds his own superlative data. Generally, body-mass exponents range from negative allometry to isometry in herbivores and primates (Table 5.4). While the data do not

Table 5.4 *Abridged data assembled by or originating from Fortelius (1985, 1990) on the scaling of postcanine tooth size[a] in herbivores with body mass*

Mammalian group (sample size)	Exponent of body mass (standard error)
Hystricomorph rodents (14)	Uppers 0.72
Primates (77)	Uppers 0.62
	Lowers 0.69
Ungulates (43)	Uppers 0.65 (0.06)
Non-selenodont artiodactyls (20)	Uppers 0.65 (0.07)
	Lowers 0.62 (0.07)
Selenodont artiodactyls (22)	Uppers 0.58 (0.07)
	Lowers 0.57 (0.07)
Bovids (13)	Uppers 0.62 (0.07)
	Lowers 0.62 (0.09)
Rhinoceroses and hyraxes (8)	Uppers 0.65 (0.04)
	Lowers 0.62 (0.04)
Bilophobunodonts (5)	Uppers 0.56 (0.14)
	Lowers 0.43 (0.13)

[a] Postcanine tooth size is estimated as the sum of the area of all premolar and molar teeth. All slopes are major axes through the data points.

really support the theory very strongly, they do suggest that exponents tend often to run significantly below isometry, a sentiment with which Martin (1990) seems to agree.

The diastema

Fracture scaling predicts that, in larger mammals, the cheek teeth will be housed in oversize jaws. A simple consequence of this is that empty spaces should appear, e.g. between the anterior and posterior teeth or behind the posterior teeth. Several herbivorous mammalian orders, such as the Lagomorpha (rabbit family), Rodentia, Artiodactyla (even-toed ungulates like cattle) and Perissodactyla (odd-toed ungulates like horses), have a gap (a diastema) between the premolars and anterior teeth (Fig. 3.10). Fracture scaling predicts that this diastema should be size-dependent, with a exponent of $M^{0.08}$ (Table 5.1). No studies have measured the size of the diastema directly. Indirect evidence comes from the study of Williams (1956) on the relationship between the length of the molar tooth row and the length of the lower jaw, but his data do not seem to show size-dependence. However, the diastema is larger in grazers than browsers (Mendoza *et al.*, 2002), which might reflect the larger body sizes in the former.

The diastema figured prominently in the theoretical work of Greaves (1978), who was concerned with the optimal location of the postcanine teeth in ungulates. The mesiodistal length of these teeth is, he says, constrained by the effectiveness of the bite force. The tooth row cannot extend too far posteriorly or else the jaw joint would dislocate because its articular surfaces would be distracted (Chapter 3). The anterior limit is not fixed, but is constrained by a diminishing bite force the further forward a bite is focussed. This theory could account for the lack of influence of diet on the mesiodistal length of the postcanine row in many mammals (Fortelius, 1990). However, even Greaves doubted that it could explain the presence of the diastema. Instead, he also considered the idea that if a long jaw is required for gape, but the length of the postcanine row is constrained by characteristics of the bite force, then a diastema might arise as a consequence. This anticipates the argument offered here, but for very different reasons.

The only other explanation for a diastema, as Greaves relates, is that it provides a mechanism for separating the function of anterior and posterior teeth, e.g. in rabbits (Ardran *et al.*, 1958) and rodents (Landry, 1970). Rodents possess soft-tissue flanges that help to seal off the posterior part of the mouth from the front of the mouth where they bite some items, like seed coverings, that they do not mean to ingest. However, there must be simpler ways of isolating the mouth from the actions of the anterior teeth – the modiolus (Chapter 2) does this, so the explanation given by fracture scaling seems more plausible.

The jaw bones

The bones that support the dentition obviously need to be built to take masticatory stresses. Fracture scaling predicts that the cross-sectional area of the jaw is related to $M^{0.5}$ while the jaw height, if this took an equal share of load-bearing responsibility, would scale as $M^{0.25}$. Smith (1983), Hylander (1985) and Bouvier (1986) discuss this in anthropoid primates, but Ravosa (1991) gives the largest dataset, finding that the height of the mandible under the second molar has an exponent of about 0.36. An obvious reason for this is the need to prevent excessive jaw deflection under load: the more that the jaw deflects, the more this impairs the alignment of upper and lower teeth during mastication. If jaw length scaled as $M^{0.33}$, while the depth of the mandible grew only as $M^{0.25}$, then the jaws of larger mammals might become unacceptably flexible.[7]

THE BEST TEST

Most size changes in evolution are enlargements. It is difficult to generalize but these changes are probably slow and thus involve subtle changes in diet in a mammalian lineage, both in the external and internal physical properties of foods. Any such dietary shift concomitant with size change reduces the value of scaling predictions. What are needed are rapid changes in which there is little time for dietary adjustment. Very much faster size changes are known in some populations of mammals trapped on islands, where larger herbivores seem to respond not by enlarging, but by reducing in size.[8] There are quite well-documented examples in living primates like this (Fooden & Albrecht, 1993), but the best examples of dwarfing are mostly from fossil evidence. The best documented are probably insular elephants, with many instances of an evolutionary reduction in body size down to just a few per cent of that of their ancestors (Roth, 1990). It is thought that such rapid change is an adaptation to the diminishing size of food patches, i.e. not to a change in diet but to a change in food distribution. If so, this provides a perfect test for the predicted negative allometry.

Maglio (1972) claimed that dwarf elephants have proportionately larger teeth, but the problem with a test on this is that elephants bring their molars into the mouth one by one. As each tooth wears out, it is shed, so making postcanine tooth size a difficult concept to quantify. Luckily, other mammals provide examples too. Gould (1975) examined data from a short lineage of four hippopotami. He showed negative allometry, i.e. that the smaller hippos had proportionately larger teeth. The slope was apparently equivalent to $M^{0.58}$, higher than predicted in Table 5.1. However, basicranial skull length was used as a measure of body size. This is likely to have inflated the slope (Fortelius, 1985) and the body-mass exponent may actually be close to 0.5. Fortelius (1985) also discusses dwarfing in several lineages of domesticated mammals where evidence for negative allometry of the teeth with dwarfing appears strong. The only objectors, in fact, appear to be Prothero & Sereno (1982), who studied dwarf rhinoceroses. They state that these mammals did not have relatively large teeth, but relatively smaller skulls instead. From the standpoint of dental function, there isn't much difference between the two interpretations, but their case needs more evidence from other groups.[9]

THE APPEARANCE OF JAWS OF DIFFERENT SIZES WITH LINEAR
ELASTIC FRACTURE SCALING

The evolutionary implications of fracture scaling are a lot easier to grasp with illustrations than with words. Figure 5.4 shows the effects of size

Getting bigger

ANCESTOR **DESCENDANT**

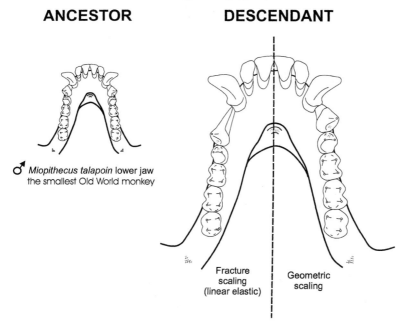

♂ *Miopithecus talapoin* lower jaw
the smallest Old World monkey

Fracture
scaling
(linear elastic)

Geometric
scaling

Fig. 5.4 A 15-fold body mass increase and its effect on jaw and tooth form in a male Old World monkey, *Miopithecus talapoin*. The 'ancestral' form at left is expanded (at right) either by linear elastic fracture (left hemi-jaw) or geometric scaling (right hemi-jaw). This fracture scaling produces lots of free spaces around the postcanine teeth, which has been filled arbitrarily by expanding the lower anterior premolar. The talapoin was chosen at random, but is the smallest known Old World monkey, including the fossil record (Delson *et al.*, 2000).

increase on the lower dentition of a male Old World monkey, the talapoin (*Miopithecus talapoin*). In Fig. 5.4, the real jaw (at left) is called the 'ancestor', while the scaled enlargement, a 15-fold enlargement in body mass, is called the 'descendant' and lies to the right. The descendant diagram shows two hemi-jaws, one geometrically scaled and the other scaled according to linear elastic fracture. The anterior dentition and the length of the jaw adapt to gape, so are always geometrically scaled and fit the jaw perfectly. However, linear elastic fracture scaling results in a proportional decrease in postcanine size. The excess jaw space that this creates could either accommodate a diastema, e.g. between the anterior premolar and the canine, or else the whole postcanine segment could be shifted forwards to leave an enlarged space behind the last molar. Nothing about fracture

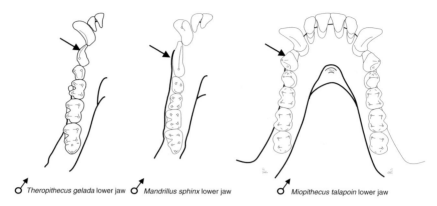

Theropithecus gelada lower jaw ♂ *Mandrillus sphinx* lower jaw   ♂ *Miopithecus talapoin* lower jaw

Fig. 5.5 A comparison between the lower jaw of the male talapoin monkey and those of two baboon species shows that the lower anterior premolar has a large extension (arrowed) in the baboons. The space for this can be created by linear elastic fracture scaling of the postcanines shown in Fig. 5.4.

scaling suggests how this arrangement should be achieved, but one possibility (shown in the figure) is that the front part of the lower anterior premolar could be expanded. This expansion is seen in larger Old World monkeys as depicted in Fig. 5.5, where the hemi-jaws of two large Old World monkey species, the gelada (*Theropithecus gelada*) and the mandrill (*Mandrillus sphinx*), are shown to the left of the talapoin jaw, all brought to the same size by simple geometric enlargement. The extended ridge of the lower anterior premolar rubs against the posterior flange of the upper canine, so sharpening it (Zingeser, 1969). It is nothing to do with diet and is much more pronounced in the baboons than in the talapoin, perhaps simply because fracture scaling creates space for this feature.

A modest size increase reveals the truth of Alexander's caveat quite quickly: jaws and teeth do not fit properly when geometric similarity is departed from. However, the consequences of size increase do not seem critical. In contrast, downsizing seems to produce far more devastating changes.

Figure 5.6 takes the identical procedure shown in Fig. 5.4, but instead reduces the body size of the talapoin by 15-fold. Linear elastic fracture scaling now produces obvious postcanine crowding and the last molar does not have enough bone to seat itself on. If this consequence is accepted, then dwarfism involves serious problems: the need for proportionately larger cheek teeth, but without the space to house them. If so, then it could only evolve under great duress. Alexander's prediction is borne out but, rather

Getting smaller

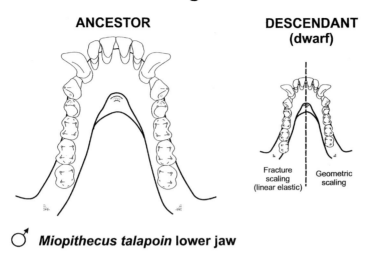

ANCESTOR

DESCENDANT
(dwarf)

Fracture
scaling
(linear elastic)

Geometric
scaling

♂ *Miopithecus talapoin* **lower jaw**

Fig. 5.6 The effect of a 15-fold body-mass decrease on jaw and tooth form in *Miopithecus talapoin*. Linear elastic fracture scaling produces dental crowding with insufficient space in the jaw for the last (third) molar. See text and compare to the size increase shown in Fig. 5.4.

than indicating that fracture scaling is impossible, suggests that it could lead to tooth loss in dwarfing mammals. As stated in Chapter 2, mammals do have variable numbers of teeth and this theory could explain it. This is returned to in Chapter 7.

CARNIVORY

Non-linear elastic behaviour

The third assumption of the fracture scaling theory was that food particles had linear elastic behaviour. This is not so for the vertebrate soft tissues that carnivores eat, which display very non-linear behaviour (Chapter 4). However, a remarkable feature of fracture scaling is that it is capable of dealing with this behaviour. The following analysis of non-linear elastic behaviour was worked out by Mai & Atkins (1975) and is only summarized here.

Linear elastic behaviour assumes that stress is in proportion to strain:

$$\sigma = \alpha \varepsilon$$

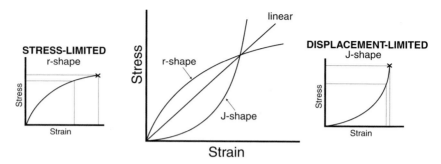

Fig. 5.7 The various types of stress–strain curves exhibited by biological tissues have been pressed into a format whereby the stress (σ) and strain (ε) are related by $\sigma = k\varepsilon^n$, with k being a constant (Purslow, 1991a). In the text, k is assumed to be 1.0. When $n < 1$, the curves are r-shaped, and when $n > 1$, J-shaped.

where $\propto = E$, the Young's modulus. Non-linear behaviour in materials such as animal soft tissues can be modelled by using a power law:

$$\sigma = k\varepsilon^n$$

where k is a constant and the exponent $n > 1.0$ for the concave J-shaped curves of most animal soft tissues, unity for most plant tissues and animal hard tissues, but <1.0 for materials like rubbers (Fig. 5.7).

Mai & Atkins (1975) show that the equivalents of Eqns 5.7–5.9 for such tissues are

$$\frac{F_x}{F_y} = \lambda^{(n+2)(n+1)} \tag{5.11}$$

$$\frac{\sigma_x}{\sigma_y} = \lambda^{-n/(n+1)} \tag{5.12}$$

$$\frac{u_x}{u_y} = \lambda^{n/(n+1)} \tag{5.13}$$

When n is set to 1.0, this trio of equations reduces to Eqns 5.7–5.9. If n gets large, then the ratio of forces approaches λ rather than $\lambda^{1.5}$, stresses move toward $\lambda^{-1.0}$ rather than $\lambda^{-0.5}$ and displacements approach λ rather than $\lambda^{0.5}$. Thus, the predicted exponents in fracture scaling vary with the deformational behaviour of the tissue.

Yamada (1970) shows many force–elongation curves for mammalian soft tissues for which $n = 3$ is a very approximate fit.[10] Assuming this to be typical, then the *length* of the blade of a carnassial tooth of a carnivore (Fig. 6.2), as an example, would be predicted to scale as $M^{0.21}$, considerably lower than for a linear elastic response (with $M^{0.25}$). However, it is important to understand exactly what Eqns 5.11–5.13 do: they predict the scaling of forces for propagating already-formed cracks in geometrically scaled particles, not the difficulty in initiating these cracks. Yet, it is crack initiation that is especially difficult in animal soft tissues (Chapter 4). The analysis may then be inappropriate. Further, it may be invalidated completely if the structure of foods is sufficiently disconnected as to be notch-insensitive (Fig. 4.12 and Appendix A). The next section discusses this aspect of food behaviour in relation to tooth size.

Structural connectivity

Disconnecting parts of a tissue results in a structure that does not respond to fractures in the way that true solids do (Chapter 4). The test of this, described in Appendix A, is called 'notch sensitivity'. Notch-insensitive structures do not transfer strain energy between structural elements and so defeat fracture mechanics by making the fracture stress independent of size. Galileo was right, but just in this special case. However, foods as a whole could be a 'special case' – are any foods actually notch-sensitive? I claim that plant foods basically are, mostly being cellular solids of a completely connected closed-cell variety (Gibson & Ashby, 1999). Examples like the seed coat and pod in Fig. 4.12 are probably exceptional. Large-scale heterogeneities such as the veins in a grass leaf can produce something close to notch insensitivity when the whole leaf is loaded in tension (Vincent, 1982), e.g. during ingestion, but this will not be the case during mastication. So fracture scaling seems to apply to plant tissues.

It is the connective tissue of vertebrates that are largely notch insensitive. In consequence, the complexities of non-linear elasticity are unlikely to apply to carnivores at all. Instead, we should expect that their teeth, such as the carnassials of cats, are not fracture scaled, but geometrically scaled. Van Valkenburgh (1990) shows that the length of the lower carnassial tooth, the first molar, of felids (members of the cat family) and mustelids (weasel family) is approximately isometric to body mass. The reduced major axis for felids gives $M^{0.35}$ (with a sample size of 16 species) while that for mustelids is $M^{0.31}$ (19 species). However, there is considerable variation between taxonomic groups, undoubtedly due to the dietary diversity of the Carnivora.[11]

Evolutionary effect on postcanine tooth size

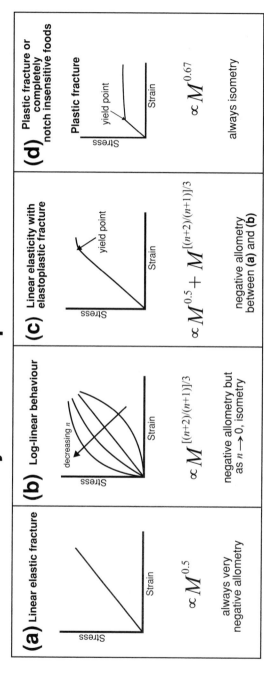

Fig. 5.8 The possible types of stress–strain curve of major foods in mammalian diets and their influence on postcanine tooth size. Mammals that eat foods with some definite elastic response and which have structural connections capable of transferring loads will show some negative allometry of the dentition. A diet of food that is completely plastic or with complete notch insensitivity (i.e. complete disconnection of cload-bearing elements) will result in dental isometry. Positive allometry is not possible.

Other mechanical behaviours exhibited by foods

Any true solid will normally have sufficient elasticity to crack. However, fruit flesh does not have to support a load or even maintain its own shape and so is relieved of the need for elastic response. In many species of fruit, the flesh can become almost completely plastic at ripeness. This tissue would not really be fractured by teeth, but rather flow away from them. The dental task may involve trapping this tissue in order to get individual cells to burst. In this way, it is possible to argue that the work done on a plastic tissue is actually proportional, not to surface areas, but to the volume of tissue being processed (Atkins, 1999). This volumetric action leads to a prediction of isometry of the postcanines of frugivores (if the fruit tissue behaves this way), as going back to Eqn 5.1 shows.

As a final possibility, foods could exhibit a combination of linear elastic behaviour, followed by yielding and a small amount of plastic deformation. This is elastoplastic fracture, for which Atkins (1999) gives effectively

$$\text{tooth size} \propto M^{0.50} + M^{[(n+2)/(n+1)]/3}. \tag{5.14}$$

OVERVIEW OF SCALING ANALYSIS AND TOOTH SIZE

Figure 5.8 sums up the various possibilities for the scaling of tooth size. They vary from strongly negative allometry to isometry.[12] Positive allometry appears to be impossible and rather than represent the norm for the scaling of tooth size in mammals, it probably indicates a dietary shift in larger mammals towards tougher food.

It might be felt by this stage though that the number of alternative scaling possibilities acts against the usefulness of such theories, making them seem infinitely flexible and capable of being twisted to provide an exponent for every circumstance. This is false: a mammal's dentition reflects what it eats. The diet dictates the dental optimum and the better that this can be characterized, the more precise will be the theoretical prediction.

Accordingly, dividing up mammals into herbivores and carnivores is obviously inadequate: tooth size predictions can only be made with confidence on species groups with homologous diets. Without knowing what diets are homologous, it is impossible to predict tooth size. Will folivores have larger teeth than carnivores or frugivores or vice versa? Right now, it is impossible to predict this. Did this analysis predict anything at all? I answer yes, but such predictions lie only in large-scale trends. The disparity between the scaling of postcanine tooth size and jaw size that follows negative allometry is a good example of this and will be returned to in Chapter 7.

THE IMPORTANCE OF SEXUAL DIMORPHISM

All the preceding arguments were derived from considerations of metabolism, in terms of the dimensions of the mouth and its component parts for processing ingested food particles at an adequate rate. Body weight entered the arguments only as a surrogate for basal metabolic rate. Its use though presents considerable statistical difficulties when analysing data for mammals with pronounced sexual dimorphism (sexual differences other than the genitalia). In some species, the body weight of males can be double that of females on the average. If the body weights of enough individuals in those species could be plotted as histograms, then there would be a bimodal distribution attributable to sex: the males on one side, females on the other. Sexual dimorphism can go either way, but in mammals the biggest differences are always associated with larger males (Ralls, 1977). The overall mean of such a bimodal distribution is meaningless because it represents the weight of very few animals – males would be larger than the 'mean' while females would be smaller. In such mammals then, the measurements of males and females have to be considered separately. This does not just apply to body mass. Bimodal distributions are a feature of the sizes of other body parts in these mammals provided that their distributions are normal. This includes the teeth (e.g. Zwell & Pilbeam, 1972; Lauer, 1975). Lauer's study includes the extreme case of sexual dimorphism in *Papio hamadryas cynocephalus*, a primate baboon species with a wide distribution in sub-Saharan Africa (and including a population of very small-bodied forms, the *kindae* baboons). However, rather than being a nuisance, sexual dimorphism can actually help establish whether there is really any critical relationship between variables like tooth size and body weight.

Inconsistencies arise almost immediately when such an analysis is performed. Females in primate species where the males are largest appear to have relatively larger postcanine teeth (Kay, 1978; Harvey *et al.*, 1978). Lucas *et al.* (1986b) and Cochard (1987) both investigated this in detail and found that the larger the male is relative to the female, the smaller is the relative size of the male's teeth. This is actually an extraordinarily regular feature that can be predicted by the degree of dimorphism in body weight (Fig. 5.9). In the extreme, males have only 60% or so of the tooth size predicted for their weights.

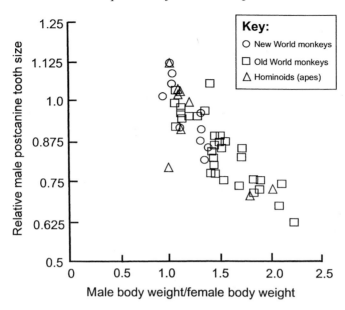

Fig. 5.9 Relative male upper postcanine tooth size (the sum of the areas of the last premolar and all the molars) plotted against the ratio of male to female body weight for 49 species of primates. The larger the male is relative to the female, the relatively smaller is the size of its postcanine teeth. Postcanine tooth size was estimated as the sum of the average area (buccolingual width × mesiodistal length) of the last maxillary premolar and the upper three molars. Data for males and females were kept separate. Postcanine tooth size in females was then regressed against mean female body weight. Relative male posatcanine tooth size was then estimated as observed tooth size for a given body size divided by that expected from the regression line for females. The relationship is highly significant ($r = -0.86$; $p < 0.001$). (From Lucas *et al.* (1986b).)

This is all very difficult to explain from an energetic perspective. In sexually dimorphic species where energy requirements have been well documented, males tend to have higher basal metabolic rates than females even after correcting for body weight differences (Benedict, 1938; Dale *et al.*, 1970). This is the opposite of the trend in relative postcanine tooth size. Pregnancy and (much more so) lactation increase the energy requirements of females. These factors could perhaps explain the trend in Fig. 5.9, but if so, then why in primates where males and females are exactly the same body size are their postcanine teeth also exactly the same size? Females of these species still get pregnant and lactate. Additionally, there are some mammals in which females are larger than males. These are not many, and the difference is not great, but it is enough to try to ascertain if in these species relative tooth size then favours males. Lucas (1980) and Fortelius (1985) assert that this seems to be so. This result suggests very clearly that there is no critical relationship between postcanine tooth size and body size and I am forced to conclude that, unless some highly systematic variation in energy requirements or diet between the sexes is found, tooth size cannot actually be dealt with accurately in the allometric format above. This does not mean that any of the above arguments are wrong, because they related mouth size to food properties, merely that the method of testing it is not correct.

To go back to the beginning, the mammalian body requires a food input delivered at a certain rate. The teeth adapt to the exact form of that input in order to deliver on that rate. What is needed now then is further documentation of that input and this is the direction to which the chapter now turns.

EVIDENCE THAT POSTCANINE TOOTH SIZE AFFECTS THE RATE OF FOOD BREAKDOWN

Before wandering there though, it seems important to hit the ground with some basic facts. Since this book is claiming that foods break down in the mouth at rates that can be influenced favourably by the right combination of tooth shape and size, this seems the right point to lay down evidence showing that tooth size really does matter. Otherwise, it might be queried whether a little cusp on a molar, like a hypocone for example, matters one way or another to the survival of a mammal.[13]

There is some physiological evidence for the importance of postcanine size in ruminants (Perez-Barberia & Gordon, 1998b) and there is a recent report claiming that molar tooth size has an effect on fitness in a wild population of mantled howler monkeys, *Alouatta palliata* (DeGusta *et al.*, 2003). However, it is a century of work on this topic in dentistry that comes to my aid and which is fairly conclusive in showing that small variations in tooth size exert a substantial influence on the rate of food breakdown. The evidence can at times be tricky to interpret because variation in the methods of measurement can introduce distortions. In particular, rather arbitrary measures of the rate of food breakdown, like masticatory efficiency and masticatory performance (Manly & Braley, 1950), exaggerate differences in food breakdown rates produced by different individuals. The following is a précis of dental findings.

The first paper to report a correlation between the rate of food breakdown and the area of the postcanine teeth (in humans) was authored in the late 1940s (Yurkstas & Manly, 1949). Their work culminated in a review by Yurkstas (1965). It is worthwhile documenting the techniques used. The state of the working surface of the postcanines was estimated by placing a sheet of opaque dental wax between upper and lower cheek teeth and asking subjects to bite on it. The wax was thinned by the bite to a degree that depended on the proximity of parts of the upper and lower teeth. This plastically distorted wax sheet was then placed under a light of fixed intensity and the proportion of light transmitted through it recorded. Then, each subject was asked to make 30 chews on a given volume of roasted peanuts and spit out the fragments. The expectorate was washed through a 2.0 mm aperture sieve and the percentage of peanut fragments passing the sieve employed as an index of food breakdown called the masticatory performance. Results for a large number of subjects showed a very strong correlation between masticatory performance and dental state (effectively, the size of the postcanine tooth surface). The relationship was non-linear, but this might simply be due to the food sizing procedure. Helkimo *et al.* (1978) and Kayser (1980) have confirmed this result with slightly different techniques, while Wilding (1993) has found that even differences in tooth size in one individual on the left- and right-hand sides of their mouths will influence breakdown rates in unilateral chewing.

All this work gives little information though on which parts of the dentition are most critical for food breakdown. Experiments by Lambrecht (1965) on the size of the 'postcanine' segment of artificial dentures were more specific, proving that a reduction of just 1.0 mm in buccolingual width of the working surface was sufficient to reduce chewing efficiency by as much as 20%. Luke & Lucas (1985) examined a range of characteristics of the oral cavity of 32 subjects in relation to the rate of breakdown of the median particle size of raw carrot. They studied the shape of the dental arcade, the height of the palate and intra-oral volume in addition to postcanine tooth size. The most important factor by far was tooth size. However, they found that females have higher rates of breakdown for their postcanine tooth areas than males (Fig. 5.10) making it important to place the teeth in their oral context (e.g as a proportion of the oral surface) and not consider them purely in isolation. In summary, all papers on this subject agree that small modifications to the working surface of the molars have a strong effect on the physiological rate of breakdown.

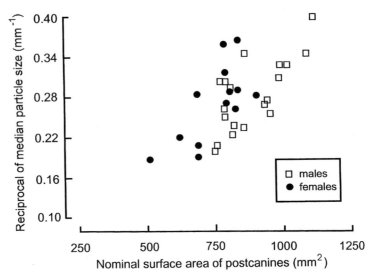

Fig. 5.10 The rate of breakdown of raw carrot particles in 32 human subjects as the reciprocal of median particle size plotted against their postcanine tooth sizes shows that females have generally higher rates of breakdown than males on this basis. (After Luke & Lucas (1985), with permission from the *British Dental Journal.*)

THE EXTERNAL PHYSICAL ATTRIBUTES OF FOODS AND TOOTH SIZE

The external physical attributes of foods, as may be recalled from Chapter 1, describe the food surface. This includes:

(1) The *extent of the food surface*, obtainable from knowledge of food particle size and shape and the total volume of food particles in the mouth, which affects the instantaneous rate of mastication.

(2) The *properties of the food surface*. This category includes stickiness (both between food particles and between a particle and the oral mucosa) and abrasiveness. These properties will both affect the rate of mastication averaged over a time period. However, stickiness and abrasiveness operate on different timescales. The stickiness of food particles changes during the time-span in which a mouthful is being processed, as saliva coats food particles and may be absorbed by them. Stickiness is important in forming a food bolus because it helps to clear the mouth of particles during swallowing (Chapter 3). In contrast, the abrasiveness of foods acts over a much longer duration – the lifespan of the individual – leading to wear that may cause the rate of mastication to decline. This is dealt with in Chapter 6.

All these food surface attributes are united by their effect on the probability of particles being broken by the teeth. Tooth size is the most important oral adaptation to these characteristics because of its dominant influence on the probability of particle fracture (the selection function). Thus, a study of the adaptation of tooth size to diet requires a systematic examination of each of these attributes. The immediate problem though is a working definition of 'relative tooth size'. Tooth size relative to body weight is clearly not adequate for anything this specific. More logically, in terms of dental function, postcanine tooth size should be considered relative to the size of the oral cavity as a whole since the probability of fracture should depend on the proportion of the oral surface given over to the teeth. However, the fracture scaling arguments above suggested that the size of the oral cavity could scale differently to tooth size, such that the postcanine teeth would not occupy the same proportion of the oral surface in all similarly adapted mammals. This is a problem to which I will offer a specific solution later on.

THE CONCEPT OF BREAKAGE SITES

The folding of tooth surfaces into cusps, fossae, crests and ridges makes some parts of these surfaces inevitably more likely to contact food particles than others. This has led to the concept of breakage sites (Chapter 3), hypothetical regions of the tooth where food particles get contacted and broken (Lucas *et al.*, 1985, 1986c; van der Glas *et al.*, 1992). Here, I describe several implications of this concept for understanding an adaptive response in terms of tooth size. In any of the following theoretical examples, only the external attribute being discussed is presumed to be varying.

Food volume (mouthful)

Whenever the number of food particles exceeds the number of breakage sites, then it is obvious that not all particles can be broken in that chew. It follows immediately that the greater the excess of food particles, the slower is the rate of food size breakdown per chew simply because fewer particles in a mouthful get broken in any chew. Accordingly, larger mouthfuls must be reduced in size at slower rates. There is good evidence in humans that this is correct (Lucas & Luke, 1984a). The probability of fracture of all particles appears to be reduced by an increase in food volume being processed. However, volume for volume, it takes a smaller number of chews (and a smaller number of swallows) to ingest food in larger mouthfuls than in smaller ones

(Lucas & Luke, 1984a). There is an upper limit, of course, when the jaws get jammed but no mammal appears to try to chew food in such quantity.

If a mammal can find food in large enough patches to ingest large mouthfuls, then it will not experience any difficulty to getting the necessary food throughput to the gut. However, if a mammal can only find food in small patches, it faces a problem. The evolutionary answer for a mammal adapting to the ingestion of small mouthfuls is to increase the number of breakage sites on its postcanine teeth, i.e. to evolve larger teeth. Even if only a limited amount of tooth surface were then utilized per chew, the benefit in increasing its food throughput would be important.

Food particle size

In general, larger particles get fractured much more commonly than small ones. We know this in humans from data on the selection function (Fig. 3.4) (Lucas & Luke, 1983; van der Glas *et al.*, 1987). This leads to the rate of food breakdown being far greater with large particles than with smaller ones. Thus, for any given food mouthful, the most important influence on the rate of food particle size reduction is food particle size. Generalizing, we can argue that the smaller the particle size in the food intake, the slower the rate of breakdown per chew.

This is not necessarily any problem though. The smaller particles are, the less mastication is needed to swallow them because ingested sizes are closer to the hypothesized particle size threshold that can be swallowed (discussed in Chapter 3). The evolutionary importance of this only arises if ingested food particles are chemically sealed. It is then necessary to ensure that all particles are broken at least once, or else they would simply become dead weight in the gut, passing the entire length of the gastrointestinal tract undigested. For two equal volumes of chemically sealed particles ingested into the mouth, one with large particles and the other small, then the larger the difference between these particle sizes, the slower would be the relative rate of processing the small particles, i.e. making sure that all are broken prior to swallowing. Thus, if a mammal habitually ingests small chemically sealed particles, then it would benefit from increase in postcanine tooth size to increase the probability of fracture.

Food particle shape

Food particles are ingested in a wide range of shapes, from relatively iso-diametric particles (i.e. with all three dimensions being more or less equal) like many seeds or fleshy fruits, to sheet forms (with one dimension being small), such as such wind-dispersed fruits or leaves, to rod-like particles (with one dimension being large), such as stalks, stems or branches. The shape of a particle must be very important initially because teeth act on food surfaces and the presentation of that food surface is very dependent on shape. The rate of size reduction also depends very greatly on shape. If a sphere, rod and sheet are compared, then the rate of production of new surface will depend really on their smallest dimension. The initial advantage that a sheet-like particle like a leaf has, with its very high specific surface, over a sphere, for example, is rapidly lost because fracture of the sphere exposes surface more quickly than fracture of the sheet. The most important effect of food particle shape on postcanine tooth size point is likely to be that sheet-like and rod-like foods are in reality small objects because their volumes are small. If they are sealed particles, then their efficient processing might require enlarged postcanines.

Stickiness

The stickiness of a food is produced by a combination of its own surface properties, any juice liberated from it during food comminution, and saliva. Particles of most fruit flesh and animal soft tissues stick together very easily and make a swallowable bolus rapidly. However, the cuticle of leaves, a hydrophobic waxy layer (Barthlott & Neinhuis, 1997), will resist the leaf surface being wetted in the mouth and inhibit bolus formation. This probably has considerable implications for tooth size, but not with equal effect along the tooth row.

During chewing, the tongue throws food particles laterally (Fig. 3.7). A sticky bolus, e.g. of fruit flesh, will not spread along the tooth row. This puts a premium on postcanine surface in the central part of the tooth row (e.g. in the first molar region) where most of the food particles would be expected to be broken. Foods like leaves or carrot, however, would disperse much more evenly along the tooth row, because they do not form a bolus easily, suggesting that a more even distribution of tooth area is then optimal. Thus, a long narrow tooth row is suited for non-bolus-forming foods while a wide short dental arch would be best for a mammal eating sticky food.

Table 5.5 tabulates these predictions while Fig. 5.11 shows them graphically. For a test, it remains just to try to put real food names to these attributes. Just a brief discussion suffices here. Leaves are small chemically sealed sheet-like particles that do not form boluses. So leaf-eaters (folivores)

Table 5.5 *The relationships between variation in the external physical (surface) attributes of food particles and tooth size*

Food attribute	Factor favouring large teeth	Factor favouring small teeth
Particle volume	Small	Large
Particle size (when chemically sealed)	Small	Large
Particle shape	Sheets/rod	Isodiametric
Bolus formation	Non-bolus-forming	Bolus-forming

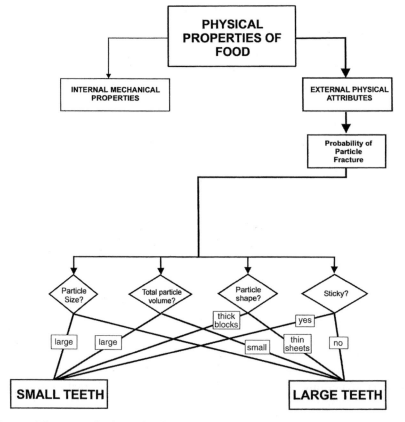

Fig. 5.11 The expected relationship between the surface attributes of food particles and postcanine tooth size. 'Large teeth' means a tooth row with equally large molars and a tendency for the premolars to be 'molarized', while 'small teeth' means a tooth row with most of its surface area located in the centre of the row around the first molar.

should have large postcanine teeth. All else being equal, the smaller the leaves that are consumed, the larger should be their teeth. Folivores that eat shoots and branches, i.e. rod-like particles, should also have the largest teeth. In contrast, fruit-eaters and carnivores should have relatively small teeth. Frugivores tend not to destroy seeds (see p. 217), but leaf-eaters commonly do. Seed-eaters should also have large teeth. Quite a lot of evidence appears to support the above. Freeman (1988) shows that many frugivorous bats have dentitions with wide arches and short tooth rows. This is also true in many primates and Kay (1978) shows that folivorous primates tend to have larger postcanines than frugivores. Carnivorous mammals tend to have reduced postcanine dentitions centred on one pair of carnassial teeth, which also appears to fit predictions.

In summary, very simple theories about the evolutionary effect of the surface attributes of foods produce predictions about tooth size in different dietary groups much more readily than the allometric theories described earlier in this chapter.

CANINE SIZE

The canines of most mammals are used to fracture the integument of a target so as to produce a deep and penetrating wound. Fracture, not fragmentation, is their major design characteristic. Early insectivorous mammals had enlarged canines in both deciduous and permanent dentitions for this purpose (Romer, 1966) and carnivores have them too. In some New World monkeys, they are used to break into the hard outer layer of fruits (*Chiropotes satanas* and *Pithecia pithecia*: Kinzey & Norconk, 1990) but this is unusual in plant feeders.

In anthropoid primates, several herbivores and some marine mammals, the canines tend to be much larger in males than in females. Even when there is no sexual difference in the permanent canines, as in gibbons (the lesser apes: genus *Hylobates*), the disparity in size between non-projecting deciduous canines and their permanent replacements betrays a role outside of feeding, in social behaviour in fact (Plavcan *et al.*, 1995). The major context for canine function in these mammals appears to be during conflicts between males (and sometimes between females) where they form potent weapons. They are rarely used to kill though because there is usually strong regulation of disputes, something that has been cogently explained by game theory (Maynard Smith & Price, 1973; Maynard Smith, 1979). There appears to be active selection for canine size enlargement in the females of some living primates (Hayes *et al.*, 1996).

Whatever their role, canines require a large gape in which to operate, a fact that relates the size of these anterior teeth back to the early part of this chapter and that provides a good test of this. Herring (1972) was the first researcher to emphasize this in a study of Old World suids (pigs) and New World tayussuids (peccaries). Later, she established that most mammals generally manage gapes of 60–70° of rotation of the mandible (Herring & Herring, 1974), which will limit canine projection to some fraction of this angle (otherwise they could not hit anything).

There are exceptions to this gape limit. The hippopotamus can produce immense gapes (Herring, 1975) and many fossil mammals, such as sabre-toothed carnivores (Emerson & Radinsky, 1980), must have been able to do so as well. This is no easy matter to arrange though. One corollary in the hippopotamus (Herring, 1975), as also in many molossid bats (Freeman, 1981a), is slack cheeks in order to allow them to stretch sufficiently during 'yawns'. When the mouth is closed, the cheeks and lips are flaccid or wrinkled, possibly impairing their use in sealing the front of the mouth during food processing. The jaw joint can also be set very low, e.g. in marmosets (Vinyard *et al.*, 2003), so as to reduce stretch in the masseter and medial pterygoid muscles (Fig. 2.16).

Canine size and gape

The upper and lower canines must clear their tips well before the maximum gape is reached. Figure 5.12 illustrates this using the skull of an Old World monkey (the gelada, *Theropithecus gelada*), where two important angles of jaw opening are distinguished, the angle at which the tips of upper and lower canines just clear each other, termed θ_c, and that at which the canines are normally used, θ_u.[14] The latter angle is likely to be at or around the maximum gape. There are no obvious adaptations to enlarged gapes in primates like this. None appears to have wrinkled lips – features that might interfere with food manipulation and certainly preclude food wadging in orang-utans and chimpanzees (Walker, 1979).

The gape angles required to just clear the tips of upper and lower canines (θ_c) can be measured on skulls. Results on a large sample of primates and a small sample of carnivores are shown in Fig. 5.13 (Lucas *et al.*, 1986b). Gapes are identical in male and female carnivores of the same species, but range from around 10° in ursid (bear) species, through intermediate values in canids (dogs) to 20° or more in felids (cats). Female primates have small canines, but male primates have a very similar range to those of carnivores, with the same maximum of 23°. Note that these gapes only clear the canine

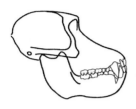 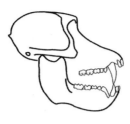 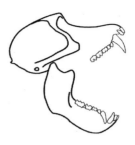

(a) Teeth together   (b) Canines tip-to-tip (θ_c) (c) Useful gape (θ_u)

Fig. 5.12 (a) The living male gelada baboon has large upper and lower canine teeth. (b) These clear each other (in this specimen) at a jaw joint rotation of approximately 17°, a gape called θ_c in the text. In this situation, the teeth are useless. (c) The gape, θ_u, at which the teeth might be used. This has been achieved in the diagram by extending the skull on the neck by 30° and by depressing the lower jaw 35°. The combination of neck and jaw opening muscles to this gape makes no difference to the actual rotation of the jaw joint (i.e. $\theta_c = 65°$), but the way that a mammal achieves it may play a substantial role in affecting the relative sizes of the upper and lower canines.

tips. They do not offer any possibility of hitting anything: that possibility only begins at this gape.

The major point to stress about these θ_c values is their variation is limited, so supporting a key contention of the earlier part of this chapter that the anterior dentition is isometric to jaw length (Lucas, 1982b). However, what explains the variability seen? One answer is the influence of variation in the height of the jaw joint, which will now be modelled.

Suppose that the jaw joint lies on the plane at which the teeth bite together and the length of the jaw between the joint and the canine is l (Fig. 5.14a). If the jaw is opened through an angle θ, then relative to its position when the jaw is closed, its vertical displacement is

$$y = l \sin \theta$$

while the horizontal retraction is given by

$$x = l(1 - \cos \theta).$$

Carnivores all essentially have their jaw joints on this plane, which is optimal for the wide gapes that they require for food acquisition. However, many herbivores have high jaw joints, which effectively restrict their gapes. Unsurprisingly, low jaw joints tend to be associated with diets requiring large gapes and with large canines and vice versa (Chapters 2 and 3). As

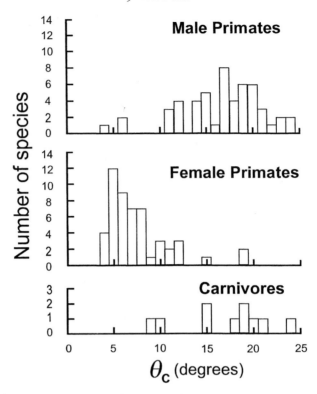

Fig. 5.13 The average gapes at which the tips of the upper and lower canines of living anthropoid primate and carnivore species are just clear of each other. Sexual dimorphism in the canines of the anthropoids is obvious: the canines of females in most of these species do not project very far. The canines of the male primates and carnivores have a similar gape distribution.

shown in Fig. 5.14b, when the jaw joint is raised above the plane of the bite, then the ratio of y/x is increased, so reducing the potency of the canines as weapons (Herring, 1972). Setting the jaw joint a distance h above the bite plane, the vertical canine displacement becomes

$$y = l \sin \theta - h(1 - \cos \theta), \qquad (5.15)$$

while the horizontal retraction is given by

$$x = l - (l \cos \theta - h \sin \theta). \qquad (5.16)$$

Even without plotting anything out, it is obvious from Fig. 5.14 that the higher the jaw joint (i.e. larger h/l), the greater the y/x ratio at any value

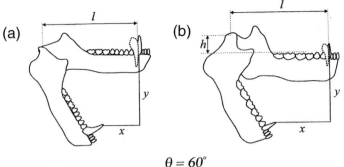

$$\theta = 60°$$

Fig. 5.14 (a) A mammal with a jaw length, l, and a jaw joint level with the bite plane of the molar teeth rotates its jaw through 60°, moving its lower canine a vertical distance y and a horizontal distance x. (b) The jaw joint is now raised by a distance h. This reduces the y/x ratio at any given gape, resulting in the lower canine moving further backwards compared to (a).

of θ. Despite large canines being apparently incompatible with high jaw joints, h/l ratios of 0.5 have been measured in male anthropoid primates. It seems likely that this would limit their canine size. Accordingly, Fig. 5.15 plots h/l ratios for male and female primates versus θ_c values. There is no relationship in females (Smith, 1984) because they usually have small canines, but there is a significant and quite strong negative relationship in males. This relationship could just be an accident. To show that it is not, the analysis must go further and predict the gape angle at which canines are used (θ_u).

I make the assumption that canine size adapts such that θ_u corresponds to a particular value of y/x, in common for all species in Fig. 5.15. As shown in this figure, the trend line through the θ_c data does not correspond with slopes of equal y/x.[15] However, correspondence can be obtained if all the θ_c values are multiplied by a constant m, such that

$$\theta_u = m\theta_c.$$

There is only one value of m that will do this. If m turned out to be <1.0, then the gape theory simply cannot hold. On the other hand, if $m \gg 1.0$, then this would provide a lot of support because obviously, $\theta_u \gg \theta_c$ (see Fig. 5.12). As shown on Fig. 5.15, the fit is $m \approx 2.8$. Taking the maximum θ_c for any living primate or carnivore, then $\theta_u = 2.8 \times 23° = 64°$, which further suggests that the theory is plausible.

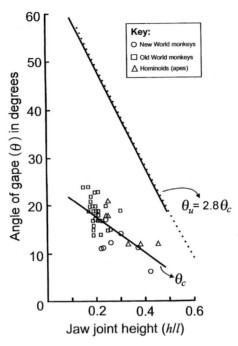

Fig. 5.15 A plot of the gape angle at which the tips of the upper and lower permanent canines of male primates just clear (θ_c) versus the height of the jaw joint, h/l. The relationship is negative ($n = 51$, $r = -0.50$; $p < 0.001$). The solid trend line was produced by multiplying values of θ_c by a constant m until it matched that of the dotted line, which is one of a family of curves of constant y/x generated from Eqns 5.15 and 5.16. The value of m turns out to be ~2.8.

Are there any relevant physiological data on maximum gapes in mammals? Carlson (1977) states that female rhesus macaque monkeys (*Macaca mulatta*) can only open their jaws to about 30°. It is possible, but unlikely, that males can open their jaws further. $\theta_c = 30°$ for male rhesus macaques should be compared to a θ_u of 42° predicted from the theory (i.e. $2.8\theta_c$). This fit is improved when an extra feature, not so far included, is considered. The head of the condylar process of the mandible of some primates, including rhesus macaques and humans, translates anteriorly during sufficiently wide jaw openings. This does not affect measured values of θ_c, but it does increase y/x ratios. The average maximum values for anterior translation in humans and rhesus macaques (Carlson, 1977) seem about 10–15% of the jaw length, *l*. Though the correction is not entirely satisfactory, all h/l ratios in Fig. 5.15 could be assumed overestimated by

15%. This reduces *m* to 2.5 and changes θ_u for male macaques from 42° to 37.5°, which again seems reasonable. Going further, could the theory explain the size of the canines in sabre-toothed carnivores? Drawings suggest that the large upper canine cleared the small lower one at about 50°. Its jaw joint was fixed and thus $\theta_c = 50° \times 2.8 = 140°$, which also seems satisfactory.

There are a considerable number of extinct primate species with high jaw joints and reduced canines, including the fossil dryopithecine ape *Sivapithecus indicus* (for which $h/l \approx 0.5$, unpublished data), *Theropithecus oswaldi* (Jablonski, 1993), particularly later members of the species and *Oreopithecus bambolii*, for which a reconstruction in Moyà Solà & Köhler (1997) gives $h/l \approx 0.60$, a record in primates. This primate probably had a very small mouthslit. It will not have escaped readers' attention that humans do not have projecting canines in either sex either. Our species has a relatively high jaw joint ($h/l \approx 0.33$), although the data shown in Fig. 5.15 suggest that this would not be sufficient to explain why our canines are very small. Humans also have very small mouthslits and consequent small maximum gapes (23°: Baragar & Osborn, 1984). This is discussed further in Chapter 7.

I have only analysed one possible influence on canine size. While it seems to predict the maximal (optimal) size of these teeth, they are of course free to vary in size below this maximum. Nothing has been said about why there is often a big disparity between the sizes of the upper and lower canines. In some mammals, such as mouse deer (chevrotains), muntjacs, Chinese water deer, sea cows and walruses, there is only an upper canine. It is possible that some of these mammals use their canines with their mouths closed. Walruses do this, for example, when they use their canines to move across ice. However, why in most sabre-toothed carnivores is the upper much longer than the lower? The answer probably depends on whether a canine bite is made more by the neck muscles than by those of the jaw.

CONCLUSIONS

Mammals are what they eat and there is no point in trying to understand tooth size without having first characterized the properties of the diet. Jaw and anterior tooth size are linked to the size of ingested food particles, but the sizes of the postcanine teeth are not. They depend on the deformation behaviour of food particles and their external physical characteristics. Kay (1978) stated that a 'coherent theory which accounts for molar size must explain why tooth dimensions are negatively allometric with respect to "metabolic body size (mass$^{3/4}$)"' and 'why all [tooth] dimensions do not follow the same regressions'. That is a pretty tall order, but I suggest that the theories contained here are indeed capable for tackling such a problem. There is a vast literature on tooth size, but little quantification of the diet. That has to change if headway is to be made.

6

Tooth wear

OVERVIEW

It will be recalled from the last chapter just how dependent the rate of food breakdown is on tooth size. However, the exposed surfaces of the teeth are subject to wear, either from food, grit or opposing tooth surfaces. Wear threatens to destroy tooth shape, decreasing the rate of food breakdown and so jeopardizing a mammal's health. Insofar as the food input is responsible for this wear, the rate must depend on the characteristics of the food surface, i.e. on the external physical attributes of foods, because wear involves small-scale events involving the interaction of surfaces. Maintaining the argument from Chapter 1, I expect the evolved response to this threat to lie in adaptation of tooth size. This chapter follows through the logic of this argument and also tries to identify the major causes of tooth wear.

INTRODUCTION

Wear is the loss of volume of an object and results from a number of processes rather than just being one process in itself. It may often involve fracture, but the chemical dissolution of stony objects like teeth also comes under the same general heading. Teeth wear mostly on their working surfaces, but any other tooth surface exposed to the oral cavity can wear too. Movement of food by the tongue and cheeks against the teeth is sufficient to cause a small amount of wear (Fox *et al.*, 1996; Ungar & Teaford, 1996), enough anyway to result in the gradual loss of perikymata (a pattern of ridges and troughs present on a newly erupted enamel surface). The most important of these other surfaces though are those where adjacent teeth in the dental arch make contact.

There have been three distinct tools for investigating wear: the naked eye, the light microscope and the scanning electron microscope. Gross wear, as observable with the naked eye, has lacked much quantification, but this

Table 6.1 *The pH of fruit flesh eaten/avoided by long-tailed macaques (* Macaca fascicularis*) in Bukit Timah, Singapore*

Species	pH[a]	Status in the diet
Calophyllum pulcherrimum (Guttiferae)	3.0–3.5	Eaten
Garcinia forbesii (Guttiferae)	2.5–3.0	Eaten
G. parvifolia (Guttiferae)	3.0–3.5	Eaten
Nephelium lappaceum (Sapindaceae)	4.6[b]	Eaten
Tinomiscium petiolare (Menispermaceae)	5.0–5.5	Eaten sporadically
Fibrauria tinctoria (Menispermaceae)	5.5–6.0	Eaten
Mezzettia parvifolia (Annonaceae)	4.0–4.5	Eaten sporadically
Tetrastigma lawsoni (Vitaceae)	3.0	Eaten
Pellacalyx saccardianus (Rhizophoraceae)	5.0	Avoided
Gnetum microcarpum (Gnetaceae)	5.0	Eaten

[a]Measured with half-unit pH paper.
[b]Data from Poon (1974).

is now changing rapidly using geographical techniques designed originally for contour mapping (Zuccotti *et al.*, 1998; Jernvall & Selänne, 1999; Ungar & Williamson, 2000). However, it is the scanning electron microscope that has yielded most insight. Initiated 25 years ago (Walker *et al.*, 1978), this 'dental microwear analysis' can resolve actual wear features (Teaford, 1994). Microwear studies have proved that mechanical processes are almost certainly the predominant cause of wear, leading to the pitting and scratching of tooth surfaces.[1] The other major possibility is erosion, the term given to the chemical dissolution of tooth tissue by acids, particularly from fruits (Table 6.1). This tends to produce a matte surface, representing the uneven loss of mineral from enamel rods, depending on their orientation. It is probable that saliva buffers acid so as to prevent much tissue loss, but I do not know.

As a topic in mechanics in general, wear remains very incompletely understood. Although understanding grew rapidly in the early to mid parts of the twentieth century due to interest in friction (Bowden & Tabor, 1950), there remains much that is not understood. Accordingly, it is difficult to predict wear rates. All this becomes obvious when you start to read around the subject. Books or articles on wear tend to be long, to subdivide the subject into categories that are not clearly connected, and to be specific rather than general. This is all very sad for dental studies because the general state of the dentition present in an individual mammal of any species, picked out at random, is worn. Unless the factors that control its wear can be understood, the mammalian dentition cannot really be understood. Mammals have tossed away the reptilian option of many tooth generations and gambled on just a couple, so the longevity of teeth is vital.

I will limit my observations here to some that could influence thinking about the causes of wear in biological terms. I cannot do any better than others in fundamental terms, but this is my brief shot at a difficult subject.

WEAR TERMINOLOGY

Dental researchers often use the term 'attrition' to describe the wear of the working surfaces against each other, while wear caused by intervening food is called 'abrasion'. With either the naked eye or a magnifying glass, abraded molar surfaces have an irregular matte surface, while attrition seems to give rise to flat glossy surfaces. The latter shiny surfaces are called 'facets' and can be seen clearly either with the naked eye or at low magnification under oblique lighting. Facets tend to develop in similar positions on the teeth of a given species because malocclusions are rare and so opposing teeth always contact in the same areas (Mills, 1955). Butler (1952) initiated the study of facets and started a system of nomenclature for them in the molar region. This work has since been extended to a wide range of mammals and their evolution in early mammals has been plotted (Crompton, 1971). The shape of wear facets provides a strong indication of whether features of the working surfaces of teeth are operating as cusps or as blades.

Under the higher magnification available with a scanning electron microscope, the difference between abraded surfaces and attrition facets begins to blur. Many facets that look featureless at lower power are actually scratched and pitted when magnified sufficiently (Teaford & Walker, 1983). Clearly, food must be entering the space between upper and lower tooth surfaces. This is only right and proper of course because this is what teeth are there for, but many questions arise from it. Why do facets and abraded surfaces look different? Are facets caused more by the contact of opposing teeth than by food, or are some of the assumptions wrong and food capable of producing glassy surfaces? Indeed, what is the predominant cause of tooth wear?

Since I want to get at causation, I will redefine the terminology slightly here. What dentists call attrition, I will call tooth–tooth wear, and what they label abrasion, I will refer to as food–tooth wear. This allows the term 'abrasion' to be specialized here to mean 'mechanical wear involving fracture'. It will not be limited in this chapter to contacts in which food intrudes. Additionally, a pit will be defined as a roughly isodiametric feature on a worn tooth, while a scratch is elongate (Teaford, 1994).

MECHANICAL WEAR

Any mechanical process whereby small fragments are lost from a large block is likely to be called abrasion. This type of wear was long thought determined by the relative hardness of contacting surfaces. Hardness itself is now known not to be the cause, but the root of much wear is still believed to lie in indentation, which is of course how hardness is measured (Appendix A). The ability of one object to indent another is determined by their relative hardness, but loss of volume depends on other properties too, such as toughness, Young's modulus and surface features that influence friction (Kendall, 2001). This is entirely logical: if wear involves loss of tissue by fracture, then toughness must play a part in this. A general theory of mechanical wear that incorporates toughness is surely necessary to explain how, for example, mammalian skin and hair can wear enamel (Aitchison, 1946; Rose *et al.*, 1981), but it has not been discovered yet.

Tooth–tooth wear

Teeth are very often the hardest substances in the mouth and are, thus, potentially serious wear agents unless their motion can be controlled precisely. This potential has been proven by examination of the wear of the molars of the guinea pig (*Cavia porcellus*). This mammal erupts its molars *in utero* and makes empty jaw movements with sufficient force before birth to produce tooth–tooth wear. Teaford & Walker (1983) show that worn molars are basically featureless flat surfaces (presumably well washed by amniotic fluid), confirming that tooth–tooth wear produces facets. Teaford & Walker (1983) and Walker (1984) have also run simple *in vitro* experiments on tooth–tooth wear, showing that the wear of enamel depends on the orientation of its component rods to the tooth surface: rods normal to the contacting surface wear faster than those that are parallel to it.

Humans also commonly wear their teeth without any food being present, but they do this after birth, not before it. Such events are called 'bruxism'. They usually take place at night, in a small proportion of the population, and are the presumptive result of psychological stress (Lavigne *et al.*, 2003). In a number of papers, Every (e.g. Every, 1970) hypothesized that all animals need to 'brux' in order to preserve the sharpness of their teeth, reasoning that if foods blunt teeth, tooth–tooth wear is there to resharpen them.[2] He called bruxism in mammals 'thegosis', a term implying that this is normal rather than pathological behaviour. His theory still attracts worldwide interest and has been critically reviewed by Murray & Sanson (1998).

I do not wish to go into the wilder aspects of Every's thinking about the direction of jaw movement in thegosis (he thought that these were opposite to those in mastication – there is no evidence for this), or his view that the dentition is basically designed as non-alimentary weaponry (a theory that could only emerge from dentistry). He was correct about the wear of some of the anterior teeth, although all that did there was to name a behaviour that had already been observed. The wear of the canines of male Old World monkeys (shown in a fantastic series of photographs in Every, 1970) is almost certainly due to specific sharpening activity. Zingeser (1969) described the facets that develop from this, but there are other undoubted instances of thegosis, such as the process of wearing of the lower incisors of rodents, which can only result from a protrusion of the jaw beyond a point where it could conceivably be used in ingestion (Osborn, 1969). However, there is no evidence though that sharpening activities are more widespread than this in mammals and they do not include the postcanines. Why then did it have a vogue? The underlying cause of its fleeting popularity seems to be that everyone believed that teeth needed to be sharp. The reasoning would be something like this: if foods break at fixed stresses, then sharp teeth will possess the smallest contact areas and so generate these stresses at the lowest forces. Accordingly, efficient teeth should be as sharp as possible. Yet, tooth sharpness cannot be maintained because contacts with food will blunt them. A search for separate sharpening activities thus seems logical.

This reasoning is flawed, as Chapter 4 should make quite clear, but there has never been a clear refutation. Instead, researchers have looked for alternatives. Osborn & Lumsden (1978) published the most famous of these, proposing that teeth are maintained sufficiently sharp by dietary use.

Food–tooth wear

The biggest threat is thought to come from quartz, which is the 'extraneous grit' of microwear studies. There is very strong evidence for its influence as a dental wear agent not just in terrestrial mammals, but also in arboreal mammals in tropical forests (Ungar *et al.*, 1995). For herbivores, a close second is the 'opaline' amorphous silica deposited in many plant tissues. This has been found on the teeth of fossils (e.g. *Gigantopithecus blacki*, a Pleistocene ape species from China: Ciochon *et al.*, 1990) and prehistoric humans (Fox *et al.*, 1996) and passes through the gut of living mammals into the faeces, e.g. of sheep (Baker *et al.*, 1959) and hyraxes (Walker *et al.*, 1978). In any plant, the shape and size of these phytoliths varies with the cells that contain them (Ball *et al.*, 1993). Some plants also contain crystals of calcium salt, such as those of calcium carbonate and calcium oxalate (Metcalfe & Chalk, 1950). The latter are more common and there is strong circumstantial evidence for them as a wear agent in prehistoric human populations (Danielson & Reinhard, 1998). A factor whose importance is completely unknown is the practice of geophagy – the consumption of 'soils' – that is widespread in mammals (e.g. in primates: Oates, 1978; Krishnamani & Mahaney, 2000; Dominy, 2001).

Additionally, woody tissues themselves may wear the teeth, certainly doing so in rodents and being the cause of the need for ever-growing incisors. Some fruit flesh contains isolated 'stone cells', which produce a gritty feel (Lucas, 1991). For carnivores, there is bone (Van Valkenburgh *et al.*, 1990) and for insectivores, the cuticle (Schofield *et al.*, 2002). Freeman (1979, 1981a) first pointed out clear differences between types of bats that eat insects with unsclerotized (e.g. moth) or sclerotized (e.g. beetle) cuticles. The latter produce many pits on teeth (Strait, 1993c), more than on primate leaf- or fruit-eaters. Similar wear is associated with fracture of bone in the mouth (Van Valkenburgh *et al.*, 1990).

INDENTATION AND WEAR

At the heart of every pit and scratch is probably an indentation process. This section discusses how indentation leads to wear and why scratches are far more probable wear features than pits. When two particles of dissimilar materials are pressed together, then the harder one will indent the softer one. Permanent indentation (due to plastic deformation) starts when the stress in the softer solid exceeds 1.1 times its yield strength, i.e. $1.1\sigma_y$, and full plasticity takes place at $2.8\sigma_y$ (Atkins, 1982). However, if the differences in hardness between the two materials are not very great, then both will be affected by the contact. To prove this, I need to distinguish between the yield stresses of the harder and softer solids. For convenience, the yield stress of the softer solid will be termed Y_s and that of the harder one, Y_h. If full plasticity in the softer solid is not achieved before the harder solid reaches $1.1Y_h$, then the harder solid will also yield. The boundary between a one-sided effect and mutual indentation is, therefore,

$$2.8Y_s = 1.1Y_h. \tag{6.1}$$

Provided that the harder solid has a yield stress $(2.8/1.1) = 2.5$ times that of the softer solid, then it will not be affected by the indentation. However, it is not very convenient to deal only in yield stresses because these can be difficult to measure. Hardness, the mean pressure that produces an indentation of unit area (Appendix A), is much simpler. The types of materials that could indent tooth tissues are all going to be true solids (i.e. with negligible airspace), so I will assume for simplicity that the hardness of potential wearing agents is always going to be a common multiple of the yield stress. I will take this multiple to be 2.8 so as to comply with Atkins (1982).[3]

Enamel microwear

I will assume initially that microwear features are plastic indentations. The force, F, required to make a pit comes directly from the definition of hardness and is

$$F = H\pi r^2 \tag{6.2}$$

where H is the hardness of enamel and r the pit radius. Taking the hardness of enamel as $H = 3700$ MPa (just an average value – it is highly variable as shown in Appendix B) and a typical pit radius as $r = 2.5$ μm (Teaford, 1988), then $F = 0.073$ N, which must be small compared to masticatory forces and very small compared to the bite forces known for any mammal (Chapter 5). Thus, many such pits could form simultaneously were suitable particles to be present in the food intake.

However, the pit might also be formed from a contact without any real indentation, the rough pit being the product of something like a cone crack (Lawn, 1993). An estimate for this force is

$$F = 6\pi r^2 \left(\frac{ER}{\pi a} \right)^{0.5} \tag{6.3}$$

where r is the radius of contact, a is the length of the crack and E and R both refer to the properties of the enamel (Sharp *et al.*, 1993). Inserting $r = 2.5$ μm and an arbitrary, but constant, value for $a = 5$ μm, and taking typical values for enamel from Appendix B, $E = 90$ GPa and $R = 13$ J m^{-2} (remembering that enamel nearly always fractures along rods), then $F = 0.033$ N, which is even smaller than for plastic indentation.

The main effect of a sliding contact against the abrading particle, so as to form a scratch, is to reduce the forces involved. These forces depend,

Table 6.2 *Frictional characteristics of tooth tissues*

Friction	Coefficient of friction (μ)	Reference
Enamel–enamel	0.10–0.42	Douglas *et al.* (1985)
Diamond–enamel	0.14 (±0.02)	Habelitz *et al.* (2001)
Diamond–dentine	0.31 (±0.05)	Habelitz *et al.* (2001)

however, very strongly on the coefficient of friction, μ, between the surfaces. Table 6.2 gives some values from the literature.

Sharp *et al.* (1993) give the force for plastic indentation as

$$F = \frac{\pi w^2 H}{(1 + 9\mu^2)^{0.5}}.$$
(6.4)

For a scratch of width $2w = 2$ μm (a typical value: Teaford, 1988), taking $\mu = 0.14$ from Table 6.1, then F ≈ 0.011 N, which is even smaller still.

Alternatively, if there is fracture without indentation, then the appropriate relation is

$$F = \frac{\pi w^2}{1.12 \left[\frac{1}{6} + \frac{13}{16}\pi\mu \right]} \left(\frac{ER}{\pi a} \right)^{0.5}.$$
(6.5)

This relation assumes Poisson's ratio to be 0.33 (Sharp *et al.*, 1993), which is probably a little high for enamel and the types of materials that can wear it (Lawn, 1993). However, for the same 2 μm wide scratch, creating a 5 μm crack, then $F \approx 0.001$ N, which is the lowest of the four force estimates.

Appendix B gives hardness values for dental tissues and various possible causative agents of pits. Few materials look capable of pitting enamel (many more could scratch it), but there are a lot of candidates lined up to damage dentine.

The biggest threat overall is from quartz with about twice the hardness of enamel, while 'opaline' amorphous silica is about 1.6 times as hard. Both quartz and opaline silica are in a range whereby they could indent enamel, but they have <2.5 times its hardness and so would themselves permanently deformed. This 'mutual indentation' has been subjected to considerable research by Atkins (1982) and results in mean pressures that are considerably below those when only one particle deforms. However, the process also depends on the geometry of loading. Most tooth-abrasive particle contacts are likely to be of a 'ball-on-flat' type whereby a piece of quartz dust or a roughly spherical phytolith presses against a larger, relatively flat, tooth surface. Assuming this geometry, then enamel that is being indented by a roughly spherical phytolith of hardness 1.6 times that of enamel, becomes fully plastic at just $2.2Y_s$ (Atkins & Felbeck, 1974). Equation 6.1 now becomes

$$(2.2/2.8)Y_s = 1.1Y_h,$$

reducing the pit-forming force to 0.057 N. Nevertheless, I can sum all this up and, while accepting that these are very rough calculations, suggest that scratches will form much more readily than pits on enamel. The difference in predicted forces is so great that this conclusion is not really affected by substituting other values for the coefficient of friction (e.g. $\mu = 0.41$; Table 6.2). The predicted 'scratch predominance' on worn enamel is supported by the literature.[4]

Teaford (1994) reported data on the enamel microwear of 20 species of primate, with widely varying diets, finding that scratches are always more frequent than pits. However, the formation of scratches requires a sliding contact, so the only occasions on which pits would be predicted to predominate would be when there is just compression. There is support for this as well. Gordon (1984) showed in the chimpanzee that pits are more frequent on the enamel of third molars than the others probably because they are closer to the jaw joint and so to the centre of rotation. There are more pits on the molars of frugivorous primates than on those of folivores, suggesting that there is less tooth sliding during mastication in the former. The extent of wear areas on the molars of Old World monkeys, a group with great dietary diversity, seems to support this. 'Crushing' areas are relatively larger in frugivorous primates, while 'shearing' facets are correspondingly enlarged in folivorous primates (Kay, 1978), and presumably covered in scratches.

Dentine microwear

Dentinal surfaces are very rarely scrutinized in microwear studies because the surface is so intrinsically heterogeneous that quantification of microwear features is made very difficult (P. S. Ungar, pers. comm.). It also collects debris, which is hard to remove (M. F. Teaford, pers. comm.). Nevertheless, the same calculations can be made as for enamel. Ignoring mutual indentation for the moment and, taking typical values from Appendix B for dentine properties ($H = 650$ MPa, $E = 20$ GPa, $R = 270$ J m^{-2}) with $\mu = 0.31$ from Table 6.2, provides a fascinating comparison with enamel (Table 6.3).

Table 6.3 *Approximate forces (in millinewtons) for the creation of microwear features, either by plastic indentation or surface fracture*

	Pit (5 μm diameter)		Scratch (2 μm width)	
	Plastic indentation (Eqn 6.2)	*Fracture* (Eqn 6.3)	*Plastic indentation* (Eqn 6.4)	*Fracture* (Eqn 6.5)
Enamel	73	33	11	1
Dentine	13	41	1.5	1

It seems that when their surfaces are compressed against an abrasive, enamel is much more likely to suffer microfractures than dentine. This seems again to be borne out by the evidence. Most published micrographs of enamel pits seem to have ragged edges. Dentine is easily indented, but this may often be just plastic modification. However, wear could result from other fracture possibilities, which depend very much on the sharpness of the abrasive particle (Lawn, 1993).

Similar to enamel, scratches on dentine are much more likely than pits. However, in a major surprise, Table 6.3 suggests that although scratches form at lower forces than those in enamel, fracture damage under sliding forces is probably equally likely in both tissues. No data in the literature appears to relate directly to this, but this result can be interpreted as showing how the material properties of enamel and dentine support the stable construction of teeth after the dentine is exposed.

Relevance to dental wear in herbivores

The dentition of most herbivores is so designed as to expose dentine at the earliest opportunity. Guinea pigs do this in the womb, while the molars of mice erupt with enamel-free areas already present on their cusp tips. Other herbivores, lacking both these characteristics, have very thin enamel at the cusp tip (e.g. in the hippopotamus: Luke & Lucas, 1983), quite unlike in humans, for example, where cuspal enamel is thickest. This also ensures that dentine will be open to wear processes as soon as possible. The need for dentine (and, in some species, cement) on the crown surface is that it allows the edges of enamel ridges, which wear more slowly, to stand proud of the rest of the crown and act as short blades on thin vegetation (Fig. 6.1). Dentine is always thought to wear away more quickly than the enamel because it is softer. However, why do enamel ridges not get progressively

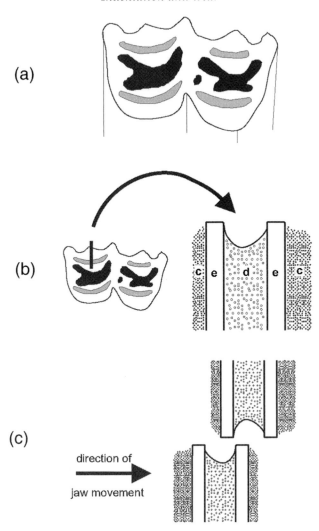

Fig. 6.1 The form of an upper molar tooth in a bovid artiodactyl, such as a cow. (a) The working surface sits on top of a tall crown. Each of the four elongate cusps can be identified by light grey areas representing half-moon shaped exposures of (grey) dentine surrounded by enamel ridges. The whole crown is originally covered with cement. The black areas represent holes. (b) This shows a schematic section through part of the tooth showing how the dentine (labelled d), and cement (c), wear down so that enamel ridges (e), stand proud of these tissues. Note that the surfaces of the dentine and cement lie just below that of the enamel. (c) A schematic diagram to show the path of movement of a lower molar across an upper, which has a strong mediolateral (or transverse) component.

higher? Surely this must follow if dentine always wears away more rapidly. One explanation is that the enamel takes the brunt of the load, so relieving dentine depressed beyond a certain distance. However, the ridges do not seem high enough for this to be plausible. How can this be explained?

When first exposed, dentine is level with enamel and subject to both compressive and sliding forces. The data on pitting suggest that dentine will wear faster than enamel in these circumstances. However, once depressed below the enamel ridges, dentine is largely immune from this compression and more subject to sliding forces. The results in Table 6.3 imply that dentine could then wear no faster than enamel. There are no data on this, although experiments and observations reported by Greaves (1973), Costa & Greaves (1983) and Walker (1984) all agree that dentine on the edge of an enamel exposure can get heavily scratched.

All this ignores mutual indentation, which is more important for dentine than enamel. Insect cuticles, bone and the woody tissues of plants have similar hardness to dentine (Appendix B) and could wear it substantially, but leave enamel unmarked. This is of some considerable interest in herbivores because, under some circumstances, the control of the enamel/dentine wear ratio goes wrong. Some diets fed to domesticated horses leave very high enamel ridges, obviously wearing the dentine but not the enamel. A veterinary surgeon has to come in and file the ridges down, a procedure often called 'floating the teeth'.

The general effects of friction on all this can be noted. For 'usual' values of μ, where $\mu \ll 1.0$, the effect is greatest on fractures during sliding (Eqn 6.5) where the force reduces. This could have important implications for browsers that ingest leaves with substantial quantities of tannins. Whether eaten because they speed up digestion (Aerts *et al.*, 1999) or because browsers cannot avoid tannins altogether, the interaction of tannins and salivary proteins, particularly those proteins that adsorb onto the enamel surface (Jensen *et al.*, 1992), is liable to increase the rate of tooth wear. True grazers encounter few such compounds (Owen-Smith *et al.*, 1993).

Unfortunately, too little is known about the properties of cement on the tooth crown in herbivores to add this into the analysis. It could act as an important test. While there is no reason to suppose that it would not behave somewhat like dentine in this regard, none of the above is anything like the last word on the mechanics of wear. Equation 6.5 has an exact look, with constants expressed as fractions, because it derives ultimately from a very complex, apparently unsurpassed, piece of applied mathematics (Hamilton & Goodman, 1966). Yet, the problem is so complex that experiments are definitely needed. Some of these have been reported on homogeneous ceramics. Lee *et al.* (2001) show on ceramic coatings that the effect of sliding contacts on indentation fractures are real – they can reduce the load, for example, in forming cone cracks – but the differences may not be anything like as great as my simplistic view suggests. Interestingly, Lee *et al.* make reference to *in vitro* wear studies on human teeth, effectively questioning whether it is necessary to model actual jaw movement patterns with quite the rigour that has marked some research on the subject (e.g. DeLong & Douglas, 1983).

The size of plant abrasives

The silica particles in plants are very small, generally being <50 μm in size (Ball *et al.*, 1993). The marks that they produce on teeth are much smaller because only a tiny part of their surface contacts the dental tissues. A lot of these particles pass through into the faeces undamaged. It is entirely possible then that their size protects them against being fractured because they lie below the 'brittle–ductile' boundary referred to Chapter 4. If material property values for quartz from Appendix B ($E = 70$ GPa, $R = 2$ J m^{-2} and $\sigma_y \approx 2500$ MPa) are inserted into Eqn 4.9, with the constant $c = 2300$ for indentation (Atkins & Mai, 1985), then the brittle–ductile boundary lies at ~ 50 μm. Rough as this calculation is, it suggests that phytolith size might be a deliberate anti-herbivore strategy by plants because it obviously enhances the wear potential of these particles if they can avoid being fractured by first contact with the dentition. Dust particles are generally an order of magnitude smaller than plant silica (Lawn, 1993) and would be expected to cause smaller wear features.[5]

THE INFLUENCE OF THE MICROSTRUCTURE OF DENTAL TISSUES ON WEAR

It has long been known that enamel structure is very variable in mammals (Boyde, 1964; Osborn, 1974) and there is increasing interest in explaining how adaptive this variation is (Janis & Fortelius, 1988; Rensberger, 2000). The decussation of enamel rods in many herbivorous mammals (e.g. in rhinoceroses: Rensberger & von Koenigswald, 1980; Fortelius, 1985; Boyde & Fortelius, 1986) produces irregular surfaces that help to grip food, not just by increased friction, but also via microindentation. In voles, the maintenance of the height of the enamel crests above the exposed dentine looks to depend on an abrupt change in decussation on a minute scale (von Koenigswald, 1982). In another rodent, the beaver (*Castor fiber*), the worn lower incisors develop a step that follows the hardness contours of the dentine very closely (Osborn, 1969). However, some microstructural features appear to make little sense (e.g. the pattern of rod decussation of canine enamel in baboons: Walker, 1984) and a coherent argument is needed that relates these features to diet and that appears to me to be lacking at the moment.

WEAR ON OTHER TOOTH SURFACES

Abrasive wear appears typical of the working surfaces of teeth, but is not limited to them. Collagen fibres, called transseptal fibres, hold adjacent teeth in a dental arch together. Should one tooth, supporting most of the load, be depressed into its socket relative to its neighbour, wear is likely to result at their point of contact. This approximal wear is not caused by transseptal fibres, which are weak and need to be remodelled by cellular activity so as to adjust their length. The attractive force between neighbours is mostly a result of a bite force that is directed mesiodistally. The classic investigation of the angulation of the bite force was by Osborn (1961) on human subjects. With the upper and lower teeth apart, Osborn found that it was possible to push thin metal strips between the approximal surfaces of teeth without effort. During a bite, these could only be removed by applying a tensile force (which Osborn measured). By going around the dental arch and performing the same test, Osborn deduced that the teeth in the arch must be forced mesially by the bite. Approximal wear is then explained by the bite being supported by particular teeth that move into their sockets, so rubbing against their neighbours.[6]

It has been calculated that Plio-Pleistocene hominins lost only 6% of their molar tooth surface at maximum to approximal wear (Wood & Abbott, 1983), but much heavier losses are found in some modern and prehistoric human populations (Murphy, 1959). Van Reenen (1982) showed that 25% of the length of a mandibular first molar could be lost to approximal wear in the lifetime of a San Bushman individual in southern Africa. His calculations seem to show that a mandibular third molar would erupt approximately in the same position in the jaw that the second molar had occupied when it erupted several years earlier. There is no logical reason why approximal wear should mirror that on the working surface of the teeth (meaning that the former could exist without the latter), but van Reenen showed an effective association between the two, with a total lifetime loss of about 60% of second molar crown volume.

Curiously, as approximal wear surfaces develop, they tend to form recip-rocal curvatures on neighbouring teeth, rather than being flat. The con-cave facet appears always to be found in the tooth that erupts first. It is probable that the tooth erupting second brushes past the first as it erupts, so spreads its wear over a larger surface area than the former, eventually invaginating it.

Loading of just one part of the postcanine dentition with food is par-ticularly likely in mammals that consume small food particles in small mouthfuls. Also, teeth within one functional group may also not come into contact with food simultaneously during a chew. Osborn (1982) showed that the helicoidal plane of occlusion in several anthropoid primates, par-ticularly some hominins, brings the molar teeth into food contact in a distal to mesial sequence (i.e. third molars, then second molars, then the first). Counter-intuitive though this may seem, the suggestion works perfectly on paper and could be explained by a dietary adaptation requiring close to the limit of bite force, a factor that C. R. Peters (e.g. Peters, 1981) has consistently suggested to explain the evolution of hominins.[7]

Other surfaces of the teeth wear, but much less so. The actions of the tongue and cheeks can push food particles against the lingual and buccal surfaces of foods with enough force to wear them (Ungar & Teaford, 1996). Buccal wear of the molars may be accentuated in mammals with cheek pouches and labial wear of the incisors by the wadging of food particles by chimpanzees and orang-utans (Walker, 1979).

WEAR AND DENTAL EFFICIENCY

In its early stages, wear imperils tooth shape. I would assume, a priori, that this would be detrimental to efficiency, but there is no need to speculate because there is evidence on this matter. Gipps & Sanson (1984) showed unequivocally in the ringtail possum (*Pseudocheirus peregrinus*) that den-tal wear decreased chewing efficiency and that this impaired digestion. Lanyon & Sanson (1986) confirmed this in the marsupial koala (*Phasco-larctos cinereus*) and Logan & Sanson (2002) have now gone further and associated dental wear and the reduction in digestive efficiency that it produces with impaired chance of a male finding a mate. This link to reproductive success represents one of the most important findings in the dental literature because it provides a direct demonstration of selective advantage.

At later stages of wear, the functioning of the entire dentition is threat-ened. There is a presumption among many dental researchers that its loss spells the end of a mammal's life. This point should probably not be pressed too hard – all mammalian physiologists are likely to feel that the particular body system that they study is the one crucial for an organism's survival. All systems – cardiovascular, neural, urinary, respiratory or whatever – show clear signs of ageing, so which of these is generally responsible for limiting a mammal's lifespan? Surely, each system is optimized to last just a lifetime and no more? The evidence that dental wear can limit lifespan in the wild though is really quite strong, involving many anecdotal reports from shrews to elephants.

The teeth of wild shrews are often worn down to the gums at an age when captives appear perfectly healthy. These wild shrews starve to death. Elephants bring their cheek teeth into action in a paired upper–lower sequence, from anterior to posterior. Once the last such tooth pair comes into occlusion, the remaining lifespan of the animal depends on these teeth or, again, they starve. Hypsodont mammals are just as vulnerable to a mismatch of wear rate. An impressive time series of zebras that used to be exhibited in the Natural History Museum in London showed that the oldest zebras generally have teeth that are very heavily worn.

Another line of evidence is provided by observations on carnivores. These mammals use one pair of teeth, called the carnassials, to fracture soft tissues. Carnivores that break bones, like dogs (Fig. 6.2) or hyenas, use molars that are posterior, or premolars anterior, to the carnassials respectively, the difference in development of these teeth being thought to be related to gape. Van Valkenburgh (1996) observed large carnivores in the wild and showed that most mammals did use the teeth that anatomists had always predicted that they would. Observations by Berkovitz & Poole (1977) on ferrets, a feral mammal closely related to polecats, suggest that the wrong choice could be fatal. Caged animals fed on rat hard pellets blunted their carnassial teeth. When the animals were subsequently fed mice, which are part of their normal diet, they were unable to bite through them easily.

ADAPTING TO WEAR

The mammalian dentition may be built to withstand wear, but it is nevertheless under undeniable threat. The principal physiological adaptations against it are extremely fine interdental sensitivity to detect abrasive particles (as discussed in Chapter 3) and the salivatory response found by Prinz (in press). The former may trigger the latter and, although the copious salivation of ruminants is supposed basically to provide buffered

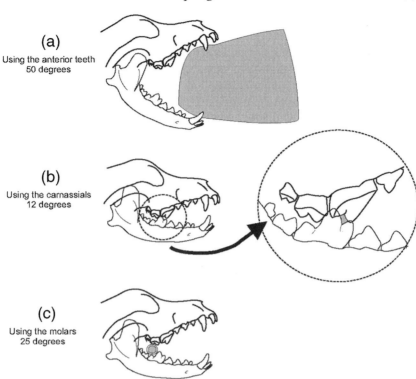

(a)
Using the anterior teeth
50 degrees

(b)
Using the carnassials
12 degrees

(c)
Using the molars
25 degrees

Fig. 6.2 The jaws of a canid (dog family) in three staged poses: (a) Gripping a prey item with its incisors and canines; (b) fracturing soft tissues with its carnassials and (c) breaking a bone with the molars lying behind the carnassials. Note the different gapes involved in each of these activities. If this mammal attempted to fracture bone with its carnassials, it would blunt the blades.

fluid for the rumen, another potential benefit may be to wash abrasives off the teeth.

The commonest evolutionary response to a wear threat is hypsodonty, particularly of the postcanines (Stirton, 1947; van Valen, 1960; Janis & Fortelius, 1988). Many herbivorous mammals show this to a greater or lesser degree (sometimes only in the last molar, on which hypsodont indices are often based: Mendoza *et al.*, 2002) and it is usually associated with grass consumption although there can be errors in blithely assuming this (Macfadden *et al.*, 1999). The family of horses, the Equidae, is a classic example of a lineage that tracks the development of hypsodonty (Simpson, 1953). The family first appeared in the fossil record of the Eocene about

55 million years ago. *Eohippus*, the first genus, had short-crowned teeth. During the Miocene, the family diversified into a large number of forms, all with high postcanine crowns. This development has consistently been associated with the development of savanna and consumption of plants, notably grasses, which contain large quantities of opaline silica. Logical though all this may be, the key experiments look still to be done.

Increase in the crown heights of incisors and canines may also represent adaptations to wear. The spatulate incisors of primates wear extensively in some species, including hominins where the cause has been the subject of considerable debate (Hylander, 1977; van Reenen, 1982; Ungar & Grine, 1991; Ungar *et al.*, 1997). The canines of mammals can be heavily worn (e.g. in great apes: St Hoyme & Koritzer, 1971; Dean *et al.*, 1992), but more usually they are subject to gross fracture, as indeed any tooth can be (e.g. Frisch, 1963; Van Valkenburgh, 1988; Van Valkenburgh & Hertel, 1993).

The other major wear adaptation that is generally recognized is the thickening of the enamel. The enamel of hominin teeth is relatively thick compared to that of the herbivores or carnivores sketched above. The thickest part is over the cusps, obviously to protect them. Other mammals with this low-cusped tooth form (bunodonty, it is called) show a similar arrangement, such as robust australopithecines (Gantt & Rafter 1998). There is some disagreement though about whether this might be for resisting wear or very high stresses (Kay, 1981).

Less well recognized is the possibility that the surface area of the postcanines could also be a response to wear (Lucas *et al.*, 1985), which DeGusta *et al.* (2003) provide some indirect evidence for. The logic is simple: abrasiveness acts at the food–tooth interface and it can, therefore, be grouped with other external physical characteristics of food particles like particle size, shape and stickiness. The only difference between them is that abrasiveness acts over a long periods of time, not within one masticatory sequence (Fig. 6.3). The argument goes like this: suppose that food particles are broken during chewing at one of a finite number of breakage sites located along the postcanine tooth row. Whenever food particles outnumber these sites, they must, in any one chew, compete to be broken, i.e. not all particles can be broken in that chew. As the sites are utilized again and again over the lifetime of the animal, they gradually wear and eventually some breakage sites may be lost because of this. It follows then that loss of breakage sites reduces the probability of particle fracture and slows the rate at which the food is broken down. This will delay bolus formation, set swallowing back and so reduce the rate of energy acquisition by the body.[8] A logical evolutionary response would be to increase the number of breakage sites by

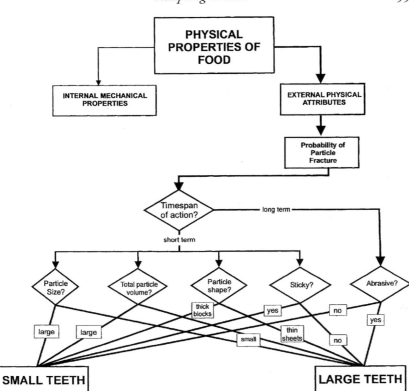

Fig. 6.3 A map of the surface attributes of foods in mammalian diets and their anticipated effect on the adaptation of tooth size. The diagram is identical to that of Fig. 5.11, except that adaptation has been expanded to include abrasion, i.e. a factor that acts over a long time-span.

increasing tooth size. For any given typical mouthful, this increase would decrease the rate at which any one of these sites is used and would, therefore, decrease their individual wear rate. The result is: larger teeth last longer.

The identification of breakage sites is not difficult: areas of wear on the tooth row show where they are. There could be many sites at any individual wear facet, for example and, in general, the larger the facet, the more breakage sites will be found there. There is no need to go further than that with defining these sites: the important point is that food–tooth contacts do have fixed locations because food is moved around the mouth. In contrast, tooth–tooth contacts are completely fixed: the same teeth will always contact each other.

There is now some evidence, from the study of DeGusta *et al.* (2003) on howler monkeys, that molar size is positively correlated with longevity in a mammal and there are also numerous examples of mammals with relatively large molars. Examples include robust australopithecines, the recently extinct ape *Gigantopithecus blacki* (Pettifor, 2000) and gomphotheres (Maglio, 1973), and the warthog, which has a last molar that probably equals the area of the rest of the postcanine tooth row. Could molar sizes like these be wear-resistant features? Why then do these mammals also have rather thick enamel? Are increases in tooth size and enamel thickness just parallel wear adaptations? I suggest not.

In the previous chapter, I argued that the type of food input that places the biggest strain on the rate of energy acquisition by the body was the ingestion of chemically sealed small particles (such as seeds) in small mouthfuls. I also suggested that the rate of breakdown in the mouth could be increased by an evolutionary enlargement of the surface area of the teeth. These other characteristics are also surface attributes and it is entirely consistent to add abrasiveness to these. What I did not mention in Chapter 5 though was the potential disadvantage of doing this. If there is little food in the mouth, but large teeth, then there is a greater chance of tooth–tooth contacts. The evidence from humans is that there would be about 50 ms in which dangerously high forces could build up between these teeth before inhibition in the jaw-closing musculature protects them. In these circumstances, it is possible to argue that an increase in enamel thickness is actually not an adaptation to food–tooth wear, but to *tooth–tooth wear* in such mammals, being the only way in which the effects of this 'auto-wear' could be countered. Teeth always contact each other in the same locations: there is no parallel to the circulation of food particles; so changing tooth size makes no difference to the rate of tooth–tooth wear. Once started, the argument must be pressed to its logical conclusion: tooth–tooth wear might be more damaging to the dentitions of some mammals than food or grit. After all, teeth are generally the hardest structures present in the mouth. They are also composed of relatively tough ceramics and are capable of inflicting heavy wear on each other. This is exactly why the dentitions of non-mammalian vertebrates are so arranged as to avoid this.

SUMMARY

While I did not say much in this chapter, the literature on tooth wear is vast. There are big problems explaining the enormous range of phenomena that has been reported because of limited theory. That given here is very basic, but to step out further does not seem justified. It is worth mentioning though that wear can be pictured very differently to the account given here. Fracture, the separation of objects, is opposed to the general trend in the universe, which is to attract objects so that they stick together (Kendall, 2001). Food–tooth or tooth–tooth wear could start by adhesion with wear only resulting when one breaks away by a fracture line that lies off the interface. Such adhesive wear is known (Ashby & Jones, 1996) and could explain the 'plucking' of rods (Walker, 1984) and why tough materials like woody tissues (e.g. in leaves) and keratin can apparently polish teeth.

7

The evolution of the mammalian dentition

OVERVIEW

This chapter is intended to be a chapter of ideas, mixing fact with suggestions that, although seemingly logical and based on the previous chapters, might require a lifetime's work to substantiate in any detail. More precisely stated then, it is speculative. When biologists do this (speculate) on the adaptive significance of changes in the form and function of plants and animals by setting these adaptations in an ecological context, they tend to refer to it as a scenario. In keeping with this theatrical jargon, this chapter goes distinctly 'off-lineage' at times, mentioning the dinosaurs even, but once the scenery is painted and the script sufficiently advanced, the mammals are swung in for their top billing.

By analogy to the sharpest way of controlling the content of television programmes, those who really do not like it can simply shut the book: there are generalizations beyond redemption here. However, all this stems from the best of intentions, which is to offer new possibilities for explaining some of the major trends in mammalian evolution, trends that are documented very largely via the dentition. I try to portray the evolution of the dentition of mammals in terms of the principles expounded in the previous chapters. However, it would be impossible to cover every lineage and foolish to claim that I can understand the range of forms that exist or have existed on the basis of current knowledge. Instead, I have divided mammals up into a small number of dietary adaptations and attempt to understand these. I finish with an overview of the dentition of primates and what this can say about the evolution of humans.

INTRODUCTION

I assume here that the dentitions of mammals adapt to the fracture properties of their diets. For this to be so, mammals must either make feeding decisions on the basis of perceived food texture or else select food on sensory grounds that correspond to a distinct range of textures. It is implicitly assumed here that one or the other is correct.

Animals and plants do not want to be eaten.[1] To prevent this, they defend themselves in various ways. Some structures are obviously defences against large predators (e.g. plant spines and thorns), while others are probably barriers against a wider range of environmental threats, such as the skin of mammals or the bark of trees. Whatever the primary role of these coverings though, they also tend to show patterns related to predator avoidance, such as spines, thorns, horns and thick dorsal (back) skin. Even the teeth themselves are weapons, at least in part (Every, 1970). Regardless of whether any feature is organized specifically as a mechanical support (like bone or wood) or primarily for defence, similar design principles are likely to apply to avoid fracture (Ashby, 1989, 1999). The construction of the basic mechanical indices was shown in Chapter 4. All the start of this chapter does is to extend their use to strategies that organisms develop to avoid being eaten.

TYPES OF MECHANICAL DEFENCE

There are two basic types of strategy to avoid mechanical damage (Ashby, 1989, 1999):

(1) A structural component could be organized such that it breaks at a stress higher than an animal could achieve without its own structure failing. These *stress-limited defences* depend on *preventing cracks from starting* and are achieved either by a high yield stress, σ_y, i.e. high hardness, or by a high value of the square root of toughness multiplied by Young's modulus, $(ER)^{0.5}$. Structures defended in this way are generally referred to by biologists as 'hard' (Lucas *et al.*, 2000).

(2) A structural component might be constructed so that it fails at such a large displacement that it limits the degree of detachment that an animal could achieve in a given time. These *displacement-limited defences* depend on *preventing cracks from growing*. The most important features of such displacement-limited defences are either the toughness of the animal or plant part alone if it is very thin, or the square root of the ratio of its toughness to Young's modulus, $(R/E)^{0.5}$, if it is fat. Biologists tend to refer to structures like this as 'tough' (Lucas *et al.*, 2000).

At first sight, stress-limited defences seem to be the best alternative because they hold out the prospect of the organism evading any damage at all. However, they are an all-or-nothing strategy. Placing all resources into the prevention of crack initiation risks catastrophe once a crack is started. A high yield stress means that a large amount of elastic strain energy can be absorbed, which is then released very easily into even the smallest crack, resulting in almost instant fragmentation. Small cracks accumulate through fatigue as well as from single bites (this is why engineers need toughness in their construction materials: Gordon, 1991), so this defence has definite risks. In contrast, a displacement-limited defence allows feeding to begin, by allowing cracks to start, but avoids detachment by frustrating crack growth. Without detachment, there can be no ingestion.

In Chapter 4, I suggested that stress-limited designs were likely to be important in deterring the first bite of an animal while displacement-limited designs would act better against repetitive fracture and fragmentation during chewing. The following (deliberately non-mammalian) example illustrates the ecological pressures that might have led to one or the other of these defences.

The big fight: dinosaur versus invertebrate

Herbivores come in all sizes, but when we think really big, we think of herbivorous dinosaurs (Sereno, 1997). Throughout most of the Mesozoic era, these reptiles were the major consumers of plants. Being very large compared to the foliage that they ate (often large compared to the size of the entire plant that they were consuming), just one feeding bout (even one bite) could be life-threatening to that plant. Additionally, dinosaurs did not chew.[2] If all this is put together, then defence against the first bite and subsequent crack initiation must have been everything to a plant's survival. Faced with such mega-herbivore pressure, it is likely that Mesozoic plants began to develop stress-limited defences. What did these defences look like?

A large dinosaur probably ingested leaves by pulling on them, but with torsion and bending as potential complications to simple tension. To prevent crack initiation under these circumstances, plant leaves need to place dense tissue as externally as possible because this 'sandwich' arrangement confers high hardness and stiffness (Gibson *et al.*, 1988). Yet this mechanical defence cannot be spread evenly in the leaf because thick-walled cells are opaque and impede the passage of light to underlying photosynthetic organelles. Since the latter provide energy keeping the plant alive, woody tissue has to be restricted in distribution, mostly being disposed as strands lining veins (hydraulic conduits) because this keeps them out of the way of photosynthesis.[3] To resist tension or bending most effectively, the optimal orientation of these woody tissues would be parallel to the direction of the bite because this would maximize $(ER)^{0.5}$. A relatively narrow leaf design reduces any chance of a bite by a big animal from all but one direction. Thus, any plant that is attacked predominantly by large herbivores would be likely to evolve long thin leaves with parallel longitudinal veins that stand proud as ribs. Thorns and spines may be mounted on veins as extra deterrents. However, these sharp structures need to be made from hard materials, and hardness requires density. Cell wall is not really very hard, so plants often employ silica, particularly in the tip of these structures, because it is very hard, usually abundant and quick and cheap to embed in tissues (Raven, 1983).

The flora of the Mesozoic is now glimpsed only in remnants, but towards the end of the Mesozoic, non-flowering plants like ferns dominated the land. The flowering plants (angiosperms) that now dominate the world's flora had begun to evolve in the Mesozoic, particularly some of the mono-cotyledonous forms (monocots), such as palms. I envisage much of this vegetation being populated by trees with a hard surface to both their trunk and branches with frequent spines. Many of these plants have little lateral expansion to their foliage and the central trunk tends to be pithy with protection confined to the outer layers. The great advantage of stress-limited defences is their self-sufficiency. Placing all investment in countering crack initiation, there is no logical point in dealing with the consequences of growing cracks, those that open cells, because these incidents trail the critical event.

The dinosaurs died out 65 million years ago and very different defoliators took their place: invertebrates. In modern floras, these animals are responsible for most of the damage to the foliage of angiosperms (Coley, 1983; Leigh, 1999). They are generally smaller than the leaves that they attack, so one bite would not be serious.[4] The most logical defence to the biting variety would be to erect displacement-limited defences in order to stop the detachment of a bitten part or simply slow the animal's progress if not. Dicotyledonous angiosperms have relatively wide leaves with a lattice-like arrangement of their veins. Woody fibres accompany these veins to defend the leaf from potential attack from all directions.[5]

Displacement-limited defences depend on toughness. As stated in Chapter 4 though, this depends on structure and structure takes time to develop. So young leaves need some other protection prior to the construction

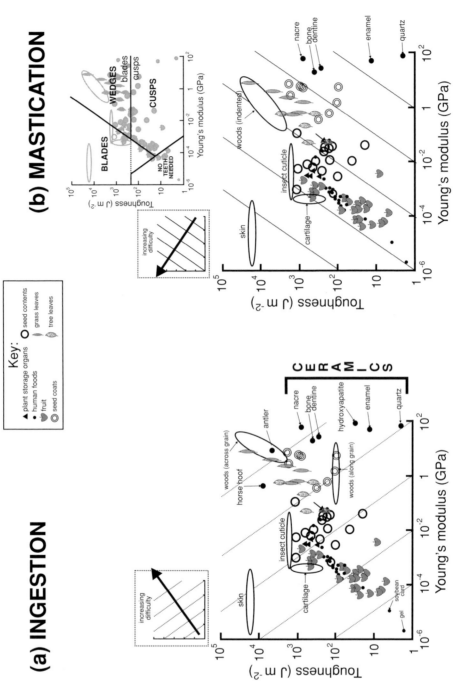

Fig. 7.1 Dental–dietary adaptation expressed in terms of plots of toughness versus Young's modulus of potential food items to express their relative resistance to (a) ingestion and (b) mastication. Data for human dental tissues and an abrasive (quartz) are included for comparison. See text for explanations.

of mechanical defences. Extensive work on leaf development (from Coley (1983) to Coley & Kursar (1996)) shows that this protection is chemical. There is an entirely logical association between the deployment of chemical defence and displacement-limited mechanical defence because the former is almost entirely intracellular while the latter 'permits' such cells to be opened. It is equally logical to expect noxious chemicals to be absent in plants with stress-limited defences because, not only are these quick to erect, but cells only open when cracks grow and this type of mechanical defence is intended to prevent that from ever happening.[6] Plants like cycads and monocots appear to contain few toxic chemicals, not I suggest because the latter are 'modern', but because they do not fit into their defensive pattern. Silica is often viewed as an archaic defence, organized before the chemical warfare of angiosperms was organized, but I do not agree.

I will now consider mammalian dentitions and diets using the same kind of arguments about defences to predation.

THE BASIS FOR DENTAL–DIETARY ADAPTATIONS

The toughness of many potential mammalian foods is plotted against their Young's moduli on logarithmic axes in Fig. 7.1. Most of the data are from Appendix B, but a substantial amount of unpublished data has been added, particularly for fruits and seeds. More plant data would make a total muddle. There are few good data on the toughness of animal soft tissues, but the latter are difficult to incorporate anyway because of the frequency with which they exhibit non-linear stress–strain relationships. For example an ellipse is shown for the rat skin data of Purslow (1983) simply to encompass the initial and final modulus, which is to say that the slope of the stress–strain curve changes by a factor of about 10^3 from the onset of loading to fracture.[7]

The use of the graphs in Fig. 7.1 is very different to that made of most bivariate graphs in biology. The points do not fall on any line, nor are they intended to. They sketch out a starry sky, the fundamental reason for this being that while Young's modulus is framed by material content and the 'rules of mixtures' that these impose, toughness is totally free from such constraints. Unless study is restricted to a small group of tissues with similar structures, when these properties *are* correlated with each other and predictable (Ashby, 1998), then the 'sky at night' is the result. This is a central fact for the thesis of this book: it is the underlying cause of dental diversity.

Lines showing equal values of $(ER)^{0.5}$ and $(R/E)^{0.5}$ have been drawn on the graphs in Fig. 7.1, denoting resistance either to ingestion (Fig. 7.1a) or mastication (Fig. 7.1b). These graphs include the same foods, but are not identical due to considerations of anisotropy. For example, woody tissue is likely to be pulled during ingestion. Thus, in Fig. 7.1a, woody tissue is represented by across-grain and along-grain ellipses with the modulus and toughness values matching that loading, i.e. the 'across-grain' ellipse consists of the modulus along the cellular axis and the toughness across the cell, vice versa for the 'along-grain' ellipse. Wood is probably mostly indented in mastication though, so the single ellipse refers to the major problem in fragmenting it – loading perpendicular to the cellular axis with toughness still being across the cell.

The smallest graphs (boxed with a dotted outline), positioned above the vertical axes of the main graphs, indicate the degree of difficulty that foods present to the dentition. This is upward and to the right for ingestion, but upward and to the left for mastication. Thus, a food that is difficult to chew may not have been so difficult to ingest and vice versa. Note, if the general theory is correct, that it would be impossible for a prey organism to be equally defended against both.

The intermediate-sized graph in Fig. 7.1b has greyed-out food symbols, superimposed on which are sectors bounded by thick solid lines that indicate dental adaptations to diets of which these foods would be typical. The most critical division is the $(R/E)^{0.5}$ line separating blades from cusps and wedges. The line is equivalent to that shown in Fig. 4.10 from experiments on humans and reflects a point where particle fragmentation seems to become impossible without blades. This must actually be a graded change rather than a threshold, but the theoretical point has to be made by imposing the latter. On the 'cusped' side of the fence, foods with higher toughness probably require marginal ridges on the cusps so as to form wedges. All the above refers only to large food particles. For the special case of thin particles (rods and sheets), the horizontal line shows that the criterion for fracture is not $(R/E)^{0.5}$, but toughness alone (Fig. 4.9). If these foods have high toughness, then blades are required to comminute them whatever their modulus, while cusps are probably satisfactory if toughness is sufficiently small (the horizontal boundary is set at $R \approx 200$ J m^{-2}, a suggestion of C. R. Peters). Finally, some foods are sufficiently low in $(R/E)^{0.5}$ or R that postcanine teeth are not necessary at all. The position of this sector on the graph is a pure guess.

I have less to say about ingestion because so little is known about it, either experimentally or observationally, making it difficult to generalize. The underlying principle is that when $(ER)^{0.5}$ is sufficiently high, ingestion gets progressively more difficult. However, the multitude of options available at the front of the mouth means that many other factors come into the picture. Experiments are currently under way to describe thresholds, but critical factors in whether food is fractured between the teeth or simply gripped and fractured against an external restraint must include friction. When a food particle is sufficiently thin, it will no longer fracture under indenting incisors, but can still be pulled to fracture by gripping it and acting against an external restraint such as (directly or indirectly) against the rooting system of a plant or against the feeder's upper limb. The reason for this is the constant in Eqn 4.9 is very much lower in tension than in indentation – thin particles may fracture when pulled, but not when indented (Atkins & Mai, 1985).

I will now consider the main dental–dietary adaptations of mammals, referring to Fig. 7.1 whenever tooth shape is being considered. It must be borne in mind though that nothing will replace the need for fieldwork that measures the mechanical properties of the actual diet. Just the age stage at which foods are eaten makes an enormous difference to these properties.

INSECTIVORY AND CARNIVORY

The evolution of early mammals and changes in tooth number

Mammals evolved from synapsid reptiles in gradual evolutionary stages. These fossil 'stages' appear to have been perfectly viable, with their lineages surviving for millions of years (Hopson, 2001). The sequence started with the synapsid reptiliomorphs (often called mammal-like reptiles) about 300 million years ago in the Late Carboniferous and culminated in the emergence of basal mammals, the Mammaliaformes, about 210 million years ago. This was not the end of the trail though. The earliest mammal that can be labelled 'eutherian' (the kin of modern mammals) emerged about 125 million years ago (Ji *et al.*, 2002). Eutherian mammals only started to diversify and become dominant land animals at the end of the Cretaceous about 65 million years ago.

Major changes in the head and neck region distinguish the extensive sequence of fossils that mark this transition. In the mouth of an early synapsid, just as in modern-day lizards, there sat a large number of teeth of simple conical form. These unicusped teeth had no roots and were anchored into position by being fused (ankylosed) to bone. The teeth were shed periodically, being replaced rapidly by another tooth in the same position. Osborn (1971b) described all the teeth occupying the same position in the mouth of a reptile as members of one tooth family. Early synapsids had many tooth generations per family, with each succeeding tooth being larger than its shed predecessor. This size increase is usually explained by the relatively constant growth rate of non-mammalian vertebrates, the size of each new tooth being proportional to the size of the growing jaw (Berkovitz, 2000). This obviously increases the chance of catching food.

Synapsid teeth sat in arches, more or less as in mammals, arranged around the margins of the tongue. However, early on, the oral cavity had no roof, allowing the tongue to gaze upwards at the upper respiratory tract. The reptilian tongue, and probably that of synapsids, contains muscles that can only act anteroposteriorly, so as to propel a food particle backwards towards the pharynx. Their pharynx has no specific musculature for swallowing, appearing to rely on superficial subcutaneous muscles to help move a single ingested food particle towards the stomach (Smith, 1992). The upper and lower teeth of a synapsid did not contact and the lower jaw moved vertically. The teeth were partly covered by cheeks on their lateral sides that hung loosely without muscle (there being little need for fine control of the buccolingual direction of food movement because synapsids did not chew) in order to accommodate a large gape.

In contrast, the mouth of the earliest mammals consisted of fewer teeth, in only two generations, housed in bony sockets. The teeth were clearly differentiated into three tooth classes: (1) peg-like incisors, (2) large canines (in both deciduous and permanent tooth generations) and (3) postcanines that possessed several cusps. The cheeks appear to have been muscular and there was a hard palate above a tongue that was capable of side-to-side food movement. Food was usually shifted to the left or right sides of the mouth by these tongue movements. The lower jaw was narrower than the upper, such that it had to move a fraction sideways in order that teeth on the non-active side would stay out of contact. The pharynx itself had a novel musculature designed to take a bolus of fragmented food particles quickly down to the oesophagus.

It is generally thought that the major selective pressure for all the above anatomical changes resulted from a dietary focus on insects. Mammals evolved alongside the angiosperms, a group of plants relying on insect pollination rather than on wind (the agent used by most other plants). In order to attract adult insects, angiosperms developed flowers containing nectar as an attractive reward for adult insects. However, for an active pollinating agent like an insect to have been superior to passive means (i.e. the fortuitous accident of wind), then all growth stages of insects would have to be supported by a food source. The nectar of angiosperm flowers could not supply nutrition to insect larvae and it seems that the latter have often favoured their young leaves as a food source.[8]

Angiosperm-linked insects offered mammals several food choices. They could, for example, have searched tree bark for adult insects running to and from flowers or instead they could have concentrated on larvae located on leaves. Tree shrews are living insectivores with dentitions somewhat similar to those of early mammals and which illustrate some of the variety of niches available to insect feeders. Pen-tailed tree shrews (*Ptilocercus lowii*) seem to catch their prey on tree bark as they rush along the trunk or large branches. Other tree shrews, e.g. in the genus *Tupaia*, are more likely to investigate foliage during day and night (Emmons, 2000).

Insectivory in early mammals would have entailed many changes to body form but a major one, apparently neglected in the literature, was un-doubtedly body size reduction. Insects probably were (as they still are) very small, so mammals would have had to reduce in size from their synapsid ancestors in order to consume them.[9] There was a range of sizes in early mammals, presumably reflecting a size range of insect prey. The small-est, *Hadroconium*, from about 195 million years ago, weighed in at only 2 g, but even a large *Sinoconodon* specimen weighed <500 g (Luo *et al.*, 2001). At sizes like that, heat loss via body surfaces is critical, thus putting a premium on rapid digestion. Also, insects are (and probably were) active mainly at night. To maintain nocturnal activity levels, mammals would have needed to control their body temperatures because locomotor stamina in the coolest part of the day requires an elevated body temperature that reptil-iomorphs lacked (Hopson, 2001). In order to supply these energy require-ments, the front end of the gut had to start processing insects into smaller pieces.

Once mammals began to masticate food, the scaling arguments of Chapter 5 suggest that the sizes of the jaw and anterior teeth were de-coupled from those of the postcanines because fracture mechanics, not gape, determines the size of the cheek teeth. The size disparity would have depended on the mechanical properties of the insects that they ate, but, almost certainly, this would have led to negative allometry of the postca-nine dentition and thus to enlargement of the posterior teeth in a dwarfing mammal. The result would have been congestion of the jaw, something that could have resulted in tooth loss. Small vertebrates do not live as long as larger ones and so a reduced lifespan in early mammals could also have been responsible for a reduction in the number of tooth generations. Thus,

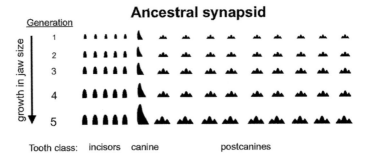

Ancestral synapsid

Generation

growth in jaw size

1
2
3
4
5

Tooth class: incisors   canine   postcanines

Descendant mammal

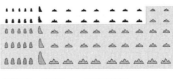

Tooth class: incisors   canine   postcanines

Fig. 7.2 A schematic view of tooth loss from reptile to mammal due to reduction in body size. In life, the jaw grows in size, but the diagram is scaled so that tooth families can be visualized. Lost teeth are shown as being greyed-out. These include tooth families (vertical in the diagram) and whole generations (horizontal). In reptiles, tooth size is proportional to jaw size, which is an adaptation to prey size. In a smaller mammalian descendant, the size of the anterior teeth and the jaw is still determined by the same selective pressures, but the size of the posterior teeth is relatively enlarged because of fracture scaling (Chapter 5). Posterior tooth loss is suggested, but this is somewhat arbitrary. The reduction in the number of tooth generations follows from a reduced lifespan.

the loss of tooth number and tooth generations in early mammals could be explained by dwarfism, as Fig. 7.2 suggests.[10]

Molar shape in early insectivorous mammals

In this book, I have consistently tied the adaptation of tooth shape to the fracture properties of foods, the requirement for particle fragmentation in mammals being a function of relatively high basal metabolic rate. The fossil evidence for the evolution of early mammals from synapsids suggests that basal metabolic rates increased only gradually from reptilian levels, with most intermediate forms only having partial control of body temperature (Hopson, 2001). It is probable that changes in tooth form during this lineage represent change, not in the mechanical properties of the diet, but in the relative amount of fracture/fragmentation required to service a particular metabolic rate.

The dental features required for fracturing insects appear to depend mostly on the mechanical properties of the cuticle.[11] Insect cuticles have complex architectures (Vincent, 1981). Though based on chitin, their properties depend very much on the degree of a process called tanning or sclerotization (Vincent & Hillerton, 1979). Tanning is an age change in most insects (Schofield *et al.*, 2002), converting a low modulus material (e.g. Young's modulus of 0.015 GPa: Vincent, 1980), probably with a J-shaped stress–strain curve, to one that is effectively linear and with a much higher modulus (Vincent, 1980, 1981). Yield strength rises dramatically with the degree of tanning (Hillerton *et al.*, 1982; Schofield *et al.*, 2002). Yet, these age changes are not inevitable. The adults of some groups, like the Lepidoptera, have very lightly tanned cuticles while parts of the cuticle of beetles can be very heavily tanned. Which type of prey did early mammals eat?

There was a consistent trend in early mammalian evolution towards the tribosphenic form where molars have sharp cusps linked by sharper concave blades (Crompton & Sita-Lumsden, 1970; Crompton, 1971). Evidence from modern insectivorous groups (Freeman, 1979, 1981b; Strait, 1993a,b) suggests that blade development is much more prominent in forms that eat softer insects. Inflexed blades were associated in Chapter 4 with foods that break at high strain and that have high Poisson's ratios. It is very likely that synapsids and early mammals tended to eat Mesozoic insects with lightly tanned cuticles – probably larvae. These cuticles are tough enough ($\sim$1.5 kJ m^{-2}; there are few published data) not to propagate cracks easily: they are a displacement-limited defence. Making a fracture in the cuticle of any size would require a sharp structure, while continuing this to obtain fragmentation would require the blades to move through the food and finally to contact. If they did not, then the crack would arrest. It is likely to have been the need to fragment insects with these properties that led to the need for tooth contact ($\equiv$ occlusion).[12]

Prinz *et al.* (2003) show that one subdivision of an insect (*Tenebrio molitor* larva – a mealworm) with a lightly tanned cuticle is sufficient for extensive digestion by gut enzymes.[13] The cusped teeth of early synapsids (Fig. 7.3a) may only have needed to make a small opening to obtain a sufficient rate of digestion from it. Such fracture does not require the teeth to contact. As metabolic requirements increased over time and body size decreased, molar shape seems to have been adapted to increase that exposure by turning fracture into fragmentation. That needed blades, which were developed between cusps. Initially, in an early mammal like *Morganucodon*, there

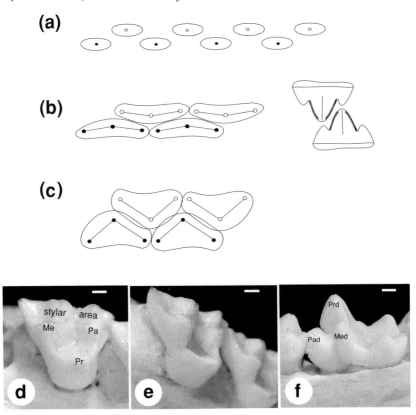

Fig. 7.3 (a, b, c) Schematic view of the evolution of molars in early mammals (see flickart for a side view). Upper teeth have unfilled cusps. (a) The spaced single cusps of a basal synapsids. (b) The occlusion of an early mammal such as *Morganucodon*, in which upper and lower molars had three cusps arranged in a mesiodistal line that had to wear against each in order to align properly and make a blade. (c) The triangulated molars of a mammal like *Kuhneotherium* fit together without wear, forming blades on mesial and distal surfaces of the teeth. (d–f) The upper and lower molars of *Didelphis marsupialis*, a polyprotodont marsupial. The stylar region of upper molars evolved very early and its relative extent on the working surface is a clear marker for insectivory in many mammals, e.g. microchiropteran bats (Freeman, 2000). Scale bars, 1 mm.

was no preformed occlusion and the upper and lower teeth, consisting essentially of a mesiodistally aligned row of three cusps (Fig. 7.3b), worked against each other to produce an edge. At best, this might have produced one subdivision of the insect. Preformed occlusion was developed in later forms and there then followed a gradual tendency to increase the number of blades that could act simultaneously so as to increase the amount of fragmentation. Presumably, higher rates of energy acquisition and diminishing size drove these trends. Triangulation of the cusps (Fig. 7.3c), such as in *Kuhneotherium*, positioned blades at angles so increasing the chance (selection function) of a blade hitting the insect and speeding up the process. The subsequent evolution of the tribosphenic form (Fig. 7.3d) has been described by Crompton & Sita-Lumsden (1970) who show how it is designed so that several blades can act almost simultaneously on either side of any given tooth with lateral jaw movement and 'waggling' of their mandibular symphyses.

The tribosphenic form has cusp-in-fossa alignments. This suggests the additional consumption of insects with more heavily tanned cuticles, probably adult insects. A tanned cuticle is more towards a stress-limited design, likely to fail by rapid crack propagation, so rendering the need for blades unnecessary for this aspect of the diet.

The need for both blades and cusps on the molars of tribosphenic mammals is explained by the position of the ellipse for insect cuticles in Fig. 7.1b, which straddles the dental divide between blades and cusps: because tanned cuticles will crack easily while untanned ones will not, so needing blades.

The great problem with all of this is that very little is still known about the diversity of insects in the diet of living mammals, let alone those in the Mesozoic, or their mechanical properties (Strait & Vincent, 1998). What information there is does not suggest that insect cuticle is anything like as tough as the veins in leaves, for example. Exactly how much fragmentation of an insect is or was necessary is difficult to estimate. Modern insectivores that have molars that approximate a tribosphenic form chew insects only briefly (e.g. the Western tarsier (*Tarsius bancanus*), a nocturnal primate observed by Jablonski & Crompton, 1994).

Just a glance at the dentitions of the living Insectivora, the ragbag 'old' order of small mammals (now disbanded), covers up dietary diversity belied by its nametag (Strait, 1993a). Little detail is yet known about the consumption of invertebrates other than insects, but this certainly seems to hold many keys to the understanding of the diversity of mammaliaforms. Some hints come from the tinted enamel referred to in Chapter 2. The Soricidae (the shrew family, which includes the smallest living mammals), have two dental types: the red-toothed Soricinae and the white-toothed Croidurinae. Red or brown pigmented enamel is known to contain iron deposits, probably always from breakdown products of haemoglobin. A group of authors including Akersten have consistently alluded to this red pigment being tied to the consumption of earth-dwelling prey like earthworms and some insect larvae (Akersten *et al.*, 2002). Siliceous grit is usually harder than dental enamel, but the red-tinted enamel of soricines contains goethite, a very hard mineral, that apparently gives the radulae of some molluscs a hardness of about 6000 MPa (Runham *et al.*, 1969). This should resist abrasion by grit better than plain white hydroxyapatite-based enamel (Akersten *et al.*, 1984). Plausible as this is, there is little support as yet from microindentation. Nanoindentation studies (Appendix A) are called for because this red layer is thin – the teeth as a whole are very small – and the hypothesis is eminently sensible. Akersten *et al.* (2002) point out that a Palaeocene multituberculate (*Lambdopsalis bulla*), in a mammalian lineage separate to therians, also had red-brown enamel, indicating at least that this terrestrial niche was available at that time.

Tooth form in carnivores

Of all the major dietary streams in mammals, carnivores (vertebrate feeders) and piscivores (fish-feeders) do the least chewing. Generally, their anterior teeth resemble those of early mammals and, in some fish-feeders, the whole dentition is similar to that of a reptile. The critical teeth for ingestion are the canines, which provide the simplest link back to reptiles because of their simple single-cusped form. They show the same kind of conical tooth form, with a recurved (actually spiralled) shape. The upper and lower canines of carnivores are generally long and projecting teeth, although their degree of projection varies greatly. The uppers are longer than the lowers. The canines act at larger gapes than the other teeth (Fig. 6.2). The essential load on them in a carnivore is not just due to indentation of the tips, but to an anteroposterior pull (Simpson, 1941; Smith & Savage, 1959). The cross-sectional shape of the canines tends to reflect this, in that they typically have oval cross-sections with the long dimension being anteroposterior (Van Valkenburgh & Ruff, 1987). Carnivores that use their canines to break bones must deal with the vagaries of the direction of the bite force depending on the orientation of the bone vis-à-vis the tooth (Rensberger, 2000). These tend to have circular cross-sections (Van Valkenburgh & Ruff, 1987).

Vertebrate soft tissues almost certainly will all lie to the 'bladed' side of Fig. 7.1b. Most carnivores have just one blade on either side of the mouth – the carnassial. They do little processing of these soft tissues because they are readily digestible even in large pieces. Generally, they only seem to manipulate one particle of food in the mouth at a time and all that the tongue has to do is organize this so that it lies between the carnassials. Very often, it appears that fracture between the carnassials is an ingestive process because the bite is made on the carcass with most of the food outside the mouth. The carnassials always have the inflexed blades referred to in Chapter 4 (Fig. 6.2). Their size indicates the probable linear dimensions of swallowed pieces and their position is presumably linked to the gape necessary for their use (Fig. 6.2). Some carnivores, such as hyaenas and canids, also fracture bones. Mineralized tissues have very low $(R/E)^{0.5}$ ratios and will fracture the carnassials if these are used to break them (Chapter 6). Hyaenas have developed blunt conical premolars in front of their carnassials, while canids use molars distal to them.

PLANT FEEDERS

Frugivory (fruit-eating)

As mentioned in Chapter 4, fruit flesh is undefended because it is intended to be eaten, being the reward offered to a frugivore for dispersing its seed(s) away from the parent plant.[14] The arrangement is the equivalent of employing animal pollinators. In the Cretaceous, before dinosaur extinction, angiosperms probably populated areas with relatively still air close to water. Some early angiosperms were actually aquatic, but by the late Cretaceous, it is clear that there were a considerable diversity of forms on land (Wing *et al.*, 1993). Being in relatively still air not only needed a pollination agent other than wind, but an alternative dispersal agent for seeds. Even today in tropical rainforests, animals disperse most of the seeds, this independence from wind having facilitated the development of such stratified forests (Eriksson *et al.*, 2000). Mammals diversified very rapidly after the dinosaurs died out 65 million years ago. Seed and fruit size records suggest that diversity peaked very early in mammalian diversification during the late Palaeocene–early Eocene epochs (Collinson & Hooker, 1991; Eriksson *et al.*, 2000), suggesting that mammals became significant consumers of fruit (frugivores) at that time.

The seeds of a plant contain its embryos – the potential members of its next generation – so these have to be heavily insured against death or else the plant species may not survive. Attack by many kinds of animals including mammals, both large and small, can devastate a seed crop at various stages of development. The attraction for animals is basically the energy reserve intended for germination, this being either carbohydrate (starch) or lipid (in the form of oil) in nature (Leighton & Leighton, 1983). Angiosperms have developed varied defences to counteract this threat. Many produce seeds in vast numbers, rather as some animals do with eggs, sacrificing

Table 7.1 *The characteristics of fruit skins and peels consumed by primates*

Species	Thickness (mm) of outer covering (type)	Fruit width (mm)
Alangium nobile (Alangiaceae)	1.8 (peel)	25
Cyathocalyx ramuliflorus (Annonaceae)	1.7 (peel)	21
Mezzettia parvifolia (Annonaceae)	0.6 (skin)	52
Willughbeia coriacea (Apocynaceae)	5.0 (peel)	30
Gnetum microcarpum (Gnetaceae)	1.5 (peel)	14
Garcinia forbesii (Guttiferae)	2.9 (peel)	27
Fibrauria tinctoria (Menispermaceae)	1.9 (peel)	20
Calamus luridus (Arecaceae)	0.6 (peel)	14
Xerospermum sp. (Sapindaceae)	2.2 (peel)	20

Source: Lucas & Corlett (1991).

the majority to ensure the survival of a few (Curran & Leighton, 2000). There is plenty of evidence that others produce seeds that contain very toxic chemicals (Bell, 1984; Waterman, 1984). Although mechanical protection is also common, it seems rare to find extensive chemical and mechanical defences together.[15] A minimal amount of mechanical protection to seeds always seems to be present, presumably in part so that mammals can distinguish seeds from flesh in the mouth by their 'hardness' (Corlett & Lucas, 1990).

I will follow a simple classification of fruits on mechanical grounds. Fleshy fruits can be distinguished from dry (fleshless) ones, while the former can be subdivided into those with a peel versus those with a skin (Janson, 1983). Dry fruits are generally either wind- or rodent-dispersed (Turner, 2001). A peel is a generally thick outer covering that separates cleanly from the underlying flesh at ripeness (Table 7.1). Primates are associated with these 'protected' fruits (Janson, 1983; Gautier-Hion *et al.*, 1985; Leighton, 1993). Familiar examples of these mechanically protected fruits are citrus fruits with their thick peels. These can be compared to unprotected fruits that have only a skin like an apple, for example, and which can be eaten by a wide variety of consumers including birds, bats and primates.

There have been many attempts since Ridley (1930) to match up the varied form of fleshy fruits with features of their dispersal agents (including mammals), but there appear to be a lot of mismatches. Some of these are probably due to inertia on the plant side of the fence (Janzen & Martin, 1982; Jordano, 1995; Herrera, 2002), but another issue is the behaviour of mammals whenever there is a fruit glut. Terborgh (1983) studied the behaviour of primates at a tropical rainforest site in Peru. Whenever there was a fruit glut, many mammals and birds would feed on these, with no apparent relationship between the anatomical form of these animals and the fruit that they ate. However, in periods when fruit was very scarce, i.e. when resources diminished and survival became difficult, each primate under study moved to a resource that distinguished it from the others and which did make considerable anatomical sense. For example, individuals of *Cebus apella*, a capuchin monkey with a large jaw and low-crowned, thick enamelled cheek teeth ate palm nuts. More lightly built squirrel monkeys (*Saimiri sciureus*) foraged for figs, while tamarins (*Saguinus* spp.) searched for nectar (Terborgh, 1983). This 'fallback' food hypothesis, also called a 'keystone food resource' hypothesis, seems to apply particularly to frugivory and could explain anatomical mismatches (Terborgh, 1986). The idea that the dentition itself was designed for such critical resources appears to be due to Rosenberger & Kinzey (1976).

Figure 7.1 shows that fruit flesh is very variable in its properties, due as much to cell size as to cell wall thinness. Its stiffness is a product of its turgidity (Chapter 4), but once cells burst, this is quickly lost. Teeth may not be needed to process many fruits because even the flat surfaces of soft tissues around the mouth can fracture cells (as suggested in Fig. 7.1b). Fruit bats do this by compressing some types of fruit flesh between the tongue and hard palate.[16] If the soft tissues of the mouth can fracture these tissues, then it is also probable that the soft tissues of the gut can too. There is little point then in the fragmentation of a lot of types of fruit flesh by chewing.[17] However, even if moisture-laden, some flesh does not give up its moisture very easily (a feature that is well known in domesticated tropical fruits too: Peleg *et al.*, 1976).

The spatulate incisors of anthropoid primates are relatively large in fruit-eating species (Hylander, 1975), making it likely that the trait evolved in association with this type of diet. One of the more plausible ecological links in fact is that between anthropoid primates and the consumption of fruits with peels (Janson, 1983; Gautier-Hion *et al.*, 1985; Leighton & Leighton, 1983; Leighton, 1993). However, the work of peeling itself (Appendix A) is generally very low and peels themselves are neither very thick (Table 7.1) nor very tough. How does a plant ward off 'fruit thieves' with fruit like this? The answer appears subtle. Successful peeling is only possible at full ripeness and requires the control of cracking such that the fracture plane remains at the junction between peel and flesh. This can only be achieved by getting a bladed tooth into the interface between the tissues to guide the crack.

Additionally, peel thickness seems generally to lie below the deformation transition described by Eqn 4.9 (Chapter 4) – in other words, cracks will not propagate within the peel when it is penetrated and so the blade of a spatulate incisor is needed to remove it.

The molar teeth of most frugivores are generally rather small (Kay, 1978; Strait, 1993b; Freeman, 2000), being low-crowned with rounded cusps (Kay, 1975), presumably designed for cellular fracture rather than fragmentation. The diets of primate fossils are commonly categorized on this basis and the introduction of fruit into the diets of insectivorous early mammals seems marked by a reduction in blade length and a rounding of cusps (e.g. Kirk & Simons, 2001). However, there can be a lot of variation in the dentitions of frugivorous primates as Fig. 7.4 indicates. A male gibbon is larger than a male long-tailed macaque, yet the latter has a larger dentition all round. How can this be understood? I suggest by looking harder at how these primates treat seeds.

Patterns of seed treatment

Seed treatment in general, whether consumption or avoidance is practiced, is of major evolutionary importance and it should be anticipated that there are considerable dental adaptations in frugivores for dealing with them. I will categorize mammals on the basis of how they treat seeds and see if this helps in understanding tooth size variations. The classification derives from the study of the behaviour of wild long-tailed macaques (*Macaca fascicularis*), a Southeast Asian Old World monkey commonly used in biomedical research (Corlett & Lucas, 1990; Lucas & Corlett, 1998). Seed(s) can be:

(1) Ingested, then swallowed almost immediately. The intact seeds pass straight through the gut and are defaecated. (*Seed swallowing*)

(2) Held by the hand while the flesh is cleaned off the seed by the incisor teeth. (*Seed cleaning*)

(3) Ingested into the mouth where the flesh is removed entirely or in part by the postcanine teeth. The seeds are then spat. (*Seed spitting*)

(4) Ingested, then chewed with the postcanine teeth until swallowed. The seeds would be destroyed by this action, rendering their content digestible. (*Seed destruction*)

These four behaviours, the *swallowing*, *cleaning*, *spitting* and *destroying* of seeds, were observed in just one social group of long-tailed macaques, but they did not occur with equal frequency (Corlett & Lucas, 1990). Clearly, individual animals have some sensory criteria for deciding how to tackle

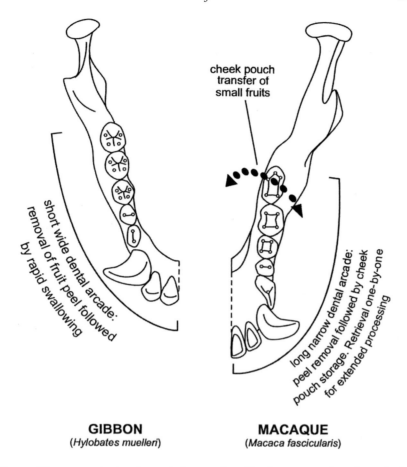

cheek pouch
transfer of
small fruits

short wide dental arcade:
removal of fruit peel followed
by rapid swallowing

long narrow dental arcade:
peel removal followed by cheek
pouch storage. Retrieval one-by-one
for extended processing

GIBBON
(*Hylobates muelleri*)

MACAQUE
(*Macaca fascicularis*)

Fig. 7.4 The contrast in dental morphology produced by the consumption of very similar fruits by sympatric primates, but which treat these foods in very different ways. The long-tailed macaque, an Old World monkey, has cheek pouches that it uses to store flesh-covered seeds, retrieving these one by one, chewing off the flesh and then spitting the seeds out.
  The gibbon tends just to swallow seeds with the flesh, later defaecating the former.

the seeds in a fruit. Long-tailed macaques destroyed the seeds of all the dry fruits that they ate, but avoided breaking down the seeds of almost all fleshy fruits.[18]

The treatment of fleshy fruits appeared to depend on fruit and seed size. Lucas (1989) postulated four size thresholds to explain the behaviour of these monkeys, which can be generalized. In Fig. 7.5, fruit size thresholds are denoted as F_1 and F_2, with $F_1 \geq F_2$, while seed size thresholds are symbolized

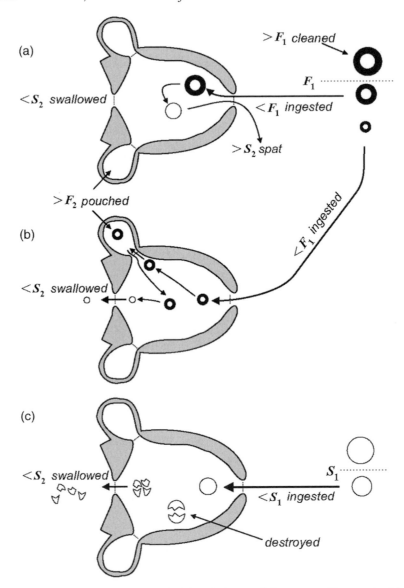

Fig. 7.5 The processing of single-seeded fruits by long-tailed macaques. The mouth of these monkeys is viewed from above, with the lips to the right and the passage leading to the pharynx shown at left. Soft oral tissues are shown in grey. These monkeys have highly sensitive cheek pouches that communicate with the vestibule through narrow openings. The flesh of a fruit is shown as a dark band around the single seed. The presence or absence of a fruit peel is ignored. (a) Shows what happens to fruits that can fit into the mouth (i.e. of a size F_1), but which have seeds that are too large to fit into a cheek pouch ($>F_2$) and too big to swallow ($>S_2$). (b) Shows the oral passage of a fruit that can fit into a cheek pouch and also be swallowed (although the two thresholds are independent). (c) Shows the thresholds that apply to a dry fruit (shown just as though it was a seed, this containing all the nutritive value).

as S_1 and S_2, with $S_1 \geq S_2$. F_1 and S_1 are both ingestion thresholds. Feeding on fruit could be understood in terms of these four thresholds.

Macaques would pluck large fruits with their hands and bring them to their mouths. The size of the opening to the mouth (mouthslit) controls the entry of large fruits. If the fruit minus any peel was too large to fit into the mouth, i.e. $>F_1$, then it would be continued to be held by the hand(s) and *cleaned* in front of the mouth. There is a relationship between seed and fruit size (Lucas & Corlett, 1991; Dominy, 2003), but it is not very strong. A fleshy fruit can contain one, several or many seeds, although of course, only large fruits can contain large seeds. If seeds within a fruit of size $>F_1$ had a size $>S_1$, then they would be cleaned outside the mouth and dropped. If however, the seeds $<S_1$, then they might enter the mouth so that more flesh could be stripped from them.

Any fruit of size $<F_1$ would be ingested. All cercopithecine Old World monkeys, including long-tailed macaques, possess muscular cheek pouches located opposite the molar teeth for storing food (Figs. 7.4 and 7.5). These pouches get distended when filled with fruits, which were the only item of natural food long-tailed macaques were seen to place in them. The pouches are probably smaller than the oral cavity proper. So, defining the maximum size of each cheek pouch as F_2, we can say that fruits $<F_2$ could be stored in them. Long-tailed macaques appear to retrieve fruits from these pouches one at a time, pushing them back into the centre of the oral cavity and removing the flesh with the teeth (Fig. 7.4) while they move to another food source. Once seeds had been cleaned, long-tailed macaques either *spat* the seeds or swallowed them. Only the smallest seeds were swallowed and found in the faeces and an intra-oral seed detection size, S_2, can be postulated, above which seeds were spat and below which they were swallowed.

The size of the fruit and seed size thresholds in macaques appear to be $F_1 \approx$ 30–40 mm, $F_2 \approx$ 10–15 mm, $S_1 \approx$ 30 mm and $S_2 \approx$ 3–4 mm, all dimensions being maximum seed width (i.e. the second largest object dimension).[19] Thus, the seed size thresholds, S_1 and S_2, are a decade apart in macaques, which explains why these monkeys spit $>$70% of the seeds that they process (Corlett & Lucas, 1990). Despite the variety of seed behaviours that they display then, long-tailed macaques are seed-spitters, probably to try to spit the seeds of all fleshy fruits that they can.[20]

The evolutionary point of this is that mammalian frugivores can be classified into *spitters*, *cleaners*, *swallowers* or *destroyers*, the basis being the fate of either the majority of seeds or those of critical ('fallback') fruits if this is thought to be more important.

What dental or oral changes are necessary for the efficient processing of seeds in these various ways? The coincidence of S_2 with the oral mucosal thresholds discussed in Chapter 3 makes it likely that all mammals possess a similar threshold. So seed spitters and seed predators are very likely to swallow very small seeds simply because they do not detect them. However, few mammals other than cercopithecines appear to use this lower detection threshold to actively eject seeds. The singular ability of cercopithecines to spit seeds appears to be connected to their possession of highly sensitive cheek pouches. Most other mammals are probably limited to either cleaning or swallowing seeds. The process of cleaning seeds requires a level of manual dexterity present only in anthropoid primates with thumbs, such as long-tailed macaques, and apes (e.g. the chimpanzee: N. J. Dominy *et al.*, unpubl. data). Anthropoids with reduced thumbs, such as spider monkeys, gibbons and colobines, have very limited ability to clean seeds. So, most mammals probably have $S_1 = S_2$, meaning that they will swallow any seed that they can ingest. This is likely to be controlled by the size of the mouthslit, which is in turn a reflection of the maximum gape of the animal. For African elephants (*Loxodonta africana*), this means that they can ingest and swallow *Balanites wilsoniana* seeds that are 88 mm in seed length and 47 mm in seed width (Chapman *et al.*, 1992).

The tooth sizes of anterior (incisor) and posterior (postcanine) teeth of frugivores can be deduced by making some simple assumptions. The time available for processing fruits is assumed to be limited, so that any behavioural changes towards one category of seed processing or another would result in changes that reduce processing time. Support for these considerations comes from Ungar's (1994) study of four Sumatran primates in which he suggests that the time spent in processing foods with the incisors may be important in determining their size.

Further, from Chapter 4, I assume that higher primates have spatulate incisors to cope with protected fruits (those with peels), such that the common ancestor of all these primates peeled these fruits efficiently. The most important dimension of a spatulate incisor is its width (i.e. its mesiodistal dimension). For the postcanines, changes in either the bucco-lingual or mesiodistal dimensions of their working surfaces could change their efficiency.

Seed swallowers peel fruit with their incisors, but do little with their posterior teeth, since flesh plus seeds are quickly swallowed. Seed swallowing is probably the ancestral behaviour pattern, requiring less in the form of manual or oral adaptations than the others. Most lineages of fleshy fruits probably evolved in the Eocene before anthropoid primates, but protected fruits (a small percentage of those available in every site yet studied) probably co-evolved with anthropoids. Seed cleaners do most of their oral processing at the front of their mouths with their anterior teeth, whereas seed spitters, perform extensive processing of fruits with both anterior and posterior teeth. Designating the size of anterior teeth as A and the overall size of the posterior teeth as P, and employing '$-$' for small and '$+$' to mean enlarged, then the following dental configurations for the three categories of primate can be anticipated:

SEED SWALLOWERS	SEED CLEANERS	SEED SPITTERS
$A-P-$	$A+P-$	$A+P+$

These represent three of the four possibilities for combining these symbols. What about $A-P+$, a primate with small anterior, but large posterior, teeth? It has been established for a long time (Hylander, 1975; Kay, 1975) that primates with this dental configuration are likely to be predominantly folivorous (leaf-eaters) rather than frugivorous. It would, of course, make a far cleaner classification if there were a way to incorporate these apparent folivores into the classification. The way to do this is to recognize that when most of these folivores eat fruits, they destroy their seeds. Many colobines are seed destroyers, eating leaves for the most part, but also consuming large quantities of seeds when these are available (McKey *et al.*, 1981; Bennett, 1983; Harrison, 1986; Davies & Baillie, 1988). The leaf-eating sifaka, *Propithecus verreauxi*, is also a seed predator (Overdorff & Strait, 1998; Yamashita, 2000).

It is possible, therefore, to finish off the classification like this:

SEED SWALLOWERS	SEED CLEANERS	SEED SPITTERS	SEED DESTROYERS
$A-P-$	$A+P-$	$A+P+$	$A-P+$

This addition makes logical sense if the processing of fruit by a seed destroyer is considered. Seed destroyers usually either concentrate on dry fruits, where the anterior dentition has little role in processing or on unripe fruit. There is no peeling involved because this does not detach from the fruit and, as pointed out by Leighton (1993), when a primate targets an unripe fruit, it does so for the nutrients contained in the seed. So, fruits are immediately thrown to the posterior teeth, which take an extra processing load, as compared to a seed swallower or seed cleaner, while the anterior teeth do not. The evidence is very strong that colobines have small anterior, but large posterior, teeth (Hylander, 1975; Kay, 1975). The gelada baboon (*Theropithecus gelada*), a cercopithecine, is another example of a folivore/seed eater (Dunbar, 1977; Iwamoto, 1979) and has an $A-P+$ dentition too (Fig. 5.5) (Jolly, 1970).

Most cercopithecines other than the gelada baboon are seed spitters. Extensive study of the redtail monkey (*Cercopithecus ascanius*) confirms this in Kibale, Uganda (Lambert, 1999). In contrast, gibbons appear to be seed swallowers, avoiding fruits that will be fit into their mouths (Whitten, 1982). These thumbless primates have an $A-P-$ dentition. New World spider monkeys (*Ateles* spp.) have small thumbs, but this is sufficient to allow them to clean some fruits (van Roosmalen, 1980). They have small posterior teeth – their last molar is occasionally absent (Zingeser, 1973) – and could be classified as $A+P-$.[21]

In contrast, the great apes seem to be seed cleaners. Certainly, chimpanzees show a much greater propensity to swallow seeds than do redtail monkeys (Lambert, 1999), but they are also equipped with hands that facilitate the processing of fruits in front of the mouth. Large fruits are processed in this way (N. J. Dominy *et al.*, unpubl. data). However, outside primates, frugivorous bats are the best example of seed cleaners and Freeman (1988) shows that the dentitions of some microchiropteran bats are very definitely $A+P-$ in form.

A note on seed treatment by humans is applicable here. Some readers may feel, from personal experience, that it is impossible to categorize seed treatment by modern-day humans – for example, some people spit grape seeds, others swallow them while still others destroy them. However, as stated in Chapter 1, populations that have control over their food supplies may not provide a good guide to the behaviour of humans with seeds in the past (any more than with diet as a whole) because the leisurely oral activities that we exhibit could be maladaptive under natural conditions. One clue to selective pressures being eased is the very fact that seed treatment is so variable. Ridley (1930) gives a very clear account of the behaviour of hunter and gatherer groups on the Malay Peninsula and states that they generally swallowed seeds, even one as large as those of the durian (*Durio zibethinus*, Bombacaceae).

Molar tooth shape in frugivores

Frugivores, whether seed destroyers or not, tend to have blunt-cusped molar teeth because most fruit flesh lies in the 'cusp' sector in Fig. 7.1b. The principal reason for this was given in Chapter 4, where it was shown that both fruit flesh and seed shells often have a very low toughness (Figs. 4.15 and 4.17). Furthermore, if the objective is to break open as many cells are possible, cusps do this much faster than blades.[22] Thus, the molars have few prominent crests ($\equiv$ 'shearing' crests analysed by Kay (1975, 1978; Kirk & Simons, 2001)) because cracks spread easily. An exception may be the noticeable marginal ridges of mammals that feed on seeds whose shells need to be wedged open (Fig. 4.18).[23] Several seed shells lie in the 'wedge' sector in Fig. 7.1b. The cusps may be more bulbous in seed destroyers because the bite force may not be aligned with the direction of jaw movement when very stiff foods are being chewed (Rensberger, 2000). The enamel of the cusp tips may also be thicker (Kay, 1981; Walker, 1981), both these features being adaptations preventing tooth fracture. Consumption of a lot of high-moisture fruit is likely to result in very loose-fitting features on upper and lower molars to allow the expressed juice to escape. As was pointed out in Chapter 2 concerning saliva flow in the mouth, this juice is probably subject to (biaxial) extensional flow, which means that its effective viscosity can be very high, so impeding tooth movement if no 'release valves' are available.[24]

Seed contents, on the other hand, do not express juice and tend to fragment rapidly. The match of cusps and fossae is also important here, but for different reasons. Cusps are only effective for the fracture of food fragments when these lie inside a certain fracture zone around those cusps (Lucas & Luke, 1984b). The size of these zones depends on the ratio of particle size in relation to cuspal dimensions. Particles lying outside these zones will be missed by cusps because the low coefficient of friction between them will cause slipping even with contact. Bulbous cusps will have larger zones of action than small ones. In fact, it is probably optimal for the cusps to be bulbous for this reason in order to avoid particle slippage. Rose & Sullivan (1961) worked this out in detail when analysing the action of certain types of industrial comminution machinery and, if their analysis can be transferred to the dentition, then the maximum size limit for comminuting seed fragments with cusps may be only 0.2–0.4 times the cuspal diameter. However, when seed contents are tough, then wedges are probably needed, as seen in colobine molars (Lucas & Teaford, 1994). In support of all this, seed contents tend to fall into the cusp or wedge sectors of Fig. 7.1b.

Seed predation is of immense ecological and evolutionary importance. The list of seed-predating mammals is long and includes primates such as the otherwise leaf-eating sifakas (Overdorff & Strait, 1998), the colobine Old World monkeys (McKey *et al.*, 1981; Bennett, 1983; Davies *et al.*, 1988) and pitheciin New World monkey genera, *Pithecia* and *Chiropotes* (Kinzey & Norconk, 1990, 1993). If the list is further extended to primates that are occasionally seed destroyers (but none the less important for that), then certain capuchin monkey species (*Cebus apella*: Terborgh, 1983), break large palm seeds while gelada baboons concentrate on grass seeds (Dunbar, 1977; Iwamoto, 1979). Seed-eating (Jolly, 1970; Peters, 1979, 1981, 1987) or hard fruit-eating (Walker, 1981) has consistently been suggested as a diet for members of the lineage of primates leading to us – the hominins – so this behaviour has deserved a lot of attention here.

Despite the above list, rodents are the most important seed predators because they often move seeds (sometimes in their cheek pouches, but this depends on seed size, a crucial factor for understanding rodent behaviour: Theimer, 2003), prior to consuming them. Though seed destroyers, they may forget to consume a significant portion, thus being seed dispersers as well (Turner, 2001). Rodents tend to feed on mechanically defended seeds.[25] However, they are extremely small animals and so cannot develop the bite forces that pigs (Curran & Leighton, 2000) and peccaries (Kiltie, 1982), important seed destroyers in Asia and Central/South America respectively, can generate. Their ever-growing incisors make sense in terms of abrading woody tissues, albeit at the cost of considerable loss of tooth tissue (Lucas & Peters, 2000).

Herbivores

Any mammal that feeds extensively on the structural, rather than reproductive, parts of plants can be called a herbivore. The category includes bark and woody tissue feeders like (some) rodents and elephants, 'root' specialists such as pigs, and leaf-eaters such as the ruminant artiodactyls (e.g. bovids – the cow family) and the 'hindgut' fermenting perissodactyls (e.g. zebras). I will deal with plant storage organs briefly in the sections on human evolution, but even without considering these, the range of foods in herbivore diets is very broad. The dentitions of the mammals that eat them are correspondingly diverse, being at a maximum between 45 and 30 million years ago (Jernvall *et al.*, 2000). To get clues that lie at the root of this diversity, which is all that can be done here, herbivore diets need to be classified in mechanical terms. Unfortunately, descriptions of plant mechanical defences have often been narrowly focussed on features like

Table 7.2 *List of jaw and dental features that differentiate grazers from browsers*

Jaw and dental features	Grazers	Browsers
Height of jaw joint	High	Low
Shape of incisal row	Straight	Curved
Size of incisal row	Large	Small
Muzzle morphology	Broad	Narrow
Width of premaxilla/width of palate in molar region	High	Low
Robusticity of mandible	High	Low
Diastema	Long	Short
Hypsodonty	High	Low

Source: Mendoza *et al.* (2002).

spines and thorns, i.e. on analogues of the 'weapons' that animals have, so the literature does not always help much. The conventional way to examine herbivorous diets has been in relation to nutritional factors like fibre content (Milton, 1979; Shipley & Spalinger, 1992; Van Soest, 1994), plant chemical defences (Janzen, 1978; Waterman, 1984) or their combination (Waterman & Kool, 1994). Body size has also figured prominently (Owen-Smith, 1988; Van Soest, 1996), although interest in this factor appears to be on the wane (Gordon & Illius, 1994). Recently, factors that influence the rate of oral processing of foods have been considered in relation to feeding preferences (e.g. Janis & Ehrhardt, 1988; Perez-Barberia & Gordon, 1998a,b, 2001), providing a link to the present discussion.

Folivores are usually separated into grazers (grass-eaters) versus browsers (consumers of other foliage). Although this distinction ignores mixed-feeders, it is very useful for defining dietary extremes. The major predators of grasses seem to be large mammals. Being small compared to these herbivores, grasses seem to show stress-related defences with relatively narrow blades, parallel venation, an often-extensive pattern of silica deposition in the tissues and little defensive chemistry (Owen-Smith *et al.*, 1993). Browse mostly consists of dicotyledonous angiosperms, which is predominantly attacked by relatively small invertebrates and so exhibit displacement-limited defences: relatively wide laminae, reticulate venous network, probably rare silica deposits and an often-extensive defence chemistry. Accordingly, the prediction is that the oral morphology of grazers should be ingestion-dominated while that of browsers should be mastication-dominated. Table 7.2 is adapted from a table in Mendoza *et al.* (2002) and indicates consistent differences between the jaws and dentitions of grazers and browsers. It would appear that grazers generally have larger squarer muzzles, paralleled with larger squarer front ends to their jaws containing larger incisors than browsers. This arrangement increases the volume of food per bite when collecting from a basically two-dimensional mat (Illius & Gordon, 1987; Janis & Ehrhardt, 1988; Solounias *et al.*, 1988; Solounias & Moelleken, 1993) and accords well with ingestion dominance.[26] Prehensile lips seem more prominent in browsers, very possibly because of greater problems in manipulating tree leaves, which often have very smooth surfaces. The coefficient of friction between such leaves and unwetted mucosa is probably much higher than that between the teeth, which may explain why many herbivores have developed horny pads for ingesting leaves to replace teeth. It is unsurprising that grazers have higher-crowned (more hypsodont) postcanines than browsers because they probably ingest more silica (Janis, 1995).

Tooth size in herbivores

Most herbivores consume foods like leaves, twigs, barks, etc. that are shaped like sheets and rods. Ingestion is still controlled basically by $(ER)^{0.5}$, but these foods do not store much strain energy when fractured across their thinnest dimension by the postcanines, so their toughness, R, is obviously crucial for understanding resistance to fragmentation. Such foods are among the most difficult items both to ingest (Fig. 7.1a) and to chew (Fig. 7.1b), (considering R, not $(R/E)^{0.5}$), but the degree of woodiness is critical to this judgement. Undoubtedly, it is because mature leaves, twigs, etc. are so tough, that herbivores deliberately select immature stages when available. Figure 7.6 indicates the problems that browsers face in finding enough of this tissue. Generally, immature tissue is located only on the ends of a branch, twig or stalk. If only a very small amount of tissue is selected, this can have a very low toughness. However, the more proximal the first bite on these structures, the tougher the food. This first bite with the anterior teeth will tend to face a tougher obstacle than the postcanines will face on average.

The most obvious repercussion of this patchy distribution of preferred tissue is on scale. Many authors have commented on the need for smaller ruminants to take immature plant parts because their digestive systems cannot handle mature parts (e.g. Van Soest, 1996). However, the toughness cline shown in Fig. 7.6 means that a larger herbivore, generally ingesting a larger mouthful, will tend to eat food that is of higher average toughness.

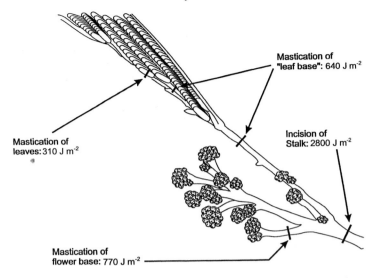

Mastication of
"leaf base": 640 J m^{-2}

Mastication of
leaves: 310 J m^{-2}

Incision of
Stalk: 2800 J m^{-2}

Mastication of
flower base: 770 J m^{-2}

Fig. 7.6 A food fragment (*Enterolobium cyclocarpum*, Leguminosae), consumed by a New World howler monkey (*Alouatta palliata*) in Costa Rica, to illustrate a general point about the problems of scale in herbivory. The larger the mammal, the more proximal on a plant part its bite point is likely to be, and the tougher the tissue will be.

This is exactly what both Bell (1971) and Jarman (1974) supposed: larger herbivores are more likely to eat food of lower quality, where quality in this sense is the inverse of toughness. The scaling arguments in Chapter 5 can be reintroduced here. Recalling Eqn. 5.6 and rearranging so that the ratio of forces of a larger mammal eating food particle x versus a smaller mammal eating food particle y gives

$$\frac{F_x}{F_y} = \lambda^{1.5} \left[\frac{R_x}{R_y} \right]^{0.5}. \tag{7.1}$$

If there were no difference in body size, then $\lambda = 1$, and the ratio of bite forces would be proportional to the square root of the ratio of the toughness of their foods, but applied to differently sized herbivores, positive allometry of postcanines results if differences in average food toughness are great enough. Much would hinge on the selectivity of the herbivore in question. Perhaps the least selective feeders are elephants and this argument might explain why elephant molars are so large that their mouth can only accommodate one of these teeth at a time. They are the only living mammals to adopt this strategy and it might be forced on them by the above arguments. However, the food data are simply not available yet, which stops this line of discussion. If grasses were tougher than browse, and grass can be very tough (Appendix B; Fig. 7.1) (Vincent, 1991; Wright & Illius, 1995), then this might generate larger molars in grazers. On the other hand, grazers are often larger than browsers (although there are exceptions), which might explain their higher jaw joints and longer diastemata in Table 7.2 (these predictions are listed in Table 5.1).

Postcanine tooth shape in herbivores

If any plant food were thin (<0.5 mm) and sufficiently tough (i.e. tougher than the dividing horizontal line in Fig. 7.1b), blades would be required on the postcanines to comminute them. These blades though are very different from those on the molars of insectivores or carnivores, being very low and produced by the exposure of at least two dental tissues. They evolved from cusps with expanded marginal ridges. These cusps generally have thin enamel (equally thin or thinner on the cusp tips than on the flanks, although some species have no enamel at all there) and are very low in height. Thus, just a small amount of wear exposes the links between cusps so to form what is called a loph (Jernvall *et al.*, 2000), an enamel crest that surrounds, often very irregularly, an enclosed pool of exposed dentine. The surface is often further complicated by the spread of cement onto the tooth crown and this together with variations in the hardness of dentine combine to form a very rough surface (Fig. 6.1) that grips the food particles. The enamel crest stands slightly proud of the other tissues (Chapter 6) and acts as a blade against similar crests on opposing teeth, being driven across them by an almost completely transverse jaw motion.[27] The key food properties for understanding this type tooth form in herbivores appear to be their sheet/rod format, their frictional properties and their toughness. Unfortunately, there are very few data yet available on any of this.

A major complicating factor for friction is the presence of tannins, which often have high concentrations in young tree leaves and unripe fruits, and which can combine with some salivary proteins to precipitate them (Chapter 3). The relative tannin consumption of mammals can often be surmised from the black stain that tannins leave on their teeth, the stain being produced presumably by the combination of tannins with salivary proteins adsorbed onto the enamel surface. The result must be a roughened enamel surface and increased friction.

HUMAN EVOLUTION AND DIET

The theory developed in this book should apply to the evolution of the dentition in any group of mammals. What limits its application mainly is the lack of quantification of diets. I have space only for one case study to try to show its power and that concerns human evolution. The major dietary difference between humans and other mammals lies in food preparation techniques and cooking which modify food properties prior to ingestion.

The classic form of explanation for human specializations is an analogy to other mammals. However, it is getting a little difficult to provide new dietary suggestions because previous efforts have spanned the whole herbivore–carnivore gamut. According to different authors, we might have been specialized herbivores (Jolly, 1970), generalized omnivores (Hatley & Kappelman, 1980), scavengers (Szalay, 1975), hunters (Ardrey, 1961) or fish/shellfish eaters (Morgan, 1972). While some may feel that these are 'old' references that have been flattened under the weight of new nutritional findings, recent literature shows no sign of consensus (e.g. the edited volume of Ungar & Teaford, 2002). It is not my intention here to find some strange new alternative diet: I have the same 'gut' feeling about this as about fads in Western societies. Instead, I am more interested in the direction that the magnetic logic of this book turns the human dietary compass. At the outset though, it must be recognized that there are intractable problems with some issues, not least the antiquity of key practices, such as cooking, which is very difficult to establish.

The face and dentition of modern humans

Both the face and teeth of modern humans (*Homo sapiens sapiens*) are much smaller than those possessed by primates of similar body size, so these structures must have been reduced in size at some point in time. The only general model for this reduction is that of Brace (1963, 1964) who argued that the most likely effect of any mutations on structures no longer being maintained by selective advantage would be to reduce their size. The problem with this general explanation is that there is no test of its applicability: it is 'last ditch'. I attempt an alternative here, but it is specific rather than general and does not apply to all aspects of structural change. The fossil record shows very clearly that some size reduction has been very recent, while part of it dates back much further. For example, while the length of the cheek tooth row is very short in modern humans, it is also short in anthropoid primates compared to others (Martin, 1990). Getting a general perspective on this should improve the chances of a viable overall explanation.

Facial and dental characteristics of hominins

Five genera of hominins[28] seem to have existed at various points in time: *Sahelanthropus, Ardipithecus, Australopithecus, Paranthropus* and *Homo*. A monospecific lineage runs from *Sahelanthropus*, the earliest known hominin, dated close to 7–6 million years ago (Brunet *et al.*, 2002) through *Ardipithecus*, living around 4.4 million years ago, putting these genera at the ends of the same line (Wong, 2003). Species of the gracile *Australopithecus* flourished between 4.2 and 2.3 million years ago while those of the robust *Paranthropus* lived between 2.3 and 1.4 million years ago. Both these australopithecine genera were not direct ancestors of modern humans. The oldest member of the genus *Homo* is *H. rudolfensis* from Koobi Fora, Kenya, known from 2.4–1.8 million years ago. Other early members of the genus were *H. habilis*, between 1.9 and 1.6 million years ago at Olduvai Gorge, Tanzania and *H. ergaster*, from 1.7 to 1.5 million years ago at Koobi Fora, Kenya. *Homo erectus*, ancestral to *H. sapiens*, is first known from about 1.8 million years ago.

I will be brief here, concentrating only on major features of these forms. The incisors of hominins are broad and spatulate, but they vary greatly in size and in orientation in the jaws. They are large and procumbent (i.e. stick out) in *Sahelanthropus* (Brunet *et al.*, 2002), but are relatively small and more vertically oriented in robust australopithecines. However, orientation is not well understood and varies with wear (Ungar *et al.*, 1997), making incisal size a better target for explanation. The canines in male hominins are always small, projecting only 4–5 degrees of gape beyond the other teeth. In this, they are unique among the apes. The postcanine teeth of all hominins are low-crowned. Their cusps are also low, blunt (e.g. in modern humans: Lucas, 1982a) and thick-enamelled, with the enamel being thickest over the cusps (Martin, 1985; Gantt & Rafter, 1998). The enamel of the molars is also quite wrinkled in appearance, somewhat resembling that of orang-utans and their fossil relatives, which are noted for this characteristic (Chaimanee *et al.*, 2003). The cusps tend to have rounded cross-sections and are very bulbous (i.e. they increase their cross-section rapidly away from the tip). The lingual cusps of the upper molars and the buccal cusps of the lowers are particularly broad. The jaw joint of hominins is situated high above the occlusal plane and jaw movements in chewing involve considerable lateral excursions.

Some hominins show these traits more prominently than others. The premolars vary in their degree of 'molarization', by which I mean that they vary in the proportion of the working surface of the postcanines that they occupy. The relative sizes of the molars also vary substantially and have variable size gradients. If the surface areas (mesiodistal length multiplied by buccolingual width) of each of the three molars, from mesial to distal, are symbolized as M1, M2 and M3, then in some populations of modern humans, the sizes of the lower molars can be expressed as $M1 > M2 > M3$, whereas in robust australopithecines, $M1 < M2 < M3$ (Robinson, 1956; Keyser, 2000). The last lower premolars of the latter have an expanded root support as compared to contemporaneous *Homo*, so adapting this tooth to take greater stresses (Wood *et al.*, 1988). Also, the posterior elements (talonids) of the mandibular molars in robust australopithecines are noticeably enlarged in these animals (Wood *et al.*, 1983), something not explained by their size (Hills *et al.*, 1983). They also seem to have made greater side-to-side excursions with their mandibles than some other hominins (Grine, 1981).

Potential explanations for some of these trends have already been presented in the theoretical sections. The ubiquitous presence of rounded cusps on the postcanine teeth implies that the hominin diet basically consisted of foods with low $(R/E)^{0.5}$. Further, even though fracture strength is a food particle size-specific trait, low cusps also indicate low σ_F/E. Bone fits these descriptions (Fig. 7.1; Appendix B) and has been suggested as a hominin specialization (Szalay, 1975), but is scarcely likely without consumption of vertebrate soft tissues, the properties of which are the antithesis of a match to hominin molar form (Fig. 7.1b). Seed shells could have been important, supporting the views of Jolly (1970) and Peters (1979, 1981, 1987). The shells shown in Fig. 7.1b are closer to those that Peters envisages for a robust australopithecine diet rather than the grass seeds that Jolly favoured. Not so far from seed shells on Fig. 7.1b are seed contents (the part of the seed that would be chewed in order to be swallowed) and plant storage organs: all these foods lie well within the $(R/E)^{0.5}$ range that Agrawal showed would fracture and break down rapidly in the mouths of modern humans (Lucas *et al.*, 2002). Some fruit flesh also resembles that of plant storage organs in texture and would also fit hominin diets.

Experiments on human chewing indicate that both jaw-closing muscle activity (Agrawal *et al.*, 1998) and the degree of lateral movement in chewing (Agrawal *et al.*, 2000) increase with food $(R/E)^{0.5}$, which is again supportive of foods with high $(R/E)^{0.5}$ being important in the diets of hominins like robust australopithecines. While bone is abrasive to teeth, seed shells are apparently not particularly so (Peters, 1982). Storage organs, particularly if these were underground (Hatley & Kappelman, 1980; Wrangham *et al.*, 1999) would have abrasives attached to them. Some of this might support a case for thick enamel on the basis of abrasion resistance.

So far, this just sketches out explanations for one food characteristic – internal mechanical factors that affect the breakage function. However, what about the effect of external physical characteristics, such as the size and shape of particles and ingest food volume (the mouthful)?

The number of teeth in the human mouth

It may be recalled from Chapter 5 that the size of the working surface of the cheek teeth has a big effect on the selection function (the probability of particle fracture). Food breakdown efficiency can be predicted in humans just by counting the number of functional cheek teeth, counting each molar as two units and a premolar as one (Helkimo *et al.*, 1978). Unfortunately, compared to early mammals or even to some primates, hominins are short on teeth, having two incisors, one canine, two premolars and three molars on each side of the upper and lower jaws – at most. This is a considerable reduction from the number in early mammals, though this did not happen recently: all anthropoid primates have far fewer teeth than early mammals. Why?

In Chapter 5, I suggested that the reduction in tooth number in early mammals was due to dwarfing. It seems reasonable to expect any re-duction in teeth in other mammalian lineages to be explained on sim-ilar grounds. An example of how this may operate is illustrated by the last instance of tooth loss prior to the emergence of the hominins. Hominins are catarrhines, a taxonomic group also containing Old World monkeys and apes. The catarrhines have a common ancestry with platyrrhines (New World monkeys), diverging something like 40–35 million years ago. It is hard to define the date – it might be much earlier (Tavaré *et al.*, 2002).

New World monkeys have an extra premolar in both upper and lower jaws compared to catarrhines. Osborn (1978) gave a highly plausible de-velopmental explanation for this premolar loss. His view is sketched in Fig. 7.7. He suggested that each tooth class (incisors, canines or post-canines) is developmentally distinct. As it grows, it competes for jaw space with other classes. In Osborn's view, whenever there is some (undefined) pressure on jaw space, the last teeth to form in each series will be the first to be lost. The upper diagram in Fig. 7.7 shows two generations of teeth

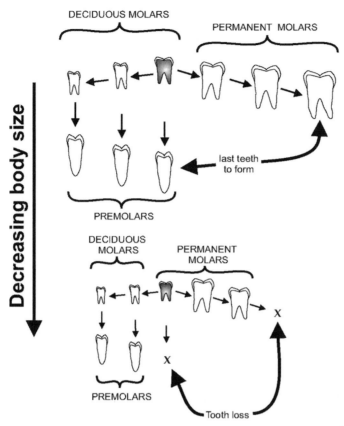

Fig. 7.7 The process of reduction in tooth number in the hypothetical ancestral catarrhine primate according to Osborn (1978). In the upper diagram, the postcanine teeth of a catarrhine ancestor are pictured as a growth series with two generations. The first generation includes both deciduous and permanent molars, the premolars being second-generation replacements of deciduous molars. The stem tooth is the first tooth to develop – that is the last deciduous molar (coloured grey). The lower diagram shows the problems that a dwarfed catarrhine would face due to fracture scaling (and, therefore, negative allometry) of the postcanine teeth. The last teeth in the series, the most posterior premolar and molar, fail to form, thus promoting the most posterior deciduous molar, the stem tooth, to permanence.

in a hypothetical catarrhine ancestor. The first tooth to form is the stem tooth of that class. The last postcanines to form are the most distal molars, premolars and permanent incisors. Under pressure for space, some of these teeth may not form. Osborn noted that there are known statistical associations like this. If the third molars in humans do not form, the next likely teeth to fail to form are the lower posterior premolars and, following this, the upper lateral incisors (Garn & Lewis, 1962). Osborn suggested that the third molars and most distal premolars were the teeth that were actually lost in the earliest catarrhine. Thus, the 'first molar' of any member of this group is actually a retained deciduous molar that has been 'promoted' to permanence because its second-generation replacement no longer forms.[29]

This is not the traditional view, which has the most anterior deciduous molar and premolar being lost simultaneously. I have always supported Osborn's view as being more likely, but what trigger precipitates tooth loss? The only one that I can suggest is dwarfing. Sadly, the fossil evidence for early anthropoid evolution is still too limited to examine this. Some early anthropoids were very small, like *Eosimias* (Beard *et al.*, 1996), which probably weighed only about 100 g, but other early Eocene finds with anthropoid affinities appear to be somewhat larger, as are the North African parapithecids (Fleagle, 1999). Yet this remains a viable possibility in my view, because it should be remembered that dwarfing involves relative changes and is not tied to any absolute size band.

A lost premolar here or here does not take tooth numbers back to those in an ancestral mammal. The overall suggestion then is that the evolution of mammals in general, and primates in particular, has been punctuated by periods when, surviving in shrinking isolated locations, populations of some species were subject to intense pressure to reduce their body size so as to cope with smaller dietary patches. This resulted in dental crowding that led to tooth loss. When conditions ameliorated, perhaps as barriers to movement were removed and the climate changed, these species could then have re-expanded slowly, both in a geographic and a somatic sense. These species were the ancestors of the major lineages that followed them (Fig. 7.8). On the direct line from the earliest mammaliaform to modern humans, some 200 million years, I hazard a guess that there might have been three or four dwarfing events. Although there has been no dwarfing within hominins, we show the signs of these past events.

Molar size gradient in primates

The important point about reduction or loss of a single tooth in a class is that it will affect the size of its neighbours because their growth is co-ordinated (Sofaer, 1973, 1977). Tooth size gradients are usually smooth (certainly so if Osborn's definitional criteria are followed), so the penultimate tooth would also be affected, and so on, leading to a change in the gradient of tooth size in

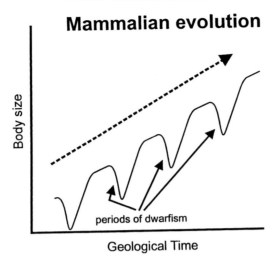

Mammalian evolution

Fig. 7.8 The suggested pattern of body size evolution in some mammalian lineages. Although the direction of body size changes has certainly been slow enlargement for most of the last 65 million years (this is called 'Cope's law': Alroy, 1998), there may have been short periods when body size had to decrease very rapidly. Dental crowding would then have resulted with the loss of some of the teeth. While some of these dwarfed mammals probably died out, others could have been ancestral to later forms. Reduction in tooth number, which the theory suggests should be abrupt and not drawn out, is the suggested evidence for this.

a series. In extreme reduction, presumably prior to loss, the final tooth may not develop to its full shape potential (Osborn, 1978), as when the upper lateral incisors in humans fail to form their typical spatulate shape (Sofaer *et al.*, 1971a,b). The size relations of the molars are remarkably variable in primates, so the size gradient seen within a dental class may reflect features of the food intake too.

The second part of Chapter 5 dealt with predictions of tooth size in response to changes in the external physical characteristics of the diet in a manner independent of any change in the size of the rest of a mammal. Specific dietary influences are unlikely to affect all tooth classes in the same way and are unlikely to result in general tooth loss – just reduction or enlargement within a tooth class. However, to keep a smooth gradient, then just as reduction proceeds in the permanent dentition from distal to mesial, enlargement should follow that trend but in reverse.

Table 5.4 suggested that an enlargement of the cheek teeth would follow a diet in which small, chemically sealed, non-sticky food particles were ingested in small mouthfuls. To this, Chapter 6 added abrasiveness.

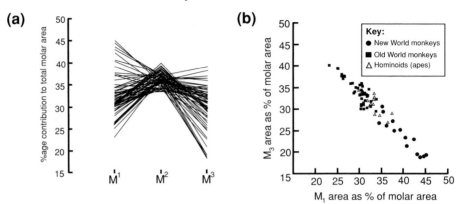

Fig. 7.9 The molar size gradients of 69 species of anthropoid primates (taken from Lucas *et al.*, 1986). The size of each tooth was calculated by summing the nominal surface areas of each molar (maximal mesiodistal length multiplied by maximum buccolingual width), in upper and lower tooth rows separately, and then expressing their percentage contribution to the total molar area. Primates can be very sexually dimorphic, so data for males and females were plotted separately (though there is no evidence of any difference between the sexes in molar gradients). (a) The contribution of the area of the upper middle molar (M^2) varies very little in these primates, while that of the first (M^1) and third (M^3) molars are inversely related, each varying three times more in contribution than the second molar. (b) Data for lower molars, plotting the percentage contributions of the first (M_1) and last (M_3) permanent molars to the total molar area. These are inversely related, strongly supporting that the molars have co-ordinated growth.

Defining food particle size in terms of particle volume, then the most common small items in primate diets are seeds and leaves. These foods are not only small in volume; they are indigestible unless they are opened.

Figure 7.9 shows that gradients in the size of molar teeth in anthropoid primates are obviously co-ordinated and that a large proportion of the variation can be expressed by the ratio of the areas of the first and third molars (Lucas *et al.*, 1986d). A high value of the M1/M3 ratio would mean 'relatively small molars' and vice versa. This ratio is strongly inversely correlated with the percentage of leaves plus seeds that have been reported in the diets of anthropoids, calculated on a cumulative annual basis (Fig. 7.10a). Sticky foods, like most fruit flesh, quickly form food boluses, leading to food being chewed on a very limited part of the cheek tooth row. Lucas *et al.* (1985, 1986a) suggested that the optimum would be a buccolingually wide tooth row with most of the working surface located in the middle of the row (i.e. with a large M1, small M3). Such an association is found

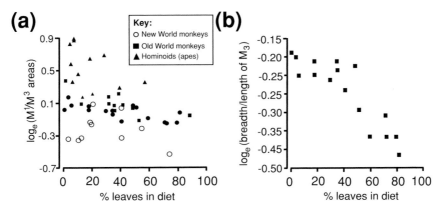

Fig. 7.10 (a) The relationship between the percentage of leaves reported in the diet of anthropoid primates and the logarithm of the ratio of areas of the upper first (M1) and third (M3) molars. There is a negative relationship that can be interpreted as suggesting large molars (i.e. low M1/M3 ratio) are required for consuming leaves. (b) The relationship between the shape (buccolingual breadth/mesiodistal length) of the lower third molar and the percentage of leaves reported in the diets of Old World monkeys. This indicates selective pressure on folivores for longer thinner molars.

in primates: the buccolingual width/mesiodistal length is positively cor-related with the M1/M3 ratio. Furthermore, the area of the last premolar contributes more to postcanine area (premolar area plus that of the summed molar areas) when the M1/M3 ratio is high. Thus, most of the working sur-face is placed in the centre of the tooth row.

Leaves do not form into boluses when they are chewed because of the hy-drophobic (and roughened) nature of their cuticular surface that dominates their exposed surface throughout mastication (Chapter 3). So leaf-eating primates will have long thin cheek tooth rows. In some Old World mon-keys, the shape of the third molar alone appears to give a good prediction of the proportion of leaves in the diet (Fig. 7.10b).[30]

This discussion suggests that there could have been great differences in the relative proportions of seeds and leaves in hominin diets. Robust australopithecines could have ingested large quantities of these items. This supports Jolly (1970) who argued for an analogy between hominins and geladas (where 'gelada' really refers also to a large number of extinct species in the genus *Theropithecus*: Jablonski, 1993). Some geladas have strong resemblances to the robust australopithecines, as also does *Gigantopithecus blacki*, an Asian ape (Pettifor, 2000). Virtually no mammal could live on a continual diet of seeds, just as they could not for fruits, because they are simply not around for long enough. Dunbar (1977) found that geladas ate young grass leaves for a considerable portion of the year. Seeds and leaves go together in the diets of leaf-eating colobine monkeys and even in some ruminants (Bodmer, 1989).

Interaction between anterior and posterior tooth sizes

The last section emphasized the cohesiveness of response of a tooth class. What about relationships between the sizes of these classes? These have long been the focus of attention in hominins, quantitative study dating back to Groves & Napier (1968). The relative sizes of the incisors and postcanines of hominins are very variable. *Sahelanthropus* has large incisors, while those of *Paranthropus* are very small. The molars of modern *Homo sapiens* look like miniatures in relation to our body size, while those of *Paranthropus* are relatively huge. One possibility for explaining this variation is by differing seed treatments among hominins. In fact, differing seed treatments are how sympatric mammalian species (i.e. those living at the same time in the same location) can share fruit resources in tropical forests. This has some particular interest for the study of hominins.

About 1.6 million years ago, in East Africa, a robust australopithecine species, *Paranthropus boisei*, was sympatric with early *Homo erectus*, a species that was (in most researchers' eyes) ancestral to *H. sapiens*. Probably also around in the same habitat was *H. habilis*, a species that probably died out soon after this date. *Homo erectus* was significantly larger in body mass than the others, but its maximum bite force was low. In contrast, the smallest hominin of the three, *P. boisei*, had the highest bite forces (Wood & Collard, 1999a). That comment represents an extrapolation from skeletal measurement, but such a reading of these measurements makes considerable sense. How could three such similar organisms have co-existed without dietary interference?

Doubtless, all hominins ate fruits. Early members of the genus *Homo* had relatively large incisors, but small postanines. Earlier in this chapter, I denoted this as $A+P-$ and as the dentition of a seed cleaner. In contrast, *P. boisei* had massive postcanines and tiny incisors. This $A-P+$ dentition is typical of a seed destroyer. Just like this, it is possible for at least two hominin species to partition the same food between themselves on the basis of seed treatment. Early *Homo* species could have eaten fruit flesh, cleaned the seeds and dropped them, while the robust australopithecine could have moved through later and consumed these cleaned seeds. Why did robust australopithecines die out?

Well, later members of the genus *Homo* reduced the size of their incisors as well, possessing a $A-P-$ dentition, i.e. that of a seed swallower. If *H. erectus* started to swallow whole seeds, then that would leave robust australopithecines without a critical resource, unless, that is, they then searched through . . . [31]

Whatever, this section has painted robust australopithecines as seed destroyers in almost exactly the way that Peters (1987) has suggested.

Canines and premolars of hominins

All hominins have small permanent canines in both sexes, an important characteristic separating *Sahelanthropus* from other hominoids (Brunet *et al.*, 2002). The only other anthropoid primate species like this are some New World monkeys. The titis (genus *Callicebus*) are very small monkeys with very small canines (Kinzey, 1972), while the largest living New World monkey, the woolly spider monkey, *Brachyteles arachnoides* (Zingeser, 1973; Lucas *et al.*, 1986b) is similar. Catarrhines (Old World monkeys, apes and hominins) other than hominins have large projecting canines in males. What precipitated their reduction in our ancestors?[32] One of the commonest methods of estimating the size of the canines in studies of the dentition of fossil hominins is to compare the size of their bases (mesiodistal length multiplied by buccolingual width) with those of the molars (Wood, 1984). As neighbours in a tightly packed jaw, with constraints over the position and length of the postcanine tooth row (Greaves, 1978), such comparisons might be influenced by competition over available jaw space (Jungers, 1978). I believe that this competition comes from 'molarized' premolars.

Any tooth used in chewing needs to be placed between the tongue and cheeks because this balance is essential in ensuring that food particles lie on the working surface of the postcanine teeth as the jaw closes. The regular and extended use of the premolars for mastication is indicated by their 'molarization' – changes to their form to resemble the molars. This increases their proportion of the postcanine working surface. These changes are seen most commonly in the most distal premolar of primates. If the premolars are used intensively for chewing, then it is probably necessary for the anterior limit of the cheek, the modiolus, to be positioned further forwards. This movement, however, jeopardizes the ability to gape widely because the mouthslit must now be narrower.

The reason for 'molarizing' premolars in primates is probably to do with ingested food particle size. The premolars in the human dentition have cusps but no fossae. These teeth normally provide preliminary breakdown from large to medium particle sizes, while molars take size reduction further once food particles have been reduced to a size where the fossae can act properly. Thus, most of the masticatory sequence is centred on the molars (Wictorin *et al.*, 1971). However, if there is something wrong with those teeth, food is placed more anteriorly on the premolars (Lundberg *et al.*, 1974). This will result in lower chewing efficiency for particles of around 5 mm or smaller and is due both to lower selection and breakage functions (van den Braber *et al.*, 2001). Molarizing the premolars will increase efficiency with small particles, but probably to the detriment of larger ones. Thus, I suggest that molarizing the premolars is an adaptation to the intake of small particles.

Modern humans are distinguished by rather limited gapes compared to other mammals, possibly managing only about 23° of condylar rotation (Baragar & Osborn, 1984). This is very small compared to the average 60–70° gape that most mammals are capable of (Herring & Herring, 1974). A smaller gape would not preclude werewolf-sized canines, but they could only project about 8–9° before the tips cleared each other or else they would not have been able to bite anything. The critical feature in humans is the soft tissue at the corner of the mouth (which can tear when the jaws are over-opened in some dental treatments: Smith, 1984).

I suggest then that the sizes of the canines are restricted by molarizing the premolars. This shifts the modiolus forwards to allow the cheeks to cover these teeth. That restricts the gape via a smaller mouthslit, and that puts pressure on reducing canine size. To address this, an index of the molarization of premolars is required. This is relatively easy: the post-canine tooth areas of the posterior premolar (conventionally termed P4) and the three molars can be summed and the proportion of the surface represented by P4 then calculated. Lucas *et al.* (1986b) correlated θ_c, the angle at which the canines just clear, with this molarized premolar index in 41–42 species of primates. Negative correlations were found in males, but not in females. Given that the restriction on canine size only applies to males, and that all basic 'size' influences have been removed in the calculation of these measurements, this is strong evidence for interaction between these teeth.

Extinct hominins are among many mammalian species with molarized premolars and, judging from modern humans, had small gapes. Accordingly, the canines had to be small. This reasoning then would have small canines in hominins being a response to the enlargement of the premolars. This section supports Jolly's (1970) general hypothesis of hominins as small-object eaters.

Food preparation and cooking

There has been such substantial reduction of the face and the dentition in the recent evolution of the genus *Homo* that it is tempting to call it 'facial dwarfing'. Rapidly dwarfing mammals seem generally to reduce their intake as a reponse to shrinking food patches. This is not what I think happened in the genus *Homo* though: like virtually all authors, I attribute this to the effect of modifying food properties prior to ingestion, by food preparation, preservation and cooking.

These practices characterize all modern human populations and change the physical properties of foods very radically. In fact, they may have almost taken the job of the dentition and masticatory apparatus away. Farrell (1956) showed that many cooked foods could be swallowed by humans and digested almost completely without any chewing.[33]

Raw vegetables in human diets are, rather unsurprisingly, an exception. Raw carrot comes through the human (Farrell, 1956) and primate (Sheine, 1979) gut into the faeces in still-recognizable pieces. In fact, there is a lot of evidence that food preparation and cooking techniques do not necessarily expose nutrients completely. For example, when peanuts are offered for consumption as whole roasted nuts, comminuted roasted meal or peanut oil, the last-named preparation is digested more completely than the other two (Levine & Silva, 1980). Levine and Silva interpreted this result as supporting the continued need for chewing in human populations. However, other factors are probably in play other than food particle sizes here. Oral stimulation influences the efficiency of fat uptake later in the gut (Mattes, 2001; Tittelbach & Mattes, 2001) and oil and comminuted meal will undoubtedly send a clearer 'fat signal' to the body than intact nuts (Chapter 3).

There is now general agreement that relative tooth size has declined for dietary reasons during at least two time periods within the lineage leading to *H. sapiens*. Early *H. erectus* is distinguished both from earlier members of the lineage *Homo* and contemporary australopithecines by virtue of a considerable reduction in the size of the postcanine teeth (Wood, 1992). However, the anterior teeth do not appear to have reduced in size at this time (Calcagno & Gibson, 1991), so giving early *H. erectus* an *A+P−* dentition (see above). Dental dimensions remained relatively stable in succeeding *H. erectus* populations, until about 300 000 years ago, when the size of the whole dentition seems to decline gradually up to the present and to an *A−P−* dentition. Brace and co-authors have documented this decline very precisely (Brace *et al.*, 1987).

What could explain these patterns of dental reduction? The general answer has always been declining use for two reasons: (1) the advent of tool use and (2) cooking. There is evidence for stone tools from about 2.5 million years ago in Ethiopia (Semaw *et al.*, 1997) and from East Africa only slightly later (Wood & Collard, 1999b). Other types of tool use may have developed much earlier (Panger *et al.*, 2002), in part because some populations of both orang-utans (van Schaik *et al.*, 1996) and chimpanzees manufacture tools, with demonstrated evidence for this dating back several hundred years (Mercader *et al.*, 2002). The first use of fire to cook is also difficult to establish. There is clear-cut evidence from about 300 000 years ago for the control of fire and so for cooking – these latter developments seem to have been a prerequisite for population movements into temperate regions – but Wrangham *et al.* (1999) have recently suggested that *H. erectus* was cooking in Africa from about 1.9 million years ago. While the details may be contentious, there is an onus on this book to see if dental reduction, indeed the reduction of the face as a whole, could be explained in terms of food preparation and cooking.

Food preparation and pre-ingestive particle size reduction

An important aspect of virtually any food preparation technique is the reduction in food particle size that it produces. A pestle and mortar is one example of this and is one of the easiest of tool sets to fashion: a crude equivalent of this is what chimpanzees use to break oil palm nuts in West Africa (Boesch & Boesch, 1990). Even now in developed human societies, fracture and fragmentation is an important and heavily overlooked aspect of our behaviour. Comminution on an industrial scale is a massive drain on energy in developed countries, accounting for perhaps 5% of the energy generated by power stations (Lowrison, 1974). I will contend here that the major selective pressure for tool development was not foraging or hunting, but the reduction of food particle size.

It will be recalled from the start of this chapter that tooth size and jaw size were proportional to each other in reptiles – as the jaw grew, teeth were replaced. In mammals, this is not so. Jaw size and the size of the anterior teeth are proportional to food particle size because of the need to open wide enough to ingest such particles. In contrast, the size of the posterior teeth is set by criteria related to the fracture and fragmentation of these particles. This does not make these structures independent of each other, but it does give them different scaling exponents. In recent human evolution, it would seem that these functions have been teased apart and a mismatch between tooth and jaw size has resulted.

A diminution in ingested food particle size due to food preparation decreases the need for gape. Better said, ingested food particle size should be proportional to linear dimensions of the face.

Raw under cooked and the problem with cube roots

Cooking is more complicated and involves the scaling arguments of Chapter 5. Suppose that it changes food toughness. Equation 5.6 can be modified to suggest the effect that this change has on tooth size. The force produced by a hominin chewing raw food, of toughness R_{raw}, will be termed F_{raw}, while the force required for cooked food, of toughness R_{cooked}, will be F_{cooked}. There is no need to invoke any difference in size of any sort, so the size ratio λ becomes 1.0 and disappears as a variable. Thus, Eqn 5.6 simplifies to

$$\frac{F_{cooked}}{F_{raw}} = \left[\frac{R_{cooked}}{R_{raw}}\right]^{0.5}. \qquad (7.2)$$

As explained in Chapter 5, the effect of change in food toughness would be first a demand on the bite force and the cross-sectional area of the jaw-closing muscles that produce it. Tooth size is responsible for the area of contact with food and should be proportional to that force. Thus, tooth size (the product of mesiodistal and buccolingual dimensions) would be proportional to the square root of the ratio of the toughness of cooked : raw food. However, instead of tooth areas, it is probably better in a comparison with the effect of food preparation to use linear dimensions: any tooth dimension should be proportional to the *cube root* of the ratio of cooked : raw food, i.e.

$$\text{any postcanine tooth dimension} \propto \left[\frac{R_{cooked}}{R_{raw}}\right]^{0.33}.$$

Now a scenario: suppose that hominins took up both food preparation and cooking, the former reducing food particle size, the latter reducing food toughness. Over time, the face and anterior teeth would reduce in proportion to the degree of comminution with tools while postcanine tooth size would reduce only as the cube root of change in toughness. I conclude that changes in toughness with cooking would have to be *colossal* to match the effect of tool use. So the face and anterior teeth would reduce in size faster than the postcanines. Over time, such a mouth would be incapable of accommodating all the postcanine teeth that tried to grow into it. This is precisely what happens in modern human populations: there is often no room for the third molar.

The effect of cooking on food

What, in a scientific sense, does cooking do to food? There are two general types of cooking: in air (e.g. roasting) or in water (boiling). It is likely that roasting preceded boiling because the latter demands a water source and receptacles. Most primates do not actually seem to use water sources very much, obtaining most of their water from food items, so it seems likely that roasting is more ancient. However, I conducted some simple experiments on both cooking methods to judge their relative effect.

Wrangham *et al.*'s (1999) target food for *H. erectus* is starchy plant storage organs. Peters *et al.* (1992) have made a compendious list of plants that almost certainly were available about 2 to 1 million years ago, while Peters & Maguire (1981) have made a specific investigation of this at a hominin fossil site. I tested several supermarket vegetables not because these were around at the time but because they might stand as examples of these food types.

Potatoes were both boiled and roasted, while turnips were just boiled. Figure 7.11 shows the results.[34] After about 4 min of boiling (shown as

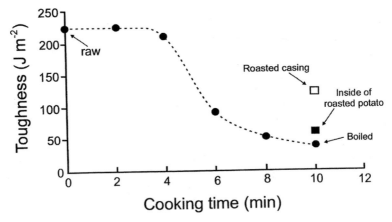

Fig. 7.11 The effects of boiling and roasting potato tubers on their toughness. Toughness reduces with boiling beyond about 4 min to levels lower than with roasting. For method, see note 34 to Chapter 7 (p. 304).

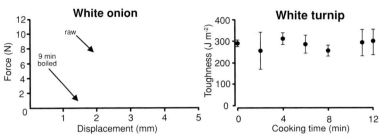

Fig. 7.12 Reduction in toughness on boiling of white onion (left), but not white turnip (right). Both were tested with a 15° wedge.

Potato

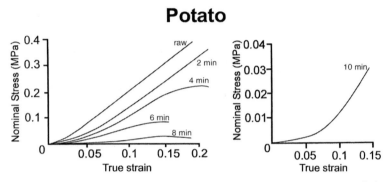

Fig. 7.13 The development of a curvilinear J-shaped stress–strain curve in potato flesh with boiling. For method, read note 34 to Chapter 7 (p. 304).

circles), the toughness of the potato parenchyma started to decay exponentially. The state after about 10 min cooking represents the usual cooked texture – its toughness was ~16% of the raw value. Roasted potatoes were tougher than boiled specimens and had a casing that still had ~50% of the raw toughness. The toughness of the internal tissue of roasted pieces was close to double that resulting from boiling. This may not be generalizable. Results of similar boiling experiments on white turnip, with white onion and Chinese leaf vegetables ('choy sim') show that turnip and leaves remain tough after boiling, while others, like onion, follow the potato example (Fig. 7.12). However, there is a second element to this: the shape of the stress–strain curve with cooking. On this, cooking has profound and general effects. Figure 7.13 show that raw potato has an essentially linear curve, which increased cooking times converts to a J-shape. The explanation is that cells lose their turgidity when they are cooked and their framework gradually collapses during compression until cell walls are being pressed against each other, when the tissue rapidly becomes much stiffer. Cracks then pass around, rather than through, cells (Lillford, 2000). This appears true of all the vegetable foods that I have tested though I hesitate in claiming that it will be universal.

Cooking also has profound effects on animal soft tissues, but they cannot be quantified as easily because the tissues are much more complex and the effects of cooking more difficult to quantify. A tough piece of meat is one that has a lot of collagenous connective tissue in it. Most of the fracture is at the level of bundles of fibres in connective tissue called perimysium, which has been the focus of a lot of studies on meat toughness (Purslow, 1985, 1991b). This can be surprisingly tough, even in cooked meat (0.4–1.8 kJ m^{-2}: Purslow, 1985), but probably less so than coagulated muscle protein. One way to estimate the toughness of meat would be to work out how much perimysium there is in it. It is much easier though just to calculate total collagen content. Chapter 5 gave data indicating that tetrapod muscle has more collagen than fish, irrespective of real differences in collagen make-up between the two vertebrate groups.

What cooking generally seems to achieve is a thwarting of the elastic crack blunting mechanism (discussed in Chapter 4) by stiffening tissue up. Uncooked skeletal muscle has a J-shaped stress–strain curve when pulled along its fibres (Fig. 7.14a–d). Yamada (1970) and colleagues tested an enormous range of raw mammalian tissues and found this type of deformational response to be by far the most common. However, after cooking, muscle stiffens and stress is approximately proportional to strain (Fig. 7.14e) (Purslow, 1991b). This will tend to connect up an otherwise purposefully 'disconnected' tissue (if this makes sense) and prevent much crack blunting. Cooking temperature is important because at temperatures where collagen starts to break down into a gel, myofibrillar proteins (those responsible for muscle contraction) start to coagulate. Overall, the toughness of any piece of meat cooked at high temperature seems to depend more on the state of these muscle proteins than on collagen (Christensen *et al.*, 2000).

The effect of cooking on evolution of teeth

The basic shape of the human dentition has remained surprisingly stable over the last few million years despite fundamental alteration in the food supply. In contrast, tooth size has varied – and varied dramatically. Simple cooking experiments suggest that cooking has a substantial influence on food properties and probably, therefore, on tooth dimensions. However,

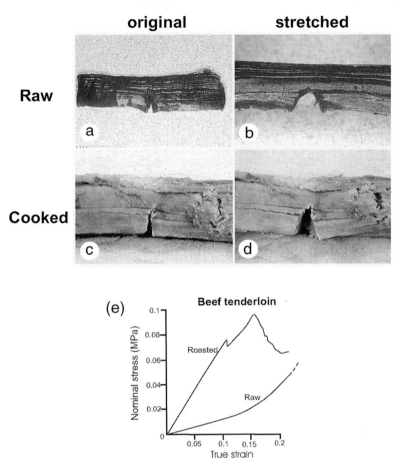

Fig. 7.14 The reduction in elastic crack blunting (see Chapter 4) in muscle tissue after roasting. Contrary to the trend in plant foods, cooking turns a J-shaped stress–strain curve into a linear one. Notch insensitivity is partially retained (Purslow, 1985).

it also looks clear that food comminution capacities of tools would far outstrip cooking in morphological effect. Tooth crowding resulting from the difference between these two effects does look like a reasonable way of explaining dental crowding in modern humans. What changes in tooth size could be anticipated though just from cooking?

With regard to the 'early cooking' theory of Wrangham *et al.* (1999) and the consumption of underground storage organs, the experiments support specific and general effects on dental dimensions. Table 7.3 indicates

Table 7.3 *The effect of cooking on potato tissue and the suggested reduction in tooth size of a hominin living on this food type*

Cooked sample	Mean (s.d.) toughness and range of cooked specimens ($J\ m^{-2}$)	Expected reduction in tooth dimensions[a] (percentage of ancestral tooth size)
Roast casing of potato	122.7 (21.7)	82%
Inside of roast potato	58.8 (13.3)	64%
Boiled (10 min)	37.9 (4.4)	56%

[a] The average toughness of the raw potato was 225 $J\ m^{-2}$, so expected tooth size is calculated as $(\text{mean cooked toughness}/225)^{0.33} \times 100$.

predictions for postcanine tooth size for a cooked potato diet. The suggested tooth size reduction is significant. It is tempting to suggest that roasting could explain the pattern of postcanine dental reduction in early *H. erectus*. The anterior teeth would probably process the tougher roasted casing while the posterior teeth would deal with both, with the greater volume of the inner tissue predominating during mastication. Just on this basis it could be predicted from Table 7.3 that anterior tooth size would not reduce as fast as that of the posterior teeth. Boiling the potato would have been dentally devastating and, if this were continued into soup, then the need for the teeth would be removed completely. The finding that cooking does not always reduce the toughness of vegetables makes investigations potentially more valuable: it may be possible to pinpoint those foods on which cooking has the greatest effect.

The changes in the stress–strain curve are also relevant. These can be modelled, very approximately, by the non-linear elasticity scaling deriving from Mai & Atkins (1975). Although the response is not actually purely elastic, provided that the load is maintained to fracture (and not cycled until eventual failure), it can be modelled in this way. Under such a monotonic loading (Atkins, 1999), sensory receptors could not judge whether part of the response is viscoelastic, plastic or elastic. Applying non-linear theory from Chapter 5 to the potato curve for 10-min boiling (Fig. 7.13) yields a stress–strain exponent of $n \approx 1.7$. This would drop the body-mass exponent of tooth size, say, to $M^{0.46}$ from $M^{0.5}$ for a diet of raw tissue when $n = 1$. Modest tooth size reduction would result.

What are the consequences of cooking meat for hominins? It will be recalled from Chapter 4 that there is a large discrepancy between the toughness of skin (or muscle) when there is a free-running crack versus the narrow radius of curvature of a crack tip kept confined by the presence of a sharp blade. A free-running crack blunts, causing toughness to soar. Cooking abolishes this and that probably allows humans with round-cusped molars to consume meat. There is still the issue of getting the skin off, both before cooking and after, because it can still be very tough due to its collagen content. I concur with many previous authors that the development of sharp stone tools was very likely to have been associated with the skinning of prey animals.

Summary of hominin diets

I do not intend to use a book account to express firm convictions on hominin diets because this was simply to indicate the depth that must be involved in substantiating claims about the dental–dietary adaptations of a mammal. A vexing issue is undoubtedly a date for the advent of cooking. Wrangham *et al.* (1999) contend that this was early and I believe too that this would be the only way to understand the decline of posterior tooth dimensions in early *Homo*. However, are starchy underground storage organs unavailable for the human gut without cooking? Modern humans have salivary amylase, but so too do most non-carnivorous mammals (Junqueira *et al.*, 1973). This suggests that some starch is available to mammals without starch granules having to be modified by heat. The most likely source of available amylose is in leaves because photosynthates are only stored there briefly before being converted to sucrose and translocated. Young leaves that are just on the verge of being productive for a plant – at the point in the tropics where yellow or reddish young foliage starts to turn green – may have been very important in the diet of early hominins, as one of the above sections indeed suggests. The roasting of storage organs must surely have preceded boiling. Remembering that primates usually do not drink, boiling may have followed a newly developed thirst on hot savannas. However, it is also possible that it was a much later development associated with life close to freshwater sources. The addition of fish to the diet, cooked or not, would probably have had a dramatic effect on tooth size.

A WORD OR TWO ABOUT LANGUAGE AND THE SENSES

The senses communicate to the body about the states of its external and internal environment. Language, both non-verbal and verbal, provides humans with the chance to convey some of this information to other members of their social group. Sometimes, there may have been an important need to share information about food, but other sensations have probably always been personal. The latter may stay below the level of consciousness and terms may never have been coined for them. The following section is intended as a short final discussion of language in relation to food texture.

The evolution of texture perception

Identification of insect prey, coupled with careful evaluation and manipulation in the mouth, must have been critical to the success of early mammals, much more so than to a reptile swallowing food particles whole. While the sense of smell and vision are most important to diurnal reptiles, the senses of smell, taste and texture, the latter two most likely first developed strongly in early mammals, were probably vital to these nocturnal insectivores. Taste release, e.g. of salts, free amino and fatty acids, would make certain that the cuticle of prey had been penetrated (Lumsden & Osborn, 1977), while texture perception would be needed to enhance food manipulation in the mouth and to assess cuticular properties. The oral processing of an adult insect might be a risky process because the cuticle of the jaws (Hillerton *et al.*, 1982) and legs can be considerably harder than the thorax and abdomen, calling for precise control of jaw movements. The hardness values reported by Schofield *et al.* (2002) for the jaws of leaf-cutter ants are probably high enough to cause the microwear seen by Strait (1993c) on modern insectivores. Thus, there should have been considerable selective pressure for an expanded sense of food texture over 200 million years ago.

Virtually all mammals seem likely to get important cues about food from texture inside the mouth. For example, herbivorous mammals are known to avoid foods with high fibre content ($\equiv$ high V_c). Yet, plant cell wall is essentially colourless, tasteless and odourless, so how can herbivores learn from their senses about it? The answer appears obviously to lie in its texture. They probably avoid toughness, not fibre (Choong *et al.*, 1992; Hill & Lucas, 1996). Seeds reveal themselves inside fruit flesh in the mouth by virtue of their extra hardness (Corlett & Lucas, 1990). This tactile-dependent distinction might be necessary from the plant side of things because the seeds of fleshy fruits will survive gut passage, sometimes with enhanced germination potential, if they are undamaged by teeth. From the mammal's side of the coin, the seeds might be toxic and important not to damage. So many frugivores may avoid hardness. They may prefer or be averse to the astringency (friction) that tannins produce and to the lipids present in some foods, such as insects or certain fruit.

The origin of language in humans

The development of spoken language seems to me to be bound up with texture, but not with texture in the mouth or even directly with food: I mean textures associated with tool-making. I envisage that language originally developed in order to transfer skills that were acquired in both the making and the using of tools to the next generation. Language was the method by which the 'handfeel' needed in making or using a tool could be communicated. The result was the development of a texture terminology.[35]

Jeannorod (1997) has pointed out that the ability to make tools depends just as much on visual as tactile information, so that an extensive language for both senses developed, generally linked to tool-making. Some auditory terms may have been useful, but this sense, plus olfaction and taste, have little value for this activity. Accordingly, the vocabulary of these senses is limited. In the case of olfaction, it is almost entirely absent, with virtually all communication requiring that partners have experienced the same smells, connotation rather than denotation predominating. Hearing, smelling and tasting were, and still are, probably of paramount importance in mammals that forage nocturnally, while vision and touch are the senses of greatest value to diurnal primates. Forelimb touch is heightened in most higher primates, probably due to use of their thumb and fingers in sensing the ripeness of fruit by palpation: fruit colour alone is an insufficient cue (N. J. Dominy *et al.*, unpubl. data). Yet the point here is that the size of the vocabulary of a sense is not related to its acuity, but to the need to communicate it. It is doubtful if there was ever a need to communicate 'mouthfeel', for example. Thus, Lucas *et al.* (2002) have argued that the texture vocabulary that is often assumed to refer to 'mouthfeel' is actually a 'hand feel' terminology that developed with tool use. This could explain why the training of human subjects in taste panels in the food industry to make ratings of various aspects of food texture has proved so difficult: it has been borrowed from a manual context.

AFTERMATH

Freed from references, my writing style can attempt a comeback.

Whirr, gurgle, gaseous snort.

Well this is what happens soon after food has been swallowed. Thank goodness that the mouth does not sound like this or it would have put me off studying it ages ago. It is difficult to see why coffee machines have to sound like it either, unless they wish to portray the noises that the flow of our digestive system is going to produce minutes after imbibing their products. The bowels just go on and on, very unlike a short snappy chew. However, ultimately, an account of the action of the mouth, at sensory, nutritive and general evolutionary levels, will only make sense in relation to those food-sinking feelings that these noises reflect.

My account of how teeth work was intended to be brief, but perhaps chewing research is rather like mammalian feeding – the getting ready for it (foraging for ideas) and the aftermath (digesting the results) take far longer than the doing of it. Even on the human side, the start (food preparation, cooking, etc.) and the end (washing up and/or culinary organization) generally take far longer than the meal. These considerations alone suggest that oral processing is extremely efficient. So why could I not match this brevity? I cannot answer this except to say that the book's length is perhaps justified by the apparent fact that there are really no other books that treat this subject in a comprehensive manner despite its importance. Some may feel offended, but I contend that this is probably so. Contrast this with locomotion, which has been much more thoroughly studied, and on which there are many good accounts. Locomotion and mastication are the two major motor activities of mammals. Both are distinctive, but . . . mastication is unique.

Mechanical properties and their measurement: material properties made easy

INTRODUCTION

Many biologists don't want to know much about the 'property testing' of materials because it sounds extraneous to many problems and something that could be handled in a routine manner by qualified technicians. Why be bothered? While it is certainly true that some biomechanical investigations do not require any detailed knowledge of this, the material properties of foods lie at the heart of the analysis of dental function. Accordingly, I suggest that the reader at least skim this appendix before tackling Chapters 4 to 6. The aim is to make material properties simple to grasp and fun to contemplate. I include basic concepts, some examples and a reference list with more for those that need it.

THE BASICS

When a force loads a solid object, it distorts. In mechanics, this distortion is called *deformation*. If the object deforms sufficiently, then it cracks (fractures). To document these behaviours, mechanical tests need to record the force applied to the object and the deformation that this produces. There are two options: controlling the force to see how the deformation varies or vice versa. Prior to fracture, there is little difference what is chosen, but to understand what happens afterwards, it is best to control deformation because this controls the rate of crack growth much more effectively.

Information about the behaviour of the deformed object can be recorded conveniently on a force–displacement graph, such as that shown in Fig. A.1. The displacement is generally a measurement of deformation in the direction of the force. The force–displacement curve of Fig. A.1 can be divided into two major parts: a deformational domain and a fracture domain. A force peak at the junction between the two indicates the instant of crack initiation. I now try to show by trial and error why mechanical properties

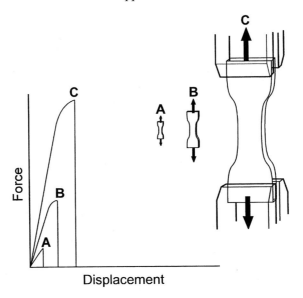

Fig. A.1 Schematic representation of force–displacement curves for geometrically scaled specimens of different sizes, all made from one material. The specimens are being pulled. The shape of the specimens is typical of those made for tensile tests, being designed such that they will not fail around the grips. It can be very difficult to shape biological tissues like this, which is one big reason for the development of several of the tests described in this appendix.

have the form that they do. Remember though that all these properties are extrapolated from simple graphs such as Fig. A.1.

THE DEFORMATIONAL DOMAIN

Imagine pulling tests on a group of regular specimens labelled A, B and C, all made of the same material and scaled replicas of each other (Fig. A.1). Pulling produces a uniform uniaxial (unidirectional) tension in each specimen. In fact, a pulling test is usually called a tensile test. The test results are the curves shown in Fig. A.1. The slopes of the initial portions of each curve are linear and define what is called the *stiffness* (force/displacement) of these specimens. Clearly though, these slopes do not describe the stiffness of the *material* that all these specimens are made from because each specimen has its own characteristic slope. The obvious reason for this is their size difference. If a way could be found to account for specimen size, then it might be possible to make the curves conform.

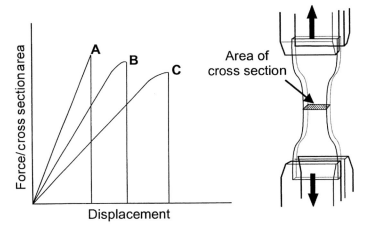

Fig. A.2 Force has now been divided by the cross-sectional area over which it acts for the three specimens in Fig. A.1. This size compensation does not make the curves on the graph coalesce in a way that could be used to characterize the material from which they are made.

If, instead of mechanics, we were dealing with chemistry, we would understand that the characteristics of a chemical reaction depend not just on the quantity of molecules that are available, but also on their concentration. Analogous to a chemical reaction, it could be hypothesized that the extent of deformation (i.e. the mechanical reaction) produced by the applied force on A, B and C was caused by differing concentrations of the force in specimens of different sizes. The force is applied uniformly to the specimens and is thus supported by their cross-sectional areas (Fig. A.2). The force divided by the cross-sectional area over which it acts is actually a basic unit in mechanics called *stress* (σ). Provided that the specimen has not been stretched too far (say less than 25% of its length), then using the original cross-sectional area for this estimate is accurate enough to obtain the stress.

However, plotting stress against deformation, as in Fig. A.2, still does not bring the loading curves for specimens A, B and C into congruence. In fact, it appears to overcompensate with the stiffest specimen (stress/displacement) now having the lowest slope. So, the displacement also needs to be scaled to specimen size. The simplest method of making this correction is to divide the displacement by specimen length. It turns out that this is again correct if the specimen has not been stretched too far: the deformation divided by the original specimen length is the *strain*, a dimensionless ratio. The symbol for strain is ε.

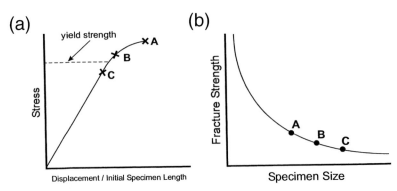

Fig. A.3 (a) The stress (force/cross-sectional area) is now plotted against displacement divided by original specimen length, which is the strain. The result is a common gradient at low stresses, which can be measured to give an estimate of Young's modulus. The crosses represent the point at which each of the specimens breaks. Note that these stresses are different. (b) If a wide enough range of specimen sizes of a material could be tested, then this graph would be the likely result, whereby the smaller the specimen, the higher the stress at which it fractures.

Stress–strain curves for the specimens are shown in Fig. A.3a. The initial slopes for all three specimens have converged, this common slope providing a measure of the deformation resistance of the material from which these specimens are made. If a test is stopped while deformation remains in the region of this common slope, then the original dimensions of the specimen are generally regained, indicating that this is an elastic property. The ratio of stress to strain in this region is called either *Young's modulus* (after its discoverer) or the *modulus of elasticity*. The symbol for Young's modulus is E.

The crosses in Fig. A.3a mark the point of fracture (crack initiation) for each of the specimens. These stresses are the *fracture strengths* (σ_F). The graph shows that these stresses are not the same for all specimens, being slightly larger for the smallest specimen, which is thus a little stronger. Choosing a much wider range of specimen sizes reveals the truth of this – fracture strength is not a true material property. Figure A.3b shows the fracture strengths of specimens A, B and C marked on a curve that also gives the loci for specimens of other sizes if these were tested in the same way: the smaller the specimen, the higher is σ_F. Below some limiting specimen size, specimens no longer crack at failure but, instead, show more extensive deformation that is not recovered on unloading, i.e. the deformation is permanent. This permanent change is called *plastic deformation*. (This phenomenon can also be called ductility – a term not used in this book.)

Plastic deformation can be induced in any material, not just by changing specimen size, but also by confining stresses to a very small part of specimen volume – such as is done when making indentations (Fig. A.9). The point of onset of plastic deformation, usually defined as a small percentage decline in the slope of the stress–strain slope, is termed the *yield strength*, σ_y, and is independent of specimen size (as indicated in Fig. A.3a).

In a pulling test, fracture results in the specimen breaking into two. If the two fragments can be refitted so as to resemble the original specimen, then this is *elastic fracture*. If the pieces are substantially longer and narrower than the original due to plastic deformation, then this is *plastic fracture*. Figure A.3a shows that the extent of plastic deformation is far greater in specimen A than specimen C. If the amount of plastic flow is relatively minor, as it usually is in biological materials, then this is called *elastoplastic fracture*.

Thus far, we have described the deformational domain as characterized by the Young's modulus, the transition from elastic (recoverable) deformation to permanent set by the yield strength and the point of crack initiation by the fracture stress.

THE FRACTURE DOMAIN

A pulling test is not a good way of observing fracture because the crack tends to rip through the specimen very rapidly. A three-point bending test is better because it produces tension only on the lower side of the specimen (Fig. A.4). The upper part of the specimen is compressed, so controlling the crack's growth. (Chapter 4 points out the stresses that can promote cracking. A compressive stress can never do this.) The specimen has been deeply notched (pre-cracked), with the notch tip made as sharp as possible.

Loading the specimen beyond the point where the crack extends from the notch gives the force–displacement curve marked ABC in Fig. A.4. The peak at B represents the point of crack growth. The force drops after this, but does not reach zero – i.e. crack growth is controlled. In fact, the test can be stopped now and the displacement put into reverse, so unloading the specimen and sending the force–displacement curve back to zero in Fig. A.4. The sequence of events is marked out by ABCA. The growth in the crack after this first experiment can be measured and the specimens reloaded. Ideally, the loading–unloading curve will then follow the curve ACDA. After unloading, we can again measure the crack growth that took place during this second loading–unloading episode. If we express

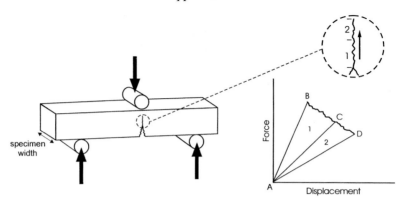

Fig. A.4 A notched specimen is being bent such that a crack extends from the sharp notch. The specimen can be loaded and unloaded repeatedly and the new length of crack measured. The graph shows two work areas, 1 and 2, that correspond to the increments of crack growth shown in the specimen inset.

crack growth as crack area, i.e. as an increase in crack length multiplied by specimen width, it will be found that the graphical areas ABCA and ACDA are in proportion to the crack areas produced in each episode. Now these graphical areas are the product of force and displacement, which is work. These areas ABCA and ACDA are therefore *work areas* and the work done in extending the crack is proportional to the new crack surface area produced. The material property that controls the fracture domain is called *toughness*, defined as the work done in producing unit area of crack. Toughness is symbolized in this text as R. Cracking is therefore an energetic process and the energy available in a cracked object controls the growth of a crack or cracks within it.

The basis of the toughness concept is easily explained. Imagine the material being made up of a lattice of atoms joined by bonds (Fig. A.5). The position of atom x lying within the material is balanced by the attraction of the atoms lying around it. Now suppose a crack (produced in the figure by a fictive blade of atomic width) ran through this material, breaking bonds on one side of x. This disturbs the equilibrium so that x moves towards the atoms to which it remains connected. Such movement does work and thus, the production of new surfaces by cracking requires work. Physical chemists refer to the cost of this surface production as the surface energy of a solid. It is identical to the surface tension of a liquid, both these quantities having the same units. Toughness has this chemical surface energy as its basis, but while surface production is its root cost, it also includes the

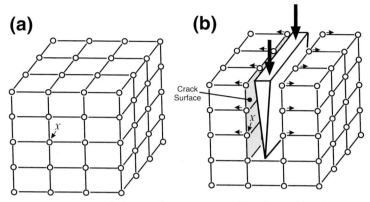

Fig. A.5 A solid consisting of a lattice of atoms connected by chemical bonds. The position of atom *x* is being disrupted by a fictive knife blade of atomic dimensions that is severing bonds on one side of this atom, so resulting in a new equilibrium position.

cost of any physical disruption around the crack in its estimation. While Young's modulus and strength were properties fully appreciated by the mid-nineteenth century, the concept of toughness was only formulated about 80 years ago. It remained unappreciated for some time after that and remains so in many areas of biology. This is ironic in that its energetic basis makes it easy to integrate into issues central to biological theory, e.g. in ecology, which deal with energetic concepts much more easily than with stresses and strains. The next sections include formal definitions and more detail.

YOUNG'S MODULUS

Young's (elastic) modulus is a measurement of the stiffness or rigidity of materials. It is measured as the force producing unit deformation of a specimen, normalized to the relevant dimensions of that specimen. These dimensional corrections convert (1) force to stress by dividing by the cross-sectional area of a specimen over which the force acts and (2) deformation to strain by dividing by the original dimension of the specimen in the direction of the force. More simply, the Young's modulus is the initial slope of a stress–strain graph. Strain has no dimensions and so the units of a modulus are force per unit area. The SI units are meganewtons per metre squared ($MN\ m^{-2}$) or giganewtons per metre squared ($GN\ m^{-2}$) where

M stands for mega (10^6) and G for giga (10^9). To improve presentation, the pascal (Pa), which equals 1 N m^{-2}, has been introduced. This is a very small unit and MPa and GPa are used here to describe the modulus.

HIGHLY EXTENSIBLE SOLIDS

When a solid is very extensible, it is no longer sufficient to use these definitions of stress and strain. Instead, account has to be taken of the progressive change of deformation as the material is distorted. It is simple to do this for strain where it can be easily shown (e.g. Ashby & Jones, 1996) that it should be calculated, not as ε, but as $\log_e(1 + \varepsilon)$, which is called the *true strain*. It is only necessary to calculate it when the change in dimensions is very great. The stress should be based on the instantaneous cross-sectional area of the specimen, not the original area. It is difficult to do this except when the volume of the material is conserved during loading when the stress can be calculated as Fl/Al_o where l_o is the original specimen length and l is its instantaneous length.

VISCOELASTICITY

Elastic behaviour is, strictly speaking, an instantaneous response. However, few biological solids are like this, usually taking time to recover their dimensions on unloading. The time lag is due mainly to their water content. This type of time-dependent elasticity is called *viscoelasticity*, the analysis of which is complex. If a viscoelastic solid is repeatedly loaded and unloaded without fracture, the result is considerable energy loss, which helps to prevent fracture.

There is little about viscoelasticity in this book or about the influence of deformation rate on the properties of materials. It is a highly complex issue on its own, without even considering fracture. There are undoubtedly areas in biology where it is extremely important, such as when there are repetitive loading–unloading cycles (e.g. the regime that the periodontal ligament endures), but under a fairly rapid masticatory movement, resulting in the fracture of food particles during that same chew, then viscoelastic theory, which specializes in the understanding of equilibria in such solids, adds inappropriate complications. Nevertheless, due to time-dependent behaviour, the values of Young's modulus obtained in tests can vary considerably with change in deformation rate. It is important to measure stress–strain behaviour at an appropriate rate. For accounts of very time-dependent foods, read Shama & Sherman (1973) and van Vliet (2002).

NON-LINEAR ELASTICITY

Many biological tissues do not have linear stress–strain curves. Most raw animal soft tissues are like this, as are many cooked plant tissues and biological gels. One way to analyse such tissues is to perform a linear regression on plots of log stress on log strain such that $\sigma = \kappa \varepsilon^n$ (as in Chapters 4 and 5), allowing the exponent n to characterize the relationship. However, many biological tissues with non-linear behaviour cannot be linearized in this way. Otherwise, depending on the purpose of the investigation, either the slope of the initial part of the stress–strain curve or that close to fracture could be used to define a modulus.

Most raw plant materials have relatively linear stress–strain curves. However, flaccidity and cooking makes them non-linear. A J-shaped concave curve is seen in most vertebrate soft tissue (e.g. Fig. 4.13), boiled vegetables and gels, while a r-shaped curve tends to be found in cooked muscle (Purslow, 1991b) and cheeses (Charalambides *et al.*, 1995).

ANISOTROPY

Many biological solids have complex structures whose mechanical response depends on the direction in which they are loaded. This direction-dependent behaviour is called *anisotropy*, the opposite of isotropy. The more complex the food behaviour, the more tests are needed to characterize it.

STRENGTH

The strength of a solid is the characteristic stress at which there is a defined change in mechanical behaviour. It either has units of kPa, where k stands for kilo (10^3), or MPa. There are two basic types of strength. The *fracture stress* is the stress at which a crack initiates while the *yield strength* is the transition from elastic to plastic behaviour in some part of the specimen.

THE DIRECTION OF STRESS

Stress is a vector quantity. Since any loading pattern can produce tension and compression within a solid, signs have to be given to indicate their differing directions. By convention, tensile stresses are positive, while those of compression are negative. When a solid is suspended in a fluid, it is subject to compressive stress from all sides. This is a state of hydrostatic pressure. The opposite, overall expansion, is not relevant to tooth function, but the overriding concept is *hydrostatic stress*, negative if it is compressive, positive if expansive.

TOUGHNESS

Toughness is a measure of the resistance to crack growth in a material. The toughness of foods is critical for analyses of tooth shape. It is measured as the work done in making new area of crack (i.e. the surface of one of the two crack faces) and its units are joules per metre squared, i.e. J m^{-2}, or kJ m^{-2}. It is an analogous quantity to the surface tension of liquids or the surface energy of solids. However, toughness differs in its calculation in that:

(1) The energy expended in the formation of new surface by cracking is divided by only one of the new surfaces formed. In contrast, in surface energy measurements in physical chemistry, the divisor is the area of both surfaces.

(2) It includes the cost of disruption of material in the vicinity of the crack, away from the crack's surfaces. Due to this, toughness values are nearly always much higher than surface energy unless the crack surface is glassy smooth. Fracture is only fully understood in such relatively homogeneous solids and even these are complicated enough. The understanding of many fracture phenomena, such as when growing cracks change direction or when they branch, is still limited.

Toughness is, at minimum, surface energy, but can be ten, one hundred, or even thousands of times greater than this because of the spread of damage around the crack. The single biggest contributor to toughening is plastic deformation.

SYMBOLS FOR TOUGHNESS

As soon as the fracture literature is entered, letters tumble out at you like some kind of kindergarten test. Generally, there are *G*s, *J*s, *K*s and *R*s to juggle with, but there are also books that could hit you with a *T* or even a *W* (Lawn, 1993). How to manoeuvre through this alphabetic nightmare? Well, the preface of Atkins & Mai (1985) indicates how a lot of these terms can be related to each other and their book describes at length the circumstances by which their separable definitions evolved. The symbol *G*, for example, has usually been restricted to solids with a linear elastic response, while the quantity embodied in *J* refers strictly to non-linear elastic situations. The term *R* is more loosely defined as the energy involved in crack resistance. In order not to lose readers, I will stick to *R* here. However, most of the other letters refer to quantities that have the same units as *R*, differing usually in the way that they are describing energy dissipation within a flawed, notched or cracked material under load. The mechanical energy stored in a material that can help pay for crack growth is called the *elastic strain energy*. However, there are many kinds of plastic or non-recoverable processes, even in apparently simple 'brittle' ceramics, into which energy can be sunk. Some of these processes obstruct crack growth and thus raise toughness. By and large, research on fracture represents the attempt to factor out that structural disruption which is intrinsic to crack growth from that which is non-essential. Much of the problem arises from the fact that the energy expended in crack growth is usually measured indirectly. While an overriding definition of the 'essential work of fracture' may eventually emerge, there appears to be no 'across-materials' consensus on this yet.

Sometimes, fracture symbols have modifying subscripts, such as K_{IC} or G_{IC}. The 'c' stands for critical value, the point at which a crack starts or starts moving. The 'I', more rarely II or III, refers to the modes of fracture. (Modes of fracture, types of failure and mechanisms of toughness are discussed in Chapter 4.) The association of energy with direction can seem strange because energy is not a vector quantity. However, K_{IC}, which nearly always has these subscripts, is very different as the following illustrates.

THE MEANING OF K_{IC}

The effect of a thin flaw on the average tensile stress, σ_∞, in a large rectangular plate of a short sharp crack, oriented at right angles to uniform tension, is shown in Fig. 4.11 where the flaw is modelled as a thin ellipse. This flaw modifies stress levels in its vicinity very greatly. At the crack tip, the tensile stress *at right angles to the long axis of the ellipse* is much higher than σ_∞, but it declines to σ_∞ well in front of the sharp edge of the flaw. Nevertheless, this flaw appears to weaken the plate so badly that if fracture stress were the criterion for fracture, the plate would break into two pieces immediately. However, experiments show that the flaw will not grow unless there is sufficient energy stored within the plate for this to happen. This is given approximately by

$$R = \frac{\sigma_\infty^2 \pi a}{E},$$

which relates the energy per unit area used up by a crack that grows from a flaw of length a to the loss in stored elastic strain energy that funded this growth. The equation can be arranged to give

$$(ER)^{0.5} = \sigma_\infty (\pi a)^{0.5}.$$

Note that the weakening effect of the flaw depends on the square root of its length. K_{IC}, sometimes called the critical stress intensity factor and sometimes fracture toughness, is simply symbolic shorthand for $\sigma_\infty(\pi\alpha)^{0.5}$. Its validity is limited to linear elastic situations where $K_{IC}^2 \approx ER$. Note that K can be measured at any stress because crack propagation does not figure in its definition. However, the critical value of K, i.e. K_C, is that when a crack propagates from the flaw in the plate. K_C will always be referred to here as K_{IC} because crack growth is usually in mode I (Lawn, 1993).

Many biological papers measure K_{IC}, but it is becoming less common to use this nomenclature because its definition excludes any plastic deformation and is difficult to employ on composite materials such as most biological tissues. However, Purslow (1991a) has suggested a form suitable for use on non-linear tissues whereby K_{IC} can be replaced by Ψ_{IC} with

$$\Psi_{IC} \propto \sigma_F a^{n/(n+1)}.$$

The term n is the exponent in 'power law' non-linear elastic equations.

OTHER QUANTITIES

POISSON'S RATIO

When most materials are pulled, they get narrower; when compressed, they get wider. Figure A.6 shows a specimen of original specimen length, l_0, and

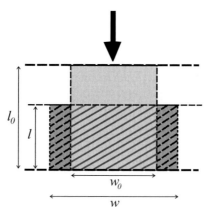

Fig. A.6 A particle is compressed, so reducing its vertical (longitudinal) dimension. However, as it does so, it spreads horizontally (laterally). The ratio of the lateral to the longitudinal strain of a material is referred to as Poisson's ratio.

width, w_0, being compressed to a new length, l, and expanded width, w, such that Poisson's ratio

$$v = \frac{(w/w_0)}{-(l/l_0)}.$$

The value of (w/w_0) is called the lateral strain and is positive because it is an expansion, while $-(l/l_0)$ is called the longitudinal strain and is negative, being a contraction. However, normal Poisson's ratios are invariably preceded by a negative sign, i.e.

$$v = -\frac{\varepsilon_{lateral}}{\varepsilon_{longitudinal}}$$

to make 'normal' deformational responses positive. For many engineering materials, Poisson's ratio is about 0.3, reflecting the fact that, when true solids are loaded, they reduce in volume. Saturated biological tissues sometimes preserve their volume when they are deformed. If they are isotropic, then $v = 0.5$. The tongue is supposed to be an example (Kier & Smith, 1985). A very high Poisson's ratio of 1.0 or more can be measured in some animal tissues while in plants, zero Poisson's ratios are possible. In fact, they might sometimes be lower than zero. When reported, a *negative* Poisson's ratio is not necessarily an error: it results from a type of cellular structure whereby cell walls collapse inwards on compressive loading (see Fig. 4.12). Some materials ('auxetic' materials) have been designed specifically for such negative ratios (Lakes, 1987).

HARDNESS

Hardness is not a property in itself but a concept derived, like many other scientific terms, from specializing a word used in everyday language. Used loosely, it means resistance to deforming under indentation (Fig. A.9). Indentation tests are the most ancient and simple of mechanical tests, wherein a blunt or sharp indenter is pressed into the flat surface of a solid. If the force on the indenter is F, while the projected area of the indentation (the area measured in the plane of the surface) is A, then hardness is defined as

$$H = \frac{F}{A}.$$

This is the correct expression for hardness values, sometimes called Meyer hardness. However, it is not the value given in manufacturers' look-up tables because these tend to divide the force by the actual area of indentation. A correction factor needs to be applied to make sense out of such data.

The units of hardness are those of stress. By itself however, hardness is an arbitrary meaningless measurement (rather like 'if I had a hammer' mechanics). Meaning began with the work of Tabor (1951) who established that hardness is an indirect measure of the yield stress within a material. For materials where there is little change of volume upon the application of load, then the hardness is three times the yield stress (Kendall, 2001). Where a solid completely collapses in itself, however, then the hardness is the same as the yield stress (Wilsea *et al.*, 1975). Low-density plant tissue obeys the latter relation as cells burst and collapse down to a pile of cell walls. In contrast, seed shells, being very high-density cellular tissue, can produce 'pile-up' around the edge of an indentation, indicating that material is being displaced. Whatever, there is always a certain amount of densification under the centre of the indenter because hydrostatic pressures are so high.

Early tests involved large indentations but these macroindentations have largely been replaced by micro-, and now nano-, indentations. Also, indenters with sharp tips, like the Vickers, Knoop and Berkovich geometries, have increasingly replaced blunt spherical types. Hardness depends to some degree on indenter geometry but understanding of what an indenter does to a surface has grown dramatically (Lawn, 1993), extending the value of the test. If resistance to indentation is totally elastic (i.e. the material springs back after indentation), then the apparent hardness is controlled by Young's modulus. If the indentation is permanent, then the amount of plastic deformation depends on the yield strength of the material. Most uses of the term 'hardness', both here and elsewhere, refer to permanent indentations. However, even if the surface appearance of an indentation remains fixed due to plastic deformation, the deepest part of a sharp-tipped indentation tends not to be so, resulting, for example, in the deepest part of an indentation formed by a sharp-tipped indenter rising up somewhat after the load is released. Measurement of this recoil can be used to estimate Young's modulus.

Indentation tests are now routinely used to measure most of the material properties of ceramics (Lawn, 1993), being attractive both for their simplicity and for their non-destructive characteristics. Many tests can be made on the same specimen simply by spacing indentations sufficiently far (more than four indentation diameters) apart. The test has always been the method of choice for investigating tooth tissues, but a rather sterile literature has been rejuvenated recently by nanoindentation, a technique developed in the last decade as an offshoot of the atomic force microscope.

This microscope uses a ceramic stylus, rather like that on the now-extinct gramophone record player, located on the end of a cantilever beam, in order to map a surface. To do this, very low forces (e.g. 10^{-8} N) are applied to the stylus, which is mounted on a relatively springy (compliant) mounting. Measurement of mechanical properties requires a much stiffer setting and a diamond stylus. Loads can range from 10^{-6} N to nearly 10^{-3} N. By swapping indenting and mapping functions, Balooch *et al.* (2001) were able to image the surface before and after indentation. The field is advancing rapidly.

It is perfectly feasible to measure indentation in mammalian field studies, preferably with a indentation of millimetre dimensions (Lucas *et al.*, 2003). Current research aims to estimate both the hardness and Young's modulus of plant materials.

NOTCH SENSITIVITY

If a notch is cut into a specimen of virtually any true solid, then its strength will decline with an increase in notch length. However, this decline is more rapid than can be predicted from the loss of cross-sectional area (Fig. A.7). The non-linear form of the curve can be predicted from arguments in fracture mechanics. Such solids are said to be *notch sensitive*. However, some biological tissues do not behave in this manner and decline in strength in simple proportion to the loss of area (Fig. A.7). Such tissues are said to be *notch insensitive*. Some food tissues are like this, a function of their structural heterogeneity and the lack of firm mechanical connection between their structural components (see Chapter 5).

ACTUAL TEST ARRANGEMENT

Before considering these, it is wise to know that there are a lot of subtleties involved in this type of work. There may be faculty in your institution that can provide practical help and there are also many standard techniques that can be looked up in materials science books. Sometimes these standards help biologists, but they can also be a hindrance. (Be aware that many of the commonest industrial standards pre-date fracture mechanics and are phrased in terms of 'strengths'. Sometimes, these 'strengths' are actually just forces.) It can be difficult to turn biological tissues into the specimen sizes and shapes demanded by some material standards. Additionally, certain conditions which these standards aim to satisfy, such as 'plane strain' (which is not discussed in this book), are not likely to be related to the way that

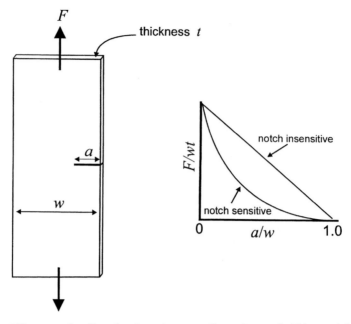

Fig. A.7 When a notch, of length a, is cut into a tensile specimen, of width w and thickness t, then the fracture stress (calculated as F/wt, i.e. force divided by cross-sectional area *without* taking the notch into account) is reduced disproportionate to the loss of supporting area represented by that notch. This phenomenon is called notch sensitivity and is central to the tenets of fracture mechanics. However, a substantial number of biological materials are not disproportionately affected by notches and show 'Galilean' strength. These are termed notch-insensitive tissues.

foods fail in the mouth. However, even if no 'standard' test is possible, all is not lost. The aim is not accuracy for the sake of it and the precision that is necessary in biology always depends in the end on the tightness of theoretical predictions. Many (but not all) estimates in biomechanics end up with an accuracy of a factor of two: the real value could be half or double that predicted. Generally, such a result is not something to be sniffed at.

Most field tests require a miniaturized version of the universal testing machines found in engineering and food science laboratories. Darvell *et al.* (1996) describe a tester that has since been expanded in range (Lucas *et al.*, 2001, 2003). The current version is shown in Fig. A.8. Together with all its accessories, it fits easily into a suitcase. By comparison, laboratory machines are often massive, the point being to make them so rigid that only the specimen under test deforms. A portable field tester obviously has a much lower stiffness. Accordingly, specimens have to be very small,

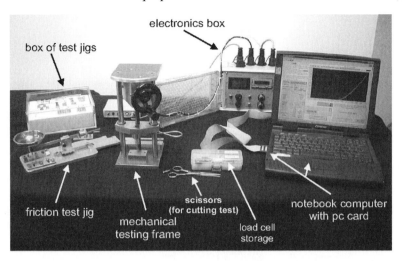

box of test jigs

electronics box

friction test jig

scissors
(for cutting test)

mechanical
testing frame

load cell
storage

notebook computer
with pc card

Fig. A.8 The 'Darvell' HKU tester and accessories for measuring the mechanical properties of mammalian foods in the field.

particularly if they are made from high-modulus materials. The other major issue with field tests is the maximum load that it can take. So far, the Darvell tester has only been fitted with 10 N and 100 N load cells.

Data can be uploaded to a notebook computer giving immediate results. I have generally interfaced a 12-bit A-to-D PCMCIA card (DAQCard 6062E, National Instruments, Austin, TX, USA), displaying and analysing the data using programs written in the Labview environment (National Instruments). A suite of programs has been written for this purpose and is available free from the author.

Really accurate work on materials demands that deformation be measured very accurately. However, deformation is constrained at the ends of specimens. Tensile grips are an obvious example because they compress the specimen locally making it very likely that it will fail around the margins of the grips. To avoid this, it is common to make specimens wider at the grips (Fig. A.1). There are a variety of grip surfaces, even involving pneumatic action, so as to provide just sufficient compression to prevent slip.

Plates for compression should allow the specimen to slide freely. This is encouraged by low friction. Some polytetrafluoroethylene (PTFE) coated tape on the plate surfaces can help this a lot. Bytac (Chemplast Inc., Wayne, NJ, USA) is such a product: a 25-μm thick layer of aluminium foil backed with adhesive and coated with a 50-μm thick layer of PTFE. Additionally, some researchers lubricate plates with oil.

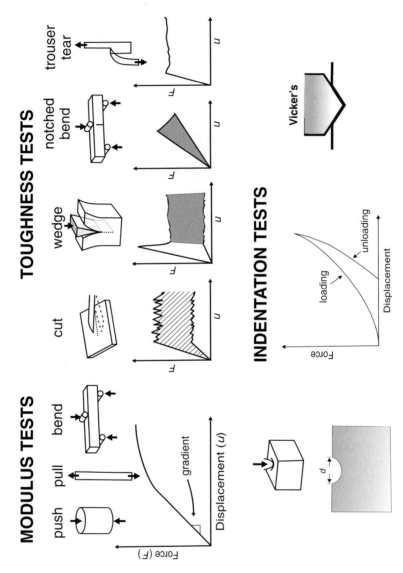

Fig. A.9 Common testing geometries for estimations of Young's modulus, toughness and indentation. The choice of test depends on the ease with which specimens can be shaped and the control that they offer over crack growth. See text for explanation.

Even with these safeguards, the only sure way to measure the strain accurately is to measure the change in specimen height or length remote from the ends of the specimen. This requires an extensometer, the most accurate of these being a laser that detects the movement of reflective markers placed on the specimen. Such techniques are impossible in the field.

The basic form of most field tests is shown in Fig. A.9. An excellent 'cook book' of methods has been produced by Vincent (1992), so this part of the appendix is kept brief. Not all tests are listed here anyway. The instrumented microtome (Atkins & Vincent, 1984) and C-ring tests (Jennings & Macmillan, 1986), to mention just two, are not included because they are not yet feasible in the field.

YOUNG'S MODULUS

COMPRESSION TESTS

All that is needed to get a value for Young's modulus for a specimen loaded in compression is to make a short cylindrical sample, making sure that the ends are trimmed squarely. The ratio of cylinder height to diameter should be low ($2:1$ is ideal) in order to eliminate any possibility of buckling.

TENSILE TESTS

Tensile specimens should be long and thin because the disturbing effect of gripping the ends of the specimen is then reduced. Ideally, test specimens, such as shown in Fig. A.1, should have a narrow waist, the strain in which can be monitored with an extensometer. This refinement is not possible in the field. If tested to failure, then a specimen should fail in the centre of the specimen, certainly not at the grips.

BENDING TESTS

There are two types of bending tests – three-point and four-point bending. The latter has many advantages, but it can be difficult to set up in the field. The length of a bending specimen is called its span. The ratio of the span to the depth (thickness) of the specimen in three-point bending should be greater than $10:1$ for relatively homogeneous materials and higher than this if there is any obvious heterogeneity.

TOUGHNESS

As related above, toughness tests usually require a pre-existing crack or notch. Rather than try to measure the flaws in specimens, which is hopeless (they are likely to be microscopic), an artificial flaw called a notch is made, from which a crack will grow during the test. The sharpness of the notch may matter: if so, it should be made 'sufficiently sharp' with a razor blade. In practice, however, this often doesn't matter because the test crack tends to extend from a small part of the microscopically irregular notch. However, test pieces generally have to conform to a certain shape to get reliable results and sometimes be above a certain size limit. These restrictions are burdensome to a biologist. Foods come in all shapes and sizes and trying to match definitions given in materials standards is often hopeless. Luckily, there is a large class of fracture tests that have been developed that do not require a notch: cutting tests. Given a choice, a materials scientist would be very unlikely to recommend them because, for one major reason, they often involve a lot of friction and for another, it can be difficult to decide how much of the work done in such tests was actually required for fracture. However, the action of teeth involves friction and, in any case, there are some ways in which much of the friction can be factored out. Most of these cutting tests derive from Atkins & Mai (1979) and involve blades and wedges (which resemble features of dentitions). Often, the shape and size of the specimen is not critical in these tests.

WEDGE

The wedge is one of the oldest fracture tests, dating from Obreimoff in the 1930s (Lawn, 1993). It is still recommended for fracture studies (Kendall, 2001) and has been introduced in a modified form into biomechanics by Vincent *et al.* (1991) and Vincent & Khan (1993).

A wedge is a single blade pressed onto a block specimen that starts a crack by bending two halves of the block apart (Fig. A.9). Crack growth is usually stable. The choice of included angle for the wedge is arbitrary, but a narrow angle (e.g. $15°$) seems to work best. The width of the block is measured (that dimension that will be cut) and then the wedge is pressed into it until a crack is started. The test can begin from this point, i.e. after fracture. The crack will normally be stable in all but the stiffest tissues and every unit movement of the edge of the wedge will result in the same unit growth of the crack just in front of it. At the end of the run, the depth needs to be measured to estimate crack area (this is generally given by the depth of wedge penetration after the start of the test). However, the work done involves not just fracture, but also friction and adhesion. To estimate the latter two factors together, the wedge can be reversed and run against the new crack surface. The work done in this second pass needs to be deducted from that in the initial run, then divided by the product of the block width and depth of wedge penetration to give an estimate of the toughness.

TROUSER-TEAR

This test is only effective for tissues that will not deflect cracks. The basic arrangement is shown in Fig. 4.1. If there is little or no stretching of the trouser legs, then with t being the specimen thickness, the toughness is calculated as

$$R = \frac{2F}{t}$$

where the force, F, should be constant during the test.

SCISSORS

Scissors provide a practical field test for fracturing thin sheets or rods, e.g. leaves and shoots (Fig. A.10). The test is not well described in textbooks and has been criticized on several occasions, so more is said about it here than other techniques. A sheet or rod, usually <1 mm in thickness, is placed between the blades of a pair of scissors. The work done during a cut can then be recorded over a given displacement. This displacement does not have to be the displacement of the scissors blade itself – it can be measured at any convenient location, such as the crosshead of a tester that is driving the blade handle provided that the load is monitored at the same place. The scissors can then returned to their original position and the specimen removed. The scissors are then driven down again over the same displacement to record the frictional work done between the blades themselves. The work done in this empty pass is then deducted from the work done in the first pass to find the work needed just to fracture the specimen. All that remains is to measure the cut length in the specimen (it is easier to cut right through it than to measure partial cuts) and its thickness. The toughness estimate is the work done in fracture divided by the area of cut. It is possible to run tests the other way around, with the

(a) SCISSORS TESTING

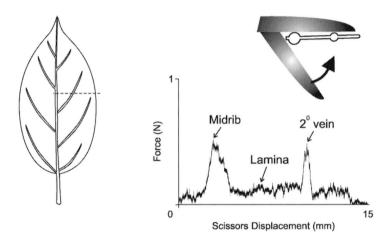

(b) PENETROMETRY

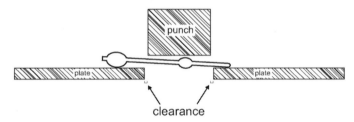

Fig. A.10 Measuring the toughness of leaves using a pair of scissors or a penetrometer. Both devices require careful mounting and attention to their condition because the tests incur a considerable amount of friction. The recording of a force–displacement curve is vital for establishing the work done in the tests. With scissors, friction is almost entirely due to metal–metal contact as the blades brush past each other in point contacts, so excluding the compression of leaf tissue between the blades. In contrast, the penetrometer resembles a punch and die separated by a small clearance in which leaf tissue gets compressed. (a) A half-leaf specimen is cut by scissors along the dotted line, with the cut running from left to right. Cutting the midrib first allows both it and any secondary vein to be identified as discrete items on a force–displacement plot. Their toughness can be obtained separately to the rest of the lamina without dissection due to characteristic force peaks. (b) A penetrometer generally involves a circular punch moving through a slightly larger hole cut in a circular plate. The edges are usually very rough, particularly around veins, and this gets dragged through the hole producing variable friction.

empty pass done first (so ignoring the friction of the specimen against the blades, which is often low). Regardless of the sequence of cuts though, the scissors blades must be cleaned before the test (never between the cut and empty pass) so as to get accurate results. A good pair of scissors is absolutely essential. I have used hairdressing scissors made from cobalt steel because plant tissues do not stain them.

When cutting young leaves the forces can be very low. Accordingly, the force between empty blades needs to be low or errors will accumulate. Whatever the scissors used, it is important to remember that these are not high-tech devices. The joint of the scissors is not stable when the blades are very wide apart and the point contact along the blades can be lost near full closure. It is sensible to work away from those limits. Small variations in blade sharpness are not critical for work on plant tissue, but the blades can be blunted quite quickly if tissues contain silica or silicates (present in some leaves and barks). The blades are otherwise harder than most tissues on which they will be used. However, the blades will still wear against each other, this blade–blade contact being essential to avoid plant material being trapped between the metal surfaces, a problem with penetrometer tests, (as discussed below). Thin specimens are likely to be 'floppy'. However, a very stiff specimen will actually drive the blades together, resulting in more work being done (and more metal being lost) than in an empty pass. A specimen this stiff should be tested another way. All in all, it is a practical test that has stood up well.

The scissors test was designed as a replacement for the popular 'penetrometer' tests used by ecologists. It is only relatively recently that actual results of penetrometer tests have been tabulated in publications, nearly always having been reported previously only as correlations. There have been many designs, but they nearly all fit the diagram in Fig. A.10 whereby a piece of biological tissue such as a leaf is punched through a hole in a plate. The results are reported as a force, a stress (force divided by the area under the punch) or a force divided by punch perimeter. However, none of these analyses can be correct and there is also friction to contend with, variously between specimen, punch and the hole (Atkins, 1980). Without a clear analysis of how to understand tissue trapped between the punch and the die at different phases of movement, the technique seems wanting (Lucas *et al.*, 1991a; Vincent, 1992; Wright & Vincent, 1996). Despite this, the technique has its proponents (Aranwela *et al.*, 1999; Edwards *et al.*, 2000), who have in turn suggested that scissoring is also flawed. I do not enter this argument here.

NOTCHED TENSION AND BENDING

These tests are very well treated in Vincent (1992) and I make no attempt to add a potted account of these tests here, particularly when examples of these tests have been shown already in this appendix and in Chapter 5.

PEELING

The bond between objects of dissimilar tissues or materials is referred to as adhesion and the work to break it as peeling. The theory and practice of peeling experiments are identical to those of fracture mechanics (Kendall, 2001). The peeling of a fruit, possible only when the outer layer strips cleanly from the underlying flesh, can be estimated by the simple geometry given in Fig. A.9 with the work of peeling given by

$$R_\mathrm{p} = \frac{2F}{w}$$

where w is the width of the peeled strip.

FRICTION

Friction is the resistance to motion between two surfaces at their interface produced by some (undefined) interaction between them. It is far from being properly understood, but it can be important to get some measurement of it for various aspects of tooth function. Figure A.11 shows the classic view of friction whereby two bodies compressive force between

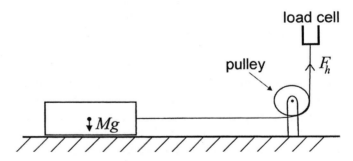

Fig. A.11 A block of mass M, so weighing Mg, is dragged across a surface by a force F_h. With the vertical force $F_\mathrm{v} = Mg$, the coefficient of friction $\mu = (F_\mathrm{h}/F_\mathrm{v})$.

them, F_v, require a frictional force, F_h, in order to slide. The coefficient of friction is

$$\mu = \frac{F_h}{F_v}.$$

Resistance to the start of movement tends to be greater than that needed to continue it, leading to two distinct concepts: static and dynamic friction. The coefficient of friction may not be constant, depending instead on the forces involved (e.g. cork: Gibson *et al.*, 1981). There are many approaches to the measurement of μ, but simple devices will usually work. The device shown in Fig. A.11 is a slider mounted on a low-friction track that can be pulled along by winding the mechanical stage.

RELATING DENTAL FORM TO DIET

Investigations of the dental–dietary adaptations of mammals can be aided on charts such as Fig. 7.1 where toughness (R) is plotted against Young's modulus (E) on logarithmic axes. This chart, based on Ashby (1999), is one of a large family of charts that could be constructed to represent the obstacles that foods present to oral processing. Measurements like this from field studies on mammals are only just beginning, examples being Yamashita (1998) on five lemur species, Dominy (2001) on four African anthropoid primates and the abstract of Strait & Overdorff (1996), who were the first researchers to take a full mechanical testing machine to the field.

HELPFUL REFERENCES

One of the biggest hurdles of entering the materials science literature is that articles often lack citations, making it difficult to know who agrees with whom. This is usually obvious in biology, but in the physical sciences, it appears often to reflect the lack of trips to libraries. Some books stress a theoretical overview and while they may emphasize the need to do things properly (i.e. in accordance with theory), they also indicate that accuracy for its own sake is nothing to aim for. Other books may just make you (have already made me) scared. I have found the following books optimistic and, in places, exciting: Cottrell (1964), Ashby & Jones (1996, 1998), Gibson & Ashby (1999), Atkins & Mai (1985), Kendall (2001) and Lawn (1993). A good introduction to viscoelasticity is by Dorrington (1980). Brennan (1980) and Bourne (2002) both describe the history of the mechanical testing of foods. Vincent (1992) explains many modern techniques in detail and is the only book that really concentrates on the mechanical properties of biological tissues and how to measure them. A general text on biomechanics that describes biomechanical techniques like electromyography and strain gauge usage is Biewener (1992).

Properties of teeth and potential foods

This appendix consists of tables listing properties mentioned in the text. As far as mechanical property measurements go, they are but a selection of those available in the literature. While some of the property values represent many years of work by investigators, others are nothing more than preliminary estimates. Where there has been a choice, I have opted for the most recent summary because some older work is probably more subject to variability due to state of preservation. Teeth provide an example. It is now established that storing teeth in deionized water loses mineral, producing a 20–30% decline in hardness inside a day (Habelitz et al., 2002). This means that it is certain that some hardness values quoted in the literature are too low. Estimates for other properties might inadvertently have been elevated: e.g. antler, which is lower in modulus and strength than limb bone, is higher in toughness (Currey & Brear, 1992), due to a slight reduction in its mineral content. Teeth extruding from the mouth may generally have lower mineralization, e.g. elephant ivory versus human dentine (Rajaram, 1986), and are undoubtedly drier. Though the toughness of ivory is unknown, a lot of fibrillar pull-out at fracture seems to imply higher toughness than that of human dentine (Rajaram, 1986).

MECHANICAL PROPERTIES OF BIOLOGICAL TISSUES

Tissue	Young's (elastic) modulus (GPa)	Toughness (J m^{-2})	Hardness (MPa)	Yield strength (MPa)	Tensile strength (MPa)	'Energy absorption' at fracture (J m^{-3} × 10^5)	Reference	Remarks
Tooth tissues								
Human (*Homo sapiens*)	Enamel 50–120 / Dentine 23–27 / Peritubular 29.8 / Intertubular 17.7–21.1	Enamel / Across rods: 200 / Between rods: 13 / Dentine / Across tubules: 270 / Between tubules: 550	Enamel 2500–6000 / Dentine 250–800 / Peritubular 2230–2540 / Intertubular 120–520		Enamel 8–35 / Dentine 31–104		Mostly compiled by Marshall et al. (2001); enamel from Cuy et al. (2002); toughness from Rasmussen et al. (1976); peri- and intertubular dentine from Kinney et al. (1996); mantle dentine much	Enamel of deciduous teeth averages 85% of that of permanent teeth (Nose, 1961; N.B. absolute hardness values reported there must be wrong). Major reviews by Braden (1976) and Waters (1980)

APPENDIX B (cont.)

Tissue	Young's (elastic) modulus (GPa)	Toughness (J m^{-2})	Hardness (MPa)	Yield strength (MPa)	Tensile strength (MPa)	'Energy absorption' at fracture (J m$^{-3} \times 10^5$)	Reference	Remarks
							softer than rest of primary dentine (Renson & Braden, 1971)	
Beaver (*Castor fiber*)			Dentine 560				Osborn (1969)	Wear patterns match hardness contours; mantle dentine is much softer, ~400 MPa
Sheep (*Ovis aries*)			Enamel 2650–3750 Dentine 324–725				Baker *et al.* (1959)	
Pigtail macaque (*Macaca nemestrina*)			Enamel 3800 Dentine 590				P. W. Lucas (unpubl. data)	Average of sub-surface enamel and primary dentine
Gibbon (*Hylobates muelleri*)			Enamel 2765 Dentine 580				P. W. Lucas (unpubl. data)	Average of sub-surface enamel and primary dentine
Orang-utan (*Pongo pygmaeus*)			Enamel 3550 Dentine 480				P. W. Lucas (unpubl. data)	Average of sub-surface enamel and primary dentine

Material						References	Notes
Indian elephant (*Elephas maximus*)	Ivory: Dry 12.5			110	8.7	Rajaram (1986)	Softer than human dentine (Rajaram 1986); probably dry in tusk; dry density 1.7 g cm^{-3}
	Wet 3.5			36	4.9		
Other tissues							
Bone (human femur unless stated)	17.7	340–400	937 (bovine tibia)	99.2	4.0	Rajaram (1986); Blackburn et al. (1992); Norman et al. (1992)	Dry density 1.94 g cm^{-3}
Antler bone	Dry 17.1		5000	188	13.5	Rajaram (1986); Currey & Brear (1992)	Dry density 1.86 g cm^{-3}
	Wet 7.5			108	14.6		
Articular cartilage	0.31– 0.08 × 10^{-3}		140–1200			Chin-Purcell & Lewis (1996); Korhonen et al. (2002)	Korhonen et al. (2002) give Poisson's ratios of 0.15–0.21
Rat skin	Initial ~10^{-6} Final 6 × 10^{-3}		Trouser-tearing: 14 000–20 000 Scissors: 590			Purslow (1983); Pereira et al. (1997)	
Horse hoof	0.18–0.56		5.5–10.7			6.5–9.5	Bertram & Gosline (1986); Kasapi & Gosline (1997)	
Fruit peel *Gnetum microcarpum* (Gnetaceae)			Scissors: across sclereids 1485 along sclereids 642			P. W. Lucas (unpubl. data)	

APPENDIX B (cont.)

Tissue	Young's (elastic) modulus (GPa)	Toughness (J m^{-2})	Hardness (MPa)	Yield strength (MPa)	Tensile strength (MPa)	'Energy absorption' at fracture (J m^{-3} × 10^5)	Reference	Remarks
Pods								
Derris thyrsifolia (Leguminosae)		Scissors: ripe brown 2000–4485 unripe green 255–469					P. W. Lucas (unpubl. data)	
Intsia palembanica (Leguminosae)		Scissors: along pod width 4906 along pod length 6950					P. W. Lucas (unpubl. data)	Pod 0.8 mm thick
Seed coverings								
Mezzettia parvifolia (Annonaceae) zone II	7	1900	205		67		Lucas *et al.* (1991b)	Woody shell: dry density 1450 kg m^{-3}
Schinziophyton rautanenii (Anacardiaceae)	Air-dry 4.96 Wet 5.23	1437 851			51.65 45.66		Williamson & Lucas (1995)	Woody shell; moisture content: air-dry 9.2% on initial weight, wet 20.5%
Cocos nucifera (Arecaceae)	2.9–4.9	1700–1900						Woody shell
Elaeis guineensis (Arecaceae)			205–235					Woody shell
Macadamia ternifolia (Proteaceae)	2–6	100–1000	180		25–80		Jennings & Macmillan (1986)	Woody shell: dry density 1300 kg m^{-3}
Callerya atropurpurea (Leguminosae)		Single edge notched tension: 330 Trouser-tearing: 355 Scissors: 3620 (±445)			5.04 (±1.12)		P. W. Lucas (unpubl. data)	Flexible covering; modulus in tension

Intsia palembanica (Leguminosae)	Scissors: 1500		P. W. Lucas (unpubl. data)	Flexible covering
Gnetum microcarpum (Gnetaceae)	Scissors: 3835			Cut made perpendicular to massive (~100 μm diameter) sclereids
Albizia splendens (Leguminosae)		267		Moisture content of whole seed 7.5%
Samanea saman (Leguminosae)		261		Moisture content of whole seed 8.6%
Adenanthera pavonina (Leguminosae)		327		Moisture content of whole seed 7.8%
Alangium ridleyi (Alangiaceae)		167		
Woody plant gall On *Distylium racemosum* (Hamamelidaceae)	C-ring: 177 C-ring (compression): 0.565	80	Hill *et al.* (1995)	
Seed contents *Callerya atropurpurea* (Leguminosae)	Wedge: 1285 (±269) Scissors: 947 (±327) 0.01	2.5	P. W. Lucas (unpubl. data)	Cotyledon
Gnetum microcarpum (Gnetaceae)	Wedge: 1143			
Brazil nut kernels	33.84×10^{-3}	160.8		
Macadamia terniifolia (Proteaceae) kernel	32.18×10^{-3}	214.3		Endosperm

APPENDIX B (cont.)

Tissue	Young's (elastic) modulus (GPa)	Toughness (J m⁻²)	Hardness (MPa)	Yield strength (MPa)	Tensile strength (MPa)	'Energy absorption' at fracture (J m⁻³ × 10⁵)	Reference
Leaves							
Lolium perenne (Graminaceae)	Longitudinal 0.55 Transverse 0.014	215–402					Vincent (1982); Greenberg *et al.* (1989)
Castanopsis fissa (Fagaceae)		Whole lamina 410 Without veins 120 Veins 2000–6000 Midrib 2000–9000					Choong (1996)
Calophyllum inophyllum (Guttiferae)	Longitudinal 0.186–0.240 Transverse 0.050–0.068	Whole lamina 724 Without major veins 220–300 Veins 6000			Without veins 2 Across veins 7–20		Lucas *et al.* (1991a)
Mollusc shell							
Nacre	60	350–1240			140		Jackson *et al.* (1988)
Insect cuticle							
Locust mandible			175–350				Hillerton *et al.* (1982)
Leaf-cutter ants (*Atta sexdens*)			Light unsclerotized: ~200 Dark sclerotized: ~300–350 Mandibular cusp (containing zinc): Light unsclerotized: ~350 Dark sclerotized: ~800				Schofield *et al.* (2002)

Isolated tissues and pure components

						Reference	
Hydroxyapatite	101		5170			Zhang et al. (1997)	
Elastin	0.0011	30		2	1.6	Gosline et al. (2002)	Extensibility 1.5
Resilin	0.002			4	4	Gosline et al. (2002)	Extensibility 1.9
Collagen	1.2			120	6	Gosline et al. (2002)	Extensibility 0.13
Periodontal ligament	0.05					Rees & Jacobsen (1997)	
Disc of temporomandibular joint	0.044					Tanne et al. (1991)	
Cell wall (of woody tissue)	Parallel to cellular axis: 25 / Perpendicular to this: 15	3450				Gibson & Ashby (1999); Lucas et al. (2000)	
Opal phytoliths			5800–6000			Baker et al. (1959)	

Soil particles

						Reference	
Quartz	70	2	7000–7750			Baker et al. (1959); Lawn (1993)	

Engineering materials

						Reference	
Diamond	1000	5	~150 000	~50 000		Ashby & Jones (1996); Atkins & Mai (1988)	
Alumina	371	100	560			Atkins & Mai (1988)	
Polymethylmethacrylate	3.4	640	65	110		Ashby & Jones (1996); Atkins & Mai (1988)	
Mild steel	210	~14 000	220	430		Ashby & Jones (1996); Atkins & Mai (1988)	

Human foods

						Reference	
Gel (mung bean starch)	Initial ~10^{-6}	0.5–22				Lucas et al. (1993); C. G. Oates et al. (unpubl. data)	
Spring roll pastry (briefly cooked)	Fresh 0.2×10^{-3} Trouser-tear: 456.9 Scissors: 139.7		0.043			Sim et al. (1993)	

APPENDIX B (cont.)

Tissue	Young's (elastic) modulus (GPa)	Toughness (J m^{-2})	Hardness (MPa)	Yield strength (MPa)	Tensile strength (MPa)	'Energy absorption' at fracture (J m^{-3} × 10^5)	Reference	Remarks
	1 day old 1.3 × 10^{-3}	Trouser-tear: 235.5 Scissors: 207.8			0.0635			
Raw carrot								
Cheese	4.57 × 10^{-3}	440					Agrawal (1999)	
Reduced-fat cheddar	0.94 × 10^{-3}	172.7					Agrawal (1999)	
Mozzarella	0.15 × 10^{-3}	70.7					Agrawal (1999)	
Parmesan	2.26 × 10^{-3}	233					Agrawal (1999)	

PROPERTIES OF NON-WOOD PLANT TISSUES USED IN FIG. 4.15

	Volume fraction occupied by cell wall (V_c)	Number of tests	Correlation coefficient r	Slope 'plastic work' ($J\ m^{-1}\ m^{-1}$) mean (standard error)	Intercept 'cell wall toughness' ($J\ m^{-2}$) mean (standard error)
Plant tissues					
Citrullus vulgaris (Cucurbitaceae) red watermelon fruit flesh	0.0031	22	0.815	21.9 (3.5)	9.5 (6.8)
Brassica rapa (Brassicaceae) green turnip	0.0192	25	0.731	341.0 (66.4)	84.5 (43.2)
Solanum tuberosum (Solanaceae) Russet Burbank potato	0.0253	35	0.467	343.5 (113.4)	89.6 (50.7)
Gossypium sp. (Malvaceae) cotton hairs	0.089	20	0.909	455.7 (49.3)	300.8 (101.2)
Callerya atropurpurea (Leguminosae) cotyledon	0.127	31	0.720	1 039.5 (186.2)	267.8 (123.5)
C. atropurpurea (Leguminosae) seed coat	0.349	27	0.633	894.6 (217.4)	1 681.3 (210.7)
Leucaena leucophala (Leguminosae) pod: inner brown layer only	0.242	25	0.822	5 775.3 (799.4)	761.3 (194.1)
Albizia splendens (Leguminosae) pod	0.349	41	0.821	3 143.2 (349.8)	970.0 (259.2)
Aleurites moluccana (Euphorbiaceae) seed shell	0.90	40	0.801	14 630 (1,778)	2 638 (515)
Schinziophyton rautenenii (Anacardiaceae) seed shell	0.94	21	0.491	11 349 (4,617)	3 767 (1 005)
Mezzettia parvifolia (Annonaceae) seed shell					
Zone I	0.95	16	0.576	24 840 (9 421)	4 558 (1 935)
Zone II		29	0.577	12 011 (2 822)	3 739 (680)
Zone III		12	0.184	2 310 (3 905)	4 192 (1 136)
Scheelea sp. (Arecaceae) palm nut endocarp	0.92	21	0.614	15 636 (4 734)	2 423 (1 532)
Cellulosic materials					
Filter paper (Whatman's, UK) No. 1	0.343	15	0.963	7 650 (597.6)	1 648.9 (217.3)
Filter paper (Whatman's, UK) No. 42	0.347	17	0.931	6 793.3 (689.5)	1 980.2 (313.8)
Filter paper (Whatman's, UK) No. 542	0.411	30	0.924	11 840 (923)	1 368.3 (218.9)
Bank paper	0.413	45	0.855	10 853 (1 005)	913.8 (108.6)
Newsprint	0.449	39	0.921	9 225 (640)	1 232 (147)

Notes

1 HOW TO GET EXCITED ABOUT TEETH

1. The most recent research, some of it still unpublished at the time of writing, suggests that teeth may be older than envisaged above, evolving as part of an adaptive explosion of body forms in the Cambrian.
2. Some ability to fracture food particles using the mouth or pharynx is not unique to mammals. Some fish (Vincent & Sibbing, 1992), insects (e.g. locusts) and birds (e.g. parrots) do it, as did an early reptile (Rybczynski & Reisz, 2001).
3. Some mammals do regurgitate food. Many of the ruminants, for example, do this, but with the intention of chewing food further rather than expelling it. Throwing up as a routine behaviour is exceptional in mammals.
4. This view has recently been challenged (Rich *et al.*, 2002).
5. Sensory feedback from the masticatory muscles is sent back to the brain via a single sensory neuron, i.e. by the fastest possible route. I do not know of any instance of this outside of the trigeminal nerve (the cranial nerve conveying these signals) in mammals. However, this will not offset massive differences in body size – an elephant could not respond at the speed that a shrew can.
6. Some taste buds are located in the inlet to the larynx in humans, which cannot have anything to do with food quality. Instead, it seems to suggest that taste sensations also help identify where food particles are, additive to the sense of touch.

2 THE BASIC STRUCTURE OF THE MAMMALIAN MOUTH

1. Some authors refer to the deciduous molars as deciduous premolars (e.g. Schwartz, 1974). I disagree, but this 1974 paper was significant as marking the first substantial break with a long-standing tradition over dental homologies.
2. The mammals with the thickest enamel appear to have been the gomphotheres (probable ancestors of elephants: Maglio, 1973) which had enamel on their molars up to 5 mm thick in places. Both gomphotheres and their teeth were very large, but it has to be remembered that the ameloblasts (cells that form enamel) of a gomphothere were probably no different in size from those of a shrew.

3. Cracks in mature enamel then probably run through a rather gel-like matrix. The toughness of a thick gel is of the same order as that of enamel (Lucas *et al.*, 1993). Averaged over the tissue though, enamel is about three times tougher than pure hydroxyapatite (White *et al.*, 2001), but about 40% less hard (Zhang *et al.*, 1997).

4. However, Fox's experiments only provide limited support of this hypothesis because the predicted effect is not large. The main problem though is that Fox tested whole teeth and, as shown later, dentine very definitely displaces fluid when it is loaded.

5. The term symphysis is the name given to a type of fibrocartilaginous joint found in the midline of the dentaries of the lower jaw and of the pubic bones of the pelvis (where it allows for substantial movement during childbirth).

6. One long-term denture-wearing subject that I measured by the same method had only 5 mm of mandibular bone height left in the molar tooth region.

7. This bone is called the lamina dura in dental radiographs, being more radio-opaque (dense) than the rest of the alveolar process.

8. An 'innocuous' result that is equation 86 of Synge (1933). Albeit a classic analysis by a famous applied mathematician, the paper is very difficult to read and the change in meaning of symbols as the paper progresses does not help. I am indebted to Waters (1975) for unravelling critical aspects of this article.

9. The periodontal ligament must provide the motive force for this, but quite how it does it is unclear.

10. Large anterior–posterior movements are also seen in elephants and, judging from tooth wear, in some extinct mammalian groups. The text sticks to rodents because these have been studied physiologically.

11. The modiolus can be palpated in humans by placing a thumb inside the corner of the mouth with the index finger over the skin. The modiolus is felt as a thickening of the soft tissue.

3 HOW THE MOUTH OPERATES

1. Some dinosaurs have been suggested to be capable of mastication and may also have had some control, albeit limited, over their body temperatures (Norman & Weishampel, 1991). This latter point has been debated for about 30 years.

2. Image analysis is now taking over from sieving.

3. Despite the apparent ease of the selection concept, there are difficulties in practical measurement. Take the largest particle sizes, the squares, shown in Fig. 3.4. Now suppose that, during any given chew in which it is being calculated, these squares break to produce the smaller-sized circles, hexagons or triangles. The selection function for the squares is easy to calculate because we know unequivocally the fraction that is broken. However, when we come to do the same with smaller sizes, then we are faced with the proportion of particles that is broken down being counterbalanced by a proportion broken into this size range from larger sizes such as the squares. If we take no notice of this, it could result in a negative selection value. This would be absurd: particles are not

getting bigger. The way around this problem has been to label particles, which we can do by dyeing them (as indicated by the different tones of the particles in Fig. 3.4) or by shaping them (as indicated by the fanciful shapes shown in the figure). By doing such experiments in humans, it has been shown that the selection function is quite stable over a huge time-span, well beyond the usual physiological timescale (van der Glas *et al.*, 1987).

4. These authors report that a solution to comminution equations evaded even a mathematician of the standing of Kolmogorov.

5. Such distributions probably do typify foods sticking to the teeth, such as biscuits and other products made from flours. Obviously, some very fine fragments are produced, but I have never seen a study of this.

6. I am not advocating here that the locomotory substrate makes no difference at all – if it did, there would be no need for special running tracks for human athletes (McMahon & Greene, 1979).

7. The space outside the dental arches is termed the vestibule, which is like a gutter. The tongue rarely needs to fish food out from here though because the control of this space by the buccinator seems to be so good that appreciable quantities of food rarely fall into it. In humans, only wearers of artificial teeth called dentures or those with paralysis of facial muscles collect food in the vestibule. However, in some mammals, the vestibule houses cheek pouches as in some rodents and in cercopithecine Old World monkeys. These act as designed food stores. In these monkeys, the mucosal lining of the pouch is wired to the brain bilaterally (Jones *et al.*, 1986; Manger *et al.*, 1996), something that is not true for other taxonomically related, but pouchless, primates (Manger *et al.*, 1995). This suggests the need for fine control of the contents of these pouches.

8. There have been other experiments in humans relevant to this sensory threshold (Öwall & Vorwerk, 1974), but with a technique that should be modified before being repeated. The swallowing of very similarly sized particles across a wide range of mammals might be related to the retentive ability of sphincters lower down in the gut: there is some evidence that very small particles in dogs slide directly through from stomach to small intestine, thus omitting a digestive step (cited in Jenkins, 1978). All this may help to explain why foods in the modern human diet, often made of extremely small particles, are not cleared effectively from the mouth by natural means. We are barely aware of their presence – until we look.

9. An analogy to mechanical testing is possible where the rate of change of either force or displacement can be controlled while the other variable is monitored. The difference between controlling and monitoring change is important because this has a great effect on crack propagation in the test specimen. Stress control is more likely to result in fast crack growth than displacement control (Atkins, 1994). However, the stiffer the material under test, the slower the rate of displacement has to be because the rate of force build-up will be proportional to specimen stiffness and sensors that could stop or reverse loading so as to avoid damage may not have time to act.

10. A large amount of research has been done on this, both theoretical and experimental, but it appears inconclusive. Early papers with novel analyses include

those by Smith & Savage (1959), Barbenel (1972), Greaves (1978), Bramble (1978) and Smith (1978). Spencer (1998) reviews and tests most of these hypotheses.

11. The joints of cercopithecine Old World monkeys have been proved to endure such forces (Hylander, 1979b; Brehnan *et al.*, 1981).

12. Nelson *et al.* (2001) did not test arabinose, which is a pity since this sugar turns up in appreciable quantities in some fruits and may be sensed by some other, as yet unknown, taste receptor. It is worth noting that the relative sweetness of sugars does not seem to tally with their calorific value.

13. Breslin *et al.* (1996) introduced the term 'monogeusia' for compounds that cannot be discriminated from each other by taste. According to these authors, sucrose, fructose and glucose are like this and are probably perceived via a single taste mechanism in the human. This conflicts with what I have written. However, these authors agree that maltose is different: it is indistinguishable from these other sugars at low concentrations, but behaves differently at higher concentrations. They suggest this is because maltose activates a second sweetness receptor.

14. The order of taste response, from strongest to weakest, appears to be cystine, alanine, glutamic acid, serine, methionine, aspartic acid, glycine, thyronine, histidine, arginine and valine. The response is heightened by the presence of purine nucleotides. Glycine triggers both the amino acid and the sweet receptor (Nelson *et al.*, 2002).

15. Critchley & Rolls (1996) have recently found neurons in the taste cortex of primates (macaques) that respond to tannic acid.

16. Recently, evidence is accumulating in humans (Hiiemae & Palmer, 1999) that some foods form a bolus in the oropharynx, behind the mouth, on the upper surface of the epiglottis.

17. Ruminants are exceptional in producing vast amounts of saliva. This is because saliva provides the fluid for the fermentation process in the first compartment of their stomachs (Kay *et al.*, 1980; Kay, 1987). This fact shows that copious salivary secretions are entirely feasible were this flow to be required for the swallowing process.

18. Cuticles pass through the mammalian gut undigested even in ruminants. However, they are apparently subject to heavy erosion in regions of high rainfall (C. Neinhuis, pers. comm.). N. J. Dominy (pers. comm.) found that few tropical rainforest leaves show the 'lotus effect'. Micro-roughening of the surface, a vital part of this effect, may also explain the self-cleaning properties of the wings of insects (Wagner *et al.*, 1996) and even perhaps the oral surface of the tongue too.

4 TOOTH SHAPE

1. There seems to be an unjustified assumption in the literature that strength is 'OK' as a fracture parameter unless massive differences in size are being considered. This is not so, and it is not something specific to compression

either, but applies to any loading regime as Griffith's (1920) original paper showed.

2. In reality, cusps may contact because of the inertia of jaw movement or because the diet of many mammals is quite varied: some foods are eaten that do not crack or they could be chewed down to the 'brittle–ductile' transition. Furthermore, if mammals swallow like most humans, then their teeth will contact at the beginning of this process. Thus, it makes sense to develop upper and lower tooth shapes that can contact each other without damage.

3. Bond (1952, 1962) based his work index on an enormous amount of data gleaned from industrial processes supporting his contention that the energy consumed in comminution (usually measured by calorimetry) is proportional to change in the reciprocal of the square root of particle size. However, he stressed that this index really did not refer to particle dimensions, but to cracks. Two nineteenth-century theories preceded Bond's. Rittinger suggested that the energy consumed was proportional to the particle volume being fractured, while Kick argued that that it was proportional to surface area. The ramifications of Rittinger's and Kick's theories are best read in Lowrison's (1974) summary of knowledge on industrial comminution, because the original accounts are nothing more than vague suggestions. Neither theory has much to back it up.

4. This secondary tensile stress is the basis of the Cook–Gordon crack-stopping effect, whereby a crack, driven by a primary stress and that is heading to a weak interface, can be arrested by a secondary stress that opens the interface before the crack reaches it (Cook & Gordon, 1964). The result is a blunted crack. The problem with emphasizing this mechanism as a major defence of foods (or teeth) against fracture is that all it really achieves is a complication in crack direction. Generally, the crack may extend along the interface for a short distance but then be driven on in the original direction as soon as favourable circumstances permit. This zigzagging crack in a food particle obviously involves a mixed mode of fracture, but again this is nothing directly to do with the shape of tooth surface that is attempting to fragment it.

5. Remember though that the stress level is not sufficient for the crack to grow. The energy balance must be in favour of it, something that generally requires that the stress field (the volume of tissue being stressed) be large enough.

6. All told, this section gives the reason why surgeons (and animal anatomists) need to use sharp scalpels. Some human foods show similar behaviours to animal soft tissues. For example, popiah skins are thin sheets of pastry used to wrap savoury foods in certain Chinese communities. (When filled and fried, these are known in the West as spring rolls.) The skins are cooked by being laid on one side against a hot metal plate for just a few seconds. A study by Sim *et al.* (1993) showed that the stress–strain curves of freshly cooked skins are completely linear up to fracture and highly extensible. They blunt cracks significantly when pulled, resulting in a toughness that is about three times that when cut with scissors. After 24 hours, this extensibility reduces and the toughness of a free-running crack is then more or less equal to that in cutting (Appendix B).

7. Other examples include muscles and tendons in which fibres commonly tear during exercise (but rarely propagate so as to split the whole structure into two, as athletes know), the periodontal ligament, an unfused mandibular symphysis and ligaments around joints.

8. There may also be a tertiary cell wall present, but this is generally irrelevant in mechanical terms.

9. There are many ways to estimate V_c. Microscopy is very effective for homogeneous tissues. When V_c is low, it is approximately equal to the thickness of the wall divided by the diameter of the cell or the average cell dimension (Ashby & Gibson, 1999). In tissues where there are elongated cells, all with the same orientation, V_c can be estimated by measuring the area occupied by cell walls in 'end-on' view. This area fraction is effectively the same as V_c. If tissues are too heterogeneous to support such laborious measurement, then as a crude calculation for living tissues, V_c is approximately equal to the chemical fibre content (neutral detergent fibre or NDF; van Soest, 1996), calculated on a wet (fresh) weight basis, multiplied by 1.5 (Choong, 1996). This multiple allows, probably too generously, for the extra density of cell wall as compared to cytoplasmic constituents. Alternatively, if the tissues are dead, as in woods or seed shells, then it is often possible to cut them into a defined shape, measure and then weigh them. The density relative to that of solid cell wall (taken to be 1500 kg m^{-3}) is the same as V_c. Lastly, if this is not possible, then as many pieces as possible can be collected together, weighed, then dropped into water in a measuring cylinder to find the displaced volume. A weight of known volume will be needed to help sink most tissue. This last Archimedes procedure is generally the least accurate.

10. There is just one direct measurement of cell wall toughness in the literature. This is on potato parenchyma and is 9.55 kJ m^{-2}, of which one-third of this, i.e. 3.18 kJ m^{-2}, was estimated to be due to elastic mechanisms acting in the wall (Hiller *et al.*, 1996). This latter value seems compatible with the 3.45 kJ m^{-2} given in the text. The propensity for plastic work is evidenced by slope/intercept ratios: these are often close to 3.0 for tissues with only primary cell walls (Appendix B), but average 10.0 or more for woods across-grain (Lucas *et al.*, 1997).

11. There is actually a negative correlation between the toughness of woods and lignin content (Lucas *et al.*, 1997).

12. Turgidity tenses the cell wall and so probably lowers the cost of fracturing it by doing some of the work for teeth. Exactly how much the toughness drops is not clear because the measurement of turgour pressure is not easy (Tomos & Leigh, 1999). Flaccid parenchyma can seem much tougher than when it is turgid, but the change in the shape of the stress–strain curve, analogous to what happens when such tissue is boiled (see Chapter 7), may mislead.

13. A bat probably fractures such fruit tissue to reduce the weight of tissue being processed in its gut. These mammals have very fast gut passage rates and the type of fruit that they concentrate on often has large quantities of nutrients (simple sugars and free amino acids) that do not require digestion. Chimpanzees

often 'wadge' foods, i.e they place plant tissues inside their lower lip and squeeze juice out of them by pressure against their lower incisons. The piths of monocots are often treated like this, most likely because these contain fibres that, although comprising only a small proportion of the tissue by volume, prevent fragmentation by the chimpanzee's blunt molars (Wrangham *et al.*, 1992).

14. J. F. V. Vincent has talked about the resistance of grasses in this way: trampling it does not result even in fraying of the blades.

15. The distal edge of a sabre-tooth canine can be sharp, but is not sharpened against another tooth. Some herbivores, such as some deer and anthropoid primates, have enlarged canines that they employ mainly in disputes among males in a species. These have sharpened blades (Zingeser, 1969; Walker, 1984), being designed for damage, but not death or ingestion.

16. Tests of any of the dental–dietary relationships suggested in this book require that a range of foods be included. If the range is too restricted, then it is likely that the various mechanical properties will be correlated with each other (Ashby, 1998), leading to spurious conclusions.

17. I have failed to locate an old note published by G. E. Hutchinson referring to the devastating effect of the canines of pet cats on hairless human skin. The note suggested canine reduction in hominins followed hairlessness.

5 TOOTH SIZE

1. One factor driving exponents up is the use of the reduced major axis, which is identical to the slope obtained from a least-squares regression, divided by the correlation coefficient. Since most biological analyses involve something less than a perfect fit, this practice always raises body-mass exponents to some degree.

2. One of these attempts was a mechanical explanation close to the theme of this book. It supposed that living organisms might be designed such that the elastic distortion of their tissues under load is everywhere identical (McMahon, 1973). This seems unlikely, which extensive tests of the hypothesis have reinforced (Alexander, 1985).

3. If engineers had to face a population of marauding King Kongs that were chewing up large buildings, then things might be rather different.

4. An exponent of 0.83 might come as a surprise to those used to allometric arguments. What other theory could result in such an odd exponent? No simple fraction gives this number.

5. The evidence for this difference in seed transport distances in mammals is not clear-cut. Many small mammals spit seeds out, e.g. bats (Phua & Corlett, 1989) and cercopithecine Old World monkeys (Corlett & Lucas, 1990; Lambert, 1999), doing this close to the parent plant. The distances from a plant that mammals defaecate swallowed seeds depend on the home range of the mammal and the patchiness of the fruit source that it is exploiting.

6. These figures are all biased towards the modern human diet. No one eats old cow or pig because their flesh is much tougher than that of younger animals (and age is the largest factor in determining meat toughness). There is no similar age bias for fish data: a large fish of any given species seems a good catch.

7. Flexible jaws would store a lot of energy. After food fracture, this storage would feed into cracks, propagating them more rapidly. Wright & Vincent (1996) discuss this in relation to a comparison of fracture tests of foods on mechanical testing machines, which are typically very stiff, and conditions in the mammalian mouth.

8. It appears that while large mammals trapped on small islands get smaller, the small mammals there may enlarge, so reducing the overall size range. Unfortunately, these rapid enlargements have not been well studied.

9. Gould (1975) and Fortelius (1985) both feel that downsizing the body would involve different patterns to the more usual upgrade. I do not agree and am unwilling to invoke developmental impediments to dwarfing, arguments that are often shuffled out to explain this phenomenon. There appear to be enormous selective pressures on mammals that reduce their body size very rapidly and I suspect a functional explanation for the patterns that they show, an explanation that fracture scaling provides.

10. The actual fit of force–elongation curves for mammalian soft tissues on a log–log plot is often poor. Certainly, many plots in Yamada (1970) remain distinctly curved after logging both stress and elongation (and after converting to true strain). Some authors claim exponents as high as 9.0 for these tissues.

11. Van Valkenburgh (1990) was concerned with the prediction of body mass from dental and skeletal measurements and thus plotted body mass on the vertical axis and carnassial length on the horizontal. The slope in the text was obtained by taking the reciprocal of the slope given in this paper, divided by the correlation coefficient.

12. Foods are more complex than this and can show a considerable amount of viscoelastic behaviour, which would be fatal in most engineering materials. However, as pointed out by Atkins (1999), this viscoelasticity would not be apparent until or unless the load was taken off. Which is to say that the shape of the stress–strain curve is the main concern, not subtle differences in the deformational processes that underlie this. This is what allows him to treat yielding as though it were equivalent to non-linear elasticity.

13. As an undergraduate obsessed with teeth, I was irked for a long time by not being able to sustain an objection to this throwaway comment from a classmate.

14. It should be noted here, for the benefit of proponents of allometry, that these variables are dimensionless and that the gape angle takes size into account. This relationship is then very close to the exploration of residuals in allometric analysis, of which examples can be found in this chapter. The gape analysis appears to me to be a more direct and elegant approach than an allometric one.

15. There is a very slight curvature to y/x lines that cannot be mimicked in the data analysis.

6 TOOTH WEAR

1. The analysis of dental microwear was developed independently for studying the wear of dentures in human subjects (Heath, 1986).

2. The triggering of this sharpening seems to be psychological stress, but is still debated. Personally, I will never forget the dreadful noise from a caged male pigtail macaque (*Macaca nemestrina*) in Singapore when sharpening its canines.

3. If work-hardening, which is not explained in this book anyway, is ignored, then the multiple would be 3.0, but the accuracy of all the following calculations is so limited that none of this really matters.

4. The picture is complicated more than a little by scratches often being variable in depth, starting very shallow, but deepening and ending up in pits (Walker, 1984).

5. The dental wear features that both particles cause are much smaller than their own dimensions due to only a small part of their surface being in contact with the tooth tissue.

6. Approximal wear is also, quite logically, referred to as interstitial wear. However, for some reason, the terms got jumbled up and many people talk about 'interproximal wear'. The editor of one dental journal used to complain rather strongly about this.

7. The occlusal plane, a plane drawn through the contact points of the postcanine dentition of mammals can be curved antero-posteriorly. This is called the curve of Spee in the human dentition. It seems associated with a high jaw joint, as explained by Osborn (1987).

8. The swallowing model of Prinz & Lucas (1997) is particularly sensitive to the rate of food breakdown. Slow breakdown is responsible for weak bolus formation in carrots. This just makes wear even more serious. Not only will wear reduce the rate of food breakdown, but it jeopardizes ever forming a bolus. If a bolus does not form, then the mouth is not properly cleared of particles.

7 THE EVOLUTION OF THE MAMMALIAN DENTITION

1. As pointed out in Chapter 4, the flesh of some fruits is an obvious exception.

2. This statement reserves the term 'chewing' for extended intra-oral processing. The 260-million-year-old Palaeozoic reptile *Suminia* (Rybczynski & Reisz, 2001) and the 230-million-year-old Mesozoic reptile *Pisanosaurus* (Sereno, 1999) could both fragment plant material to some extent and there is also some evidence that some late Cretaceous dinosaurs developed this ability (Norman & Weishampel, 1991). However, none appears to have chewed as mammals do. A major general trend in dinosaur evolution is increase in body size (Sereno, 1997), but some of the later herbivores were as small as 30 kg, possibly consuming a class of angiosperms (flowering plants) called monocots (Norman & Weishampel, 1991; Crane *et al.*, 1995) – although Barrett & Willis (2001) find no evidence for this. These newly targeted angiosperms may also have defended themselves by developing thorns and spines.

3. Young leaves in the tropics tend not to photosynthesize immediately, delaying their greening while they develop. They can be covered entirely with small spines or hairs, but these are shed when the leaf starts to function (Coley, 1983). Relatively few plants seem to have this defence, even though herbivory statistics support its efficacy against herbivores.

4. Many invertebrates bite, but some suck cell contents (Leigh, 1999).

5. This is where woody tissue excels, in fact, because its structural design sacrifices the highest possible modulus for a massive increase in toughness (Jeronimidis, 1980), so maximizing $(R/E)^{0.5}$.

6. Young leaves protected by hairs or small spines do not seem to possess much defensive chemistry (Coley, 1983), as predicted from this stress-limited strategy. The assumption that I make here is that plant chemical defences are not a furtive attempt to poison animals without their sensing it, but are intended to provide clear clues via taste receptors that poisoning would result if feeding continued. I suspect that, as is the case in living birds, dinosaurs had few taste buds.

7. These texture maps will eventually have to be converted to non-linear deformation (along the lines of 'power law' relationships between stress and strain) rather than modulus in order to reflect the resistance of many raw animal (and cooked vegetable) foods in mammalian (and human) diets more accurately. Chapter 5 and Appendix A give an inkling of how this could be done. Other important food properties can be mapped (Ashby, 1999), but no single chart can show all relevant features of either dentitions or diets. The aim here is to highlight general patterns of adaptation.

8. Taking this argument to its logical conclusion results in a viewpoint which seems not to have any precedent: some leaves could have been sacrificed by plants to nurture the growth of insect larvae that would subsequently mature to form part of the population of pollinators. In other words, angiosperms needed to breed their (guild of) pollinators. This might explain why many angiosperm families appear to have very close ecological relationships with those of Lepidoptera (Ehrlich & Raven, 1965, 1967). This view explains little about extant ecosystems because butterflies and moths are not now the major pollinators of plants, but perhaps they once were. In contrast, ecological accounts focus purely on angiosperm *defences* against herbivory (e.g. Coley & Kursar, 1996). Why though are leaf defences so often faulty?

9. Some very small synapsids have recently been discovered, but their overall size range appears to remain well above that of the earliest mammals.

10. It is possible that birds, which are basically dwarfed dinosaurs (Sereno, 1999), sacrificed the chance of developing mastication purely by enormous weight reduction during their evolution. This would not allow the space necessary for 'fracture-scaled' postcanines.

11. The internal tissues of insects are often thought to have little fracture resistance, but Evans & Sanson (1998) provide evidence from experiments that disagrees with this. There appears to be no information in the literature on the mechanical properties of the internal parts of insects.

12. It is unclear whether a heavily tanned cuticle is really so very different a mechanical obstacle to a lightly tanned one. Moore & Sanson (1995) compared the ability of two marsupial species, the northern quoll (*Dasyurus hallacatus*) and the large short-nosed bandicoot (*Isoodon macrourus*), to break down beetles, finding that the former, which has the sharper teeth, was more effective.

13. Digestion of the cuticle itself requires the presence of a gut enzyme called chitinase (Jeuniaux, 1961), the antiquity of which appears unclear.

14. I define seeds in this section as the unit that separates from the fruit during oral processing. It is the 'seed' that a layman would recognize. However, it may actually possess an outer covering that is derived not from seed tissue, but from the fruit (from the ovary). However, there is no point to get into this. The embryology of seeds is exceedingly complex (Corner, 1976) and mechanically protective layers, when found, may actually develop either from the seed coat (Corner, 1949) or fruit wall.

15. Janzen (1978) suggests in effect that these defences are parsimoniously organized in that heavily mechanically protected seeds will be free from toxic chemicals and vice versa.

16. The chimpanzee often compresses the pith of monocots between the lower lips and incisors, most likely because it contains fibres that, although comprising only a small proportion of the tissue by volume, prevent fragmentation with the molars (Wrangham *et al.*, 1992).

17. A bat probably juices fruit to reduce the weight of tissue being processed in its gut. These mammals have very fast gut passage rates and the type of fruit that they concentrate on often has large quantities of nutrients (simple sugars and free amino acids) that do not require digestion (Wendeln *et al.*, 2000).

18. From an examination of stomach contents, Gautier-Hion (1980) concluded that some African cercopithecine monkeys partially damage a considerable proportion of the seeds that they swallow.

19. There could be differences due to the age of individuals and because this is learnt behaviour.

20. Some other variables do confound this pattern, but the overall dominance of seed size is clear in the data (Corlett & Lucas, 1990).

21. Eaglen (1984) demonstrates that New World spider monkeys have smaller incisors than Old World colobine monkeys compared to body weight. He argues that this reflects phylogeny because otherwise highly frugivorous spider monkeys must surely be anticipated to possess larger incisors than folivorous leaf monkeys. The arguments given in this section suggest that this is not always expected. The incisal size of colobines vary. According to Ungar's (1996) interpretation of Hylander (1975), the *Presbytis* species, which are heavy seed destroyers, may have smaller incisors than *Trachypithecus* species, which are closely related but which consume more leaves.

22. Dominy *et al.* (2001) offer a calculation of how few cells a blade breaks, using a leaf tissue as an example.

23. Some New World monkeys open fruits with their canines (Kinzey & Norconk, 1990, 1993). These teeth have obvious ridges on them.

24. It has been suggested to me that the crenulated (wrinkled) enamel seen in many frugivorous mammals can be explained in this way.

25. Seeds are heavily protected, very often by chemistry rather than mechanics. The 'choice' of defence may be constrained by seed size. A seedling has to break out of its container. Nearly all seeds display a suture or some other weak point that allows a germinating seed, turgid from imbibing water, to produce a crack from such a 'notch' (e.g. Lucas *et al.*, 1991b). Given the type of woody tissue from which seed coverings are made, it is relatively simple to show that very small seeds cannot have substantial mechanical protection because the weak element would have to be so large that a seed predator could easily utilize it, rendering the investment in the rest of the seed covering redundant (Lucas & Corlett, 1998). The critical size range of seeds appears to be of the order of 1–2 mm in diameter (or maximum seed width), below which any protection is likely to be chemical. This size range coincides with the mucosal sensitivity threshold described in Chapter 3, which turns out to have important repercussions for the oral treatment of seeds. So mechanical protection of seeds is probably size-limited. However, although mechanical defence is 'permitted' in large seeds, many of them do not have this. Circumstantial evidence links the presence of hard shells mainly to an oil-based food reserve (Peters, 1987). It is unclear why seed starch needs to be defended at all if it is generally indigestible to mammals.

26. The lower incisors here include the most lateral tooth, conventionally termed a canine.

27. We know that the enamel crests are important because a rare condition called amelogenesis imperfecta causes the enamel of inflicted herbivores to wear as rapidly as its other dental tissues. These individuals cannot chew properly. The length of these crests can remain relatively constant despite considerable wear (Fortelius, 1985).

28. Hominins used to be hominids: the old use of the word 'hominid' is equivalent to the new term 'hominin'. The change is purely taxonomic and reflects recent agreement that our line from the apes does not deserve familial rank. The reason to have taxonomy as part of biological science is that it is difficult to get very far without being able to put a name to an organism. The problem with hominin taxonomy is that it seems to be professional death to agree with anyone else. Thus, while sycophancy is rare, nomenclatural stability is not. Luckily, even though we can't always put a name to a hominin fossil, we can usually put a face to it.

29. Better said, the second-generation replacement 'no longer forms to the point of mineralization' because there are many instances in mammals of tooth germs that initiate but are resorbed very early in their development. Many insectivores only have one generation of teeth due to this, including shrews (Kindahl, 1959). Why? I suggest an extension of the same type of explanation as for early mammals: a very short lifespan. Some modern primates seem to be dwarfs. The marmosets and tamarins are the clearest example (Ford, 1980), these primates having lost their third molars.

30. Taking this further to involve the other premolars is more difficult. As mentioned above, there are only two premolars in catarrhine primates, but three in cebids. The lower anterior premolar in Old World monkeys is enlarged for sharpening against the upper canine.

31. To forestall the development of a new insult, no one is descended from a robust australopithecine.

32. There have been many explanations for canine reduction in males during the early evolution of hominins. I cannot review them all here, but few have been refuted. Holloway (1967) felt that canine reduction was related to tool use. Kinzey (1971) supposed that the canines might never have been large. There is still no convincing fossil evidence to support this. Nothing has changed since Delson & Andrews' (1975) view that the most parsimonious suggestion is that the last common ancestor of catarrhines had large canines in males. However, the presence of small canines in *Sahelanthropus* certainly means that reduction was very early.

33. Farrell's work seems to be the principal reason why British dentists have little interest in understanding how teeth work.

34. As an indication of the methods involved in these experiments, two large Washington baking potatoes were cut into standardized 20-mm thick slices. All the parts from one potato, except two reserved for raw comparison, were placed into boiling water and slices recovered at 2-min intervals. The pieces of the other potato were roasted for 50 min in a 200 °C oven. All particles were cooled rapidly after cooking and then placed in plastic bags to conserve their moisture prior to testing. Their toughness and stiffness were then measured by wedge and cylindrical compression tests respectively. It is well known that potato tissue is not homogeneous, so all test specimens were extracted from the same subsurface region of the tuber. Tissue that had been in contact with the boiling water was excluded. Roasting produced a rubbery outer casing, so this was tested separately from deeper tissue. The degree of reduction in toughness from both cooking methods is discussed in the text.

35. The term 'texture' here is not restricted to surface feel, as is often the case, but includes deformational and fracture terms.

References

Abd-el-Malek, S. (1955) The part played by the tongue in mastication and deglutition. *Journal of Anatomy* **89**: 250–254.

Abler, W. L. (1992) The serrated teeth of tyrannosaurid dinosaurs, and biting structures in other animals. *Paleobiology* **18**: 161–183.

Aerts, R. J., Barry, T. N. & McNabb, W. C. (1999) Polyphenols and agriculture: beneficial effects of proanthocyanidins in forages. *Agriculture, Ecosystems and Environment* **75**: 1–12.

Agrawal, K. R. (1999) The effect of food texture on chewing patterns in human subjects. Ph.D. thesis. Hong Kong: University of Hong Kong.

Agrawal, K. R. & Lucas, P. W. (2003) The mechanics of the first bite. *Proceedings of the Royal Society London series B* **270**: 1277–1282.

Agrawal, K. R., Lucas, P. W., Prinz, J. F. *et al.* (1997) Mechanical properties of foods responsible for resisting food breakdown in the human mouth. *Archives of Oral Biology* **42**: 1–9.

Agrawal, K. R., Lucas, P. W., Bruce, I. C. *et al.* (1998) Food properties that influence neuromuscular activity during human mastication. *Journal of Dental Research* **77**: 1931–1938.

Agrawal, K. R., Lucas, P. W. & Bruce, I. C. (2000) The effect of food fragmentation index on mandibular closing angle in human mastication. *Archives of Oral Biology* **45**: 577–584.

Aidos, I., Lie, O. & Espe, M. (1999) Collagen content in farmed Atlantic salmon (*Salmo salar* L.). *Journal of Agricultural and Food Chemistry* **47**: 1440–1444.

Aiello, L. & Dean, C. (1990) *An Introduction to Human Evolutionary Anatomy*. London: Academic Press.

Aitchison, J. (1946) Hinged teeth in mammals: a study of the tusks of muntjacs (*Muntiacus*) and Chinese water deer (*Hydropotes inermis*). *Proceedings of the Zoological Society London* **116**: 329–338.

Akersten, W. A., Lowenstam, H. & Walker, A. (1984) 'Pigmentation' of soricine teeth: composition, ultrastructure, and function. Abstract, *American Society of Mammalogists, 64th Annual Meeting* no. 153, 40.

Akersten, W. A., Lowenstam, H., Walker, A. *et al.* (2002) How and why do some shrews have red teeth? Abstract, *Society of Vertebrate Paleontology, 62nd Annual Meeting* p. 31A.

305

Alexander, R. M. (1985) Body support, scaling, and allometry. In: *Functional Vertebrate Morphology* (eds. M. Hildebrand, D. M. Bramble, K. F. Liem & D. B. Wake), pp. 26–37. Cambridge, MA: The Belknap Press of Harvard University.

(1991) Optimization of gut structure and diet for higher vertebrate herbivores. *Philosophical Transactions of the Royal Society London series B* **333**: 249–255.

(1994) Optimum gut structure for specified diets. In: *The Digestive System in Mammals* (eds. D. J. Chivers & P. Langer), pp. 54–62. Cambridge: Cambridge University Press.

(1996) *Optima for Animals*, 2nd edn. Princeton, NJ: Princeton University Press.

(1998) News of chews: the optimization of mastication. *Nature* **391**: 329.

(2001) Design by numbers. *Nature* **412**: 591.

Alroy, J. (1998) Cope's rule and the dynamics of body mass evolution in North American fossil mammals. *Science* **280**: 731–734.

Anapol, F. & Herring, S. W. (2000) Ontogeny of histochemical fiber types and muscle function in the masseter muscle of miniature swine. *American Journal of Physical Anthropology* **112**: 595–613.

Anderson, D. J. (1976) The incidence of tooth contacts in normal mastication and the part they play in guiding the final stage of mandibular closure. In: *Mastication* (eds. D. J. Anderson & B. Matthews), pp. 237–241. Bristol: Wright.

Anderson, D. J., Hannam, A. G. & Matthews, B. (1970) Sensory mechanisms in mammalian teeth and their supporting structures. *Physiological Review* **50**: 171–195.

Anderson, D. J., Hector, M. P. & Linden, R. W. A. (1985) The possible relation between mastication and parotid secretion in the rabbit. *Journal of Physiology* **364**: 19–29.

Andreasen, J. O. (1972) *Traumatic Injuries of the Teeth*. Copenhagen: Munksgaard.

Aranwela, N., Sanson, G. & Read, J. (1999) Methods of assessing leaf-fracture properties. *New Phytologist* **144**: 369–393.

Ardran, G. M., Kemp, F. H. & Ride, W. D. L. (1958) A radiographic analysis of mastication and swallowing in the domestic rabbit *Oryctolagus cuniculus*. *Proceedings of the Zoological Society London* **130**: 257–274.

Ardrey, R. (1961) *African Genesis*. New York: Atheneum.

Ashby, M. F. (1989) On the engineering properties of materials. *Acta Metallurgica* **37**: 1273–1293.

(1998) Checks and estimates for material properties. I. Ranges and simple correlations. *Proceedings of the Royal Society London series A* **454**: 1301–1321.

(1999) *Materials Selection in Mechanical Design*, 2nd edn. Oxford: Butterworth Heinemann.

Ashby, M. F. & Jones, D. R. H. (1996) *Engineering Materials*, vol. 1, 2nd edn. Oxford: Butterworth Heinemann.

(1998) *Engineering Materials*, vol. 2, 2nd edn. Oxford: Butterworth Heinemann.

Ashby, M. F., Easterling, K. E., Harryson, R. *et al.* (1985) The fracture and toughness of woods. *Proceedings of the Royal Society London series A* **398**: 261–280.

Atkins, A. G. (1974) Imparting strength and toughness to brittle composites. *Nature* **252**: 116–118.

(1980) On cropping and related processes. *International Journal of Mechanical Science* **22**: 215–231.

(1982) Topics in indentation hardness. *Metal Science* **16**: 127–137.

(1994) Scale effects in engineering failures. *Engineering Failure Analysis* **1**: 201–214.

(1999) Scaling laws for elastoplastic fracture. *International Journal of Fracture* **95**: 51–65.

Atkins, A. G. & Felbeck, D. K. (1974) Applying mutual indentation hardness phenomena to service failures. *Metal Engineering Quarterly* **14**: 55–61.

Atkins, A. G. & Mai, Y.-W. (1979) On the guillotining of materials. *Journal of Materials Science* **14**: 2747–2754.

(1985) *Elastic and Plastic Fracture*. Chichester: Ellis Horwood.

Atkins, A. G. & Vincent, J. F. V. (1984) An instrumented microtome for improved histological sections and the measurement of fracture toughness. *Journal of Materials Science Letters* **3**: 310–312.

Baker, G., Jones, L. H. P. & Wardrop, I. D. (1959) Cause of wear in sheeps' teeth. *Nature* **184**: 1583–1584.

Ball, T. B., Brotherson, J. D. & Gardner, J. S. (1993) A typologic and morphometric study of variation in phytoliths from einkorn wheat (*Triticum monococcum*). *Canadian Journal of Botany* **71**: 1182–1192.

Balooch, M., Demos, S. G., Kinney, J. H. *et al.* (2001) Local mechanical and optical properties of normal and transparent root dentin. *Journal of Materials Science: Materials in Medicine* **12**: 507–514.

Baragar, F. A. & Osborn, J. W. (1984) A model relating patterns of human jaw movement to biomechanical constraints. *Journal of Biomechanics* **17**: 757–767.

Baragar, F. A., van der Glas, H. W. & van der Bilt, A. (1996) An analytic probability density for particle-size in human mastication. *Journal of Theoretical Biology* **181**: 169–178.

Barbenel, J. C. (1972) The biomechanics of the human temporomandibular joint: a theoretical study. *Journal of Biomechanics* **5**: 251–256.

Barnes, J. (1987) *Early Greek Philosophy*. London: Penguin.

Barrett, P. M. & Willis, K. J. (2001) Did dinosaurs invent flowers? Dinosaur–angiosperm coevolution revisited. *Biological Reviews* **76**: 411–447.

Barthlott, W. & Neinhuis, C. (1997) Purity of the sacred lotus or escape from contamination in biological surfaces. *Planta* **202**: 1–8.

Basmajian, J. V. & De Luca, C. J. (1985) *Muscles Alive*, 5th edn. Baltimore, MD: Williams & Wilkins.

Beard, K. C., Tong, Y., Dawson, M. *et al.* (1996) Earliest complete dentition of an anthropoid primate from the late Middle Eocene of Shanxi province, China. *Science* **272**: 82–85.

Beecher, R. M. (1977) Function and fusion at the mandibular symphysis. *American Journal of Physical Anthropology* **47**: 325–336.

(1979) Functional significance of the mandibular symphysis. *Journal of Morphology* **159**: 117–130.

Bell, E. A. (1984) Toxic compounds in seeds. In: *Seed Physiology*, vol. 1, *Development* (ed. D. R. Murray), pp. 245–264. Sydney: Academic Press.

Bell, R. H. V. (1971) A grazing ecosystem in the Serengeti. *Scientific American* **225**: 86–89.

Bemis, W. E. (1984) Morphology and growth of lepidoserenid lungfish tooth plates (Pisces: Dipnoi). *Journal of Morphology* **179**: 73–93.

Benedict, F. G. (1938) *Vital Energetics: A Study in Comparative Basal Metabolism.* Washington, DC: Carnegie Institute of Washington.

Bennett, E. L. (1983) The banded langur: ecology of a colobine in West Malaysia. Ph.D. thesis. Cambridge: University of Cambridge.

Berkovitz, B. K. B. (2000) Tooth replacement patterns in non-mammalian vertebrates. In: *Development, Function and Evolution of Teeth* (eds. M. F. Teaford, M. M. Smith & M. W. J. Ferguson), pp. 186–200. Cambridge: Cambridge University Press.

Berkovitz, B. K. B. & Poole, D. F. G. (1977) Attrition of the teeth in ferrets. *Journal of Zoology* **183**: 411–418.

Berkovitz, B. K. B., Moxham, B. J. & Newman, H. N. (1995) *The Periodontal Ligament in Health and Disease*, 2nd edn. London: Mosby-Wolfe.

Bertram, J. E. A. & Gosline, J. M. (1986) Fracture toughness design in horse hoof keratin. *Journal of Experimental Biology* **125**: 29–47.

Bialek, W. (1987) Physical limits to sensation and perception. *Annual Review of Biophysics and Biophysical Chemistry* **16**: 455–478.

Biewener, A. (1992) *Biomechanics.* Oxford: IRL Press.

Blackburn, J., Hodgskinson, R., Currey, J. D. *et al.* (1992) Mechanical properties of microcallus in human cancellous bone. *Journal of Orthopaedic Research* **10**: 237–246.

Blanton, P. L., Biggs, N. L. & Perkins, R. C. (1970) Electromyographic analysis of the buccinator muscle. *Journal of Dental Research* **49**: 389–394.

Bodmer, R. E. (1989) Frugivory in Amazonian Artiodactyla: evidence of the evolution of the ruminant stomach. *Journal of Zoology* **219**: 457–467.

Boesch, C. & Boesch, H. (1990) Tool use and tool making in wild chimpanzees. *Folia Primatologica* **54**: 86–99.

Bond, F. C. (1952) Third theory of comminution. *American Institute of Mining Engineers Transactions* **193**: 484–494.

(1962) The laws of rock breakage. In: *Zerkleinern Symposion* (ed. H. Rumpf), pp. 194–202. Düsseldorf: Verlag Chemie.

Boughter, J. D. & Gilbertson, T. A. (1999) From channels to behavior: an integrative model of NaCl taste. *Neuron* **22**: 213–215.

Bourne, M. C. (1976) Compression rates in the mouth. *Journal of Texture Studies* **8**: 373–376.

(2002) *Food Texture and Viscosity*, 2nd edn. New York: Academic Press.

Bouvier, M. (1986) A biomechanical analysis of mandibular scaling in Old World monkeys. *American Journal of Physical Anthropology* **69**: 473–482.

Bowden, F. P. & Tabor, D. (1950) *The Friction and Lubrication of Solids.* Oxford: Oxford University Press.

Boyde, A. (1964) The structure and development of enamel. Ph.D. thesis. London: University of London.

Boyde, A. & Fortelius, M. (1986) Development, structure and function of rhinoceros enamel. *Zoological Journal of the Linnean Society* **87**: 181–214.

Brace, C. L. (1963) Structural reduction in evolution. *American Naturalist* **97**: 39–49.

(1964) The probable mutation effect. *American Naturalist* **98**: 453–455.

Brace, C. L., Rosenberg, K. & Hunt, K. D. (1987) Gradual change in human tooth size in the late Pleistocene and post-Pleistocene. *Evolution* **41**: 705–720.

Braden, M. (1976) Biophysics of the tooth. In: *Frontiers of Oral Physiology*, vol. 2 (ed. Y. Kawamura), pp. 1–37. Basel: Karger.

Bramble, D. M. (1978) Origin of the mammalian feeding complex: models and mechanisms. *Paleobiology* **4**: 271–301.

Brehnan, K., Boyd, R. L., Gibbs, C. H. *et al.* (1981) Direct measurement of loads at the temporomandibular joint in *Macaca arctoides*. *Journal of Dental Research* **60**: 1820–1824.

Brennan, J. G. (1980) Food texture measurement. In: *Food Analysis Techniques* (ed. R. D. King), pp. 1–18. London: Applied Science.

Breslin, P. A. S., Beauchamp, G. K. & Pugh, E. N. Jr (1996) Monogeusia for fructose, glucose, sucrose, and maltose. *Perception and Psychophysics* **58**: 327–341.

Brochu, C. (1999) Phylogenetics, taxonomy, and historical biogeography of Alligatoroidea. *Journal of Vertebrate Paleontology* **19** (supplement to No. 2): 9–100.

Brunet, M., Guy, F., Pilbeam, D. *et al.* (2002) A new hominid from the Upper Miocene of Chad, Central Africa. *Nature* **418**: 145–151.

Buckland-Wright, J. C. (1975) The structure and function of cat skull bones in relation to the transmission of biting forces. Ph.D. thesis. London: University of London.

Butler, P. M. (1952) The milk molars of the Perissodactyla, with remarks on molar occlusion. *Proceedings of the Zoological Society London* **121**: 777–817.

(1978) Molar cusp nomenclature and homology. In: *Development, Function and Evolution of Teeth* (eds. P. M. Butler & K. A. Joysey), pp. 439–453. New York: Academic Press.

Cachel, S. (1984) Growth and allometry in primate masticatory muscles. *Archives of Oral Biology* **29**: 287–293.

Calcagno, J. M. & Gibson, K. R. (1991) Selective compromise: evolutionary trends and mechanisms in hominid tooth size. In: *Advances in Dental Anthropology* (eds. M. A. Kelley & C. S. Larsen), pp. 59–76. New York: Wiley.

Calder, W. A. (1984) *Size, Function, and Life History*. Cambridge, MA: Harvard University Press.

Carlson, D. S. (1977) Condylar translation and the function of the superficial masseter in the rhesus monkey (*M. mulatta*). *American Journal of Physical Anthropology* **47**: 53–64.

Carlsson, G. E. (1974) Bite force and masticatory efficiency. In: *Frontiers of Oral Physiology*, vol. 2 (ed. Y. Kawamura), pp. 265–292. Basel: Karger.

Carpita, N. C. & Gibeaut, D. M. (1993) Structural models of primary cell walls in flowering plants: consistency of molecular structure with the physical properties of the walls during growth. *Plant Journal* **3**: 1–30.

Chaimanee, Y., Jolly, D., Benammi, M. *et al.* (2003) A Middle Miocene hominoid from Thailand and orangutan origins. *Nature* **422**: 61–65.

Chandrashekar, J., Mueller, K. L., Hoon, M. A. (2000) T2Rs function as bitter taste receptors. *Cell* **100**: 703–711.

Chapman, L. J., Chapman, C. A. & Wrangham, R. W. (1992) *Balanites wilsoniana*: elephant dependent dispersal? *Journal of Tropical Ecology* **8**: 275–283.

Charalambides, M. N., Williams, J. G. & Chakrabarti, S. (1995) A study of the influence of ageing on the mechanical properties of Cheddar cheese. *Journal of Materials Science* **30**: 3959–3967.

Chin-Purcell, M. V. & Lewis, J. L. (1996) Fracture of articular cartilage. *Journal of Biomechanical Engineering* **118**: 545–556.

Choong, M. F. (1996) What makes a leaf tough and how this affects the pattern of *Castanopsis fissa* leaf consumption by caterpillars. *Functional Ecology* **10**: 668–674.

   (1997) Patterns of herbivory in tropical Fagaceae. Ph.D. thesis. Hong Kong: University of Hong Kong.

Choong, M. F., Lucas, P. W., Ong, J. Y. S. *et al.* (1992) Leaf fracture toughness and sclerophylly: their correlations and ecological implications. *New Phytologist* **121**: 597–610.

Christensen, M., Purslow, P. P. & Larsen, L. M. (2000) The effect of cooking temperature on mechanical properties of whole meat, single muscle fibres and perimysial connective tissue. *Meat Science* **55**: 301–307.

Ciochon, R. L., Piperno, D. R. & Thompson, R. G. (1990) Opal phytoliths found on the teeth of the extinct ape *Gigantopithecus blacki*: implications for paleodietary studies. *Proceedings of the National Academy of Science, USA* **87**: 8120–8124.

Cipollini, M. L. & Levey, D. J. (1997) Secondary metabolites of fleshy vertebrate-dispersed fruits: adaptive hypotheses and implications for seed dispersal. *American Naturalist* **150**: 346–372.

Cochard, L. R. (1987) Postcanine tooth size in female primates. *American Journal of Physical Anthropology* **74**: 47–54.

Coley, P. D. (1983) Herbivory and defensive characteristics of tree species in a lowland tropical rainforest. *Ecological Monographs* **53**: 209–233.

Coley, P. D. & Kursar, T. A. (1996) Anti-herbivore defenses of young tropical leaves: physiological constraints and ecological trade-offs. In: *Tropical Forest Plant Ecophysiology* (eds. S. S. Mulkey, R. L. Chazdon & A. P. Smith), pp. 305–336. New York: Chapman & Hall.

Collinson, M. E. & Hooker, J. J. (1991) Fossil evidence of interactions between plants and plant-eating mammals. *Philosophical Transactions of the Royal Society London series B* **333**: 197–208.

Cook, J. & Gordon, J. E. (1964) A mechanism for the control of crack propagation in all brittle systems. *Proceedings of the Royal Society London series A* **282**: 508–518.

Corlett, R. T. (1996) Characteristics of vertebrate-dispersed fruits in Hong Kong. *Journal of Tropical Ecology* **12**: 819–833.

Corlett, R. T. & Lucas, P. W. (1990) Alternative seed-handling strategies in primates: seed-spitting by long-tailed macaques. *Oecologia* **82**: 166–171.

Cornelissen, J. H. C. (1999) A triangular relationship between leaf size and seed size among woody species: allometry, ontogeny, ecology and taxonomy. *Oecologia* **118**: 248–255.

Corner, E. J. H. (1949) The annonaceous seed and its four integuments. *New Phytologist* **48**: 332–364.

  (1976) *The Seeds of Dicotyledons*. Cambridge: Cambridge University Press.

Costa, R. L. & Greaves, W. S. (1983) The pattern of wear responsible for the formation of enamel ridges on teeth with exposed dentin. *American Journal of Physical Anthropology* **60**: 185.

Cottrell, A. H. (1964) *The Mechanical Properties of Matter*. New York: Wiley.

Cox, H. L. (1952) The elasticity and strength of paper and other fibrous materials. *British Journal of Physics* **3**: 72–79.

Crane, P. R., Friis, E. M. & Pederson, K. R. (1995) The origin and early diversification of angiosperms. *Nature* **363**: 342–344.

Creighton, G. K. (1980) Static allometry of mammalian teeth and the correlation of tooth size and body size in contemporary mammals. *Journal of Zoology* **191**: 235–243.

Critchley, H. D. & Rolls, E. T. (1996) Responses of primate taste cortex neurons to the astringent tastant tannic acid. *Chemical Senses* **21**: 135–145.

Crompton, A. W. (1963) On the lower jaw of *Diarthrognathus* and the evolution of the mammalian lower jaw. *Proceedings of the Zoological Society London* **140**: 697–753.

  (1971) The origin of the tribosphenic molar. In: *Early Mammals* (eds. D. M. Kermack & K. A. Kermack), pp. 65–87. Supplement no. 1 to *Zoological Journal of the Linnean Society*, vol. 50.

Crompton, A. W. & Hiiemae, K. M. (1970) Molar occlusion and mandibular movements during occlusion in the American opossum (*Didelphis marsupialis*). *Zoological Journal of the Linnean Society* **49**: 21–47.

Crompton, A. W. & Kielan-Jaworowska, Z. A. (1978) Molar structure and occlusion in Cretaceous therian mammals. In: *Development, Function and Evolution of Teeth* (eds. P. M. Butler & K. A. Joysey), pp. 249–288. New York: Academic Press.

Crompton, A. W. & Sita-Lumsden, A. (1970) Functional significance of the therian molar pattern. *Nature* **222**: 678–679.

Curran, L. M. & Leighton, M. (2000) Vertebrate responses to spatio-temporal variation in seed production by mast-fruiting Bornean Dipterocarpaceae. *Ecological Monographs* **70**: 121–150.

Currey, J. D. (1967) The failure of exoskeletons and endoskeletons. *Journal of Morphology* **123**: 1–16.

(1980) Mechanical properties of mollusc shell. In: *The Mechanical Properties of Biological Materials* (eds. J. F. V. Vincent & J. D. Currey), pp. 75–97. Cambridge: Cambridge University Press.

Currey, J. D. & Brear, K. (1992) Fractal analysis of compact bone and antler fracture surfaces. *Biomimetics* **1**: 102–118.

Cuy, J. L., Mann, A. B., Livi, K. J. *et al.* (2002) Nanoindentation mapping of the mechanical properties of human molar enamel. *Archives of Oral Biology* **47**: 281–291.

D'Arcy Thompson, W. (1961) *On Growth and Form*. Cambridge: Cambridge University Press.

Dale, H. E., Shanklin, M. D., Johnson, H. D. *et al.* (1970) Energy metabolism of the chimpanzee. In: *The Chimpanzee*, vol. 2 (ed. G. H. Bourne), pp. 100–122. Basel: Karger.

Damuth, J. (1990) Problems in estimating body masses of archaic ungulates using dental measurements. In: *Body Size in Mammalian Paleobiology* (ed. J. Damuth & B. J. MacFadden), pp. 229–253. Cambridge: Cambridge University Press.

Danielson, D. R. & Reinhard, K. J. (1998) Human dental microwear caused by calcium oxalate phytoliths in prehistoric diet of the Lower Pecos region, Texas. *American Journal of Physical Anthropology* **107**: 297–304.

Darvell, B. W. (1990) On uniaxial compression tests and the validity of indirect tensile strength. *Journal of Materials Science* **25**: 757–780.

Darvell, B. W., Lee, P. K. D., Yuen, T. D. B. *et al.* (1996) A portable fracture toughness tester for biological materials. *Measurement Science and Technolology* **7**: 954–962.

Davies, A. G. & Baillie, I. C. (1988) Soil-eating by red-leaf monkeys (*Presbytis rubicunda*) in Sabah, northern Borneo. *Biotropica* **20**: 252–258.

Davies, A. G., Bennett, E. L. & Waterman, P. G. (1988) Food selection by two South-east Asian colobine monkeys (*Presbytis rubicunda* and *Presbytis melalophos*) in relation to plant chemistry. *Biological Journal of the Linnean Society* **34**: 33–56.

de Bruijne, D. W., Hendricks, H. A. C. M., Anderliesten, L. *et al.* (1993) Mouthfeel of foods. In: *Food Colloids and Polymers: Stability and Mechanical Properties* (eds. E. Dickinson & P. Walstra), pp. 204–213. Cambridge: Royal Society of Chemistry.

Dean, M. C. (2000) Incremental markings in enamel and dentine: what they can tell us about the way teeth grow. In: *Development, Function and Evolution of Teeth* (eds. M. F. Teaford, M. M. Smith & M. W. J. Ferguson), pp. 119–130. Cambridge: Cambridge University Press.

Dean, M. C., Jones, M. E. & Pilley, J. R. (1992) The natural history of tooth wear, continuous eruption and periodontal disease in wild shot great apes. *Journal of Human Evolution* **16**: 23–29.

Dechow, P. C. & Carlson, D. S. (1990) Occlusal force and craniofacial biomechanics during growth in rhesus monkeys. *American Journal of Physical Anthropology* **83**: 219–237.

Dechow, P. C. & Hylander, W. L. (2000) Elastic properties and masticatory bone stress in the macaque mandible. *American Journal of Physical Anthropology* **112**: 553–574.

DeGusta, D., Everett, M. A. & Milton, K. (2003) Natural selection on molar size in a wild population of howler monkeys (*Alouatta palliata*). *Proceedings of the Royal Society London series B* (Supplement: *Biological Letters*). 0b10005.S1.

Delgado, S., Casane, D., Bonnaud, L. *et al.* (2001) Molecular evidence for pre-Cambrian origin of amelogenin, the major protein of enamel. *Molecular Biology and Evolution* **18**: 2146–2153.

DeLong, R. & Douglas, W. H. (1983) Development of an artificial oral environment for the testings of dental restorations: bi-axial and movement control. *Journal of Dental Research* **62**: 32–36.

Delson, E. G. & Andrews, P. J. (1975) Evolution and interrelationships of the catarrhine primates. In: *Phylogeny of the Primates: A Multidisciplinary Approach* (eds. W. P. Luckett & F. S. Szalay), pp. 405–446. New York: Plenum.

Delson, E., Terranova, C. J., Jungers, W. J. *et al.* (2000) Body mass in Cercopithecidae (Primates, Mammalia): estimation and scaling in extinct and extant taxa. *American Museum of Natural History, Anthropological Papers* **83**: 1–159.

Demes, B. & Creel, N. (1988) Bite force, diet, and cranial morphology of fossil hominids. *Journal of Human Evolution* **17**: 657–670.

Demment, M. W. & Van Soest, P. J. (1985) A nutritional explanation for body-size patterns of ruminant and nonruminant herbivores. *American Naturalist* **125**: 641–672.

Deutsch, D., Palmon, A., Dafni, L. *et al.* (1995) The enamelin (tuftelin) gene. *International Journal of Developmental Biology* **39**: 135–143.

Diekwisch, T. G. H. (1998) Subunit compartments of secretory stage enamel matrix. *Connective Tissue Research* **38**: 101–111.

Diekwisch, T. G. H., Berman, B. S., Gentner, S. *et al.* (1995) Initial enamel crystals are spatially not associated with mineralised dentine. *Cell and Tissue* **279**: 149–167.

Dixon, A. D. (1963) Nerve plexuses in the oral mucosae. *Archives of Oral Biology* **8**: 435–447.

Dobrin, P. B. & Doyle, J. M. (1970) Vascular smooth muscle and the anisotropy of dog carotid artery. *Circulation Research* **27**: 105–119.

Dominy, N. J. (2001) Trichromacy and the ecology of food selection in four African primates. Ph.D. thesis. Hong Kong: University of Hong Kong.
 (2003) Color as an indicator of food quality to anthropoid primates: ecological evidence and an evolutionary scenario. In: *Anthropoid Origins: New Visions* (eds. C. Ross & R. F. Kay), pp. 599–628. New York: Kluwer.

Dominy, N. J., Lucas, P. W., Osorio, D. *et al.* (2001) The sensory ecology of primate food perception. *Evolutionary Anthropology* **10**: 171–186.

Dorrington, K. L. (1980) The theory of viscoelasticity in biomaterials. In: *The Mechanical Properties of Biomaterials* (eds. J. F. V. Vincent & J. D. Currey), pp. 289–314. Cambridge: Cambridge University Press.

Douglas, W. H., Sakaguchi, R. L. & DeLong, R. (1985) Frictional effects between natural teeth in an artificial mouth. *Dental Materials* 1: 115–119.

Drake, B. K. (1963) Food crushing sounds: an introductory study. *Journal of Food Science* 28: 233–241.

  (1965) Food crushing sounds: comparisons of objective and subjective data. *Journal of Food Science* 30: 556–559.

Druzinsky, R. E. (1993) The time allometry of mammalian chewing movements: chewing frequency scales with body mass in mammals. *Journal of Theoretical Biology* 160: 427–440.

Dunbar, R. I. M. (1977) Feeding ecology of gelada baboons: a preliminary report. In: *Primate Ecology* (ed. T. H. Clutton-Brock), pp. 251–273. London: Academic Press.

Dyment, M. L. & Synge, J. L. (1935) The elasticity of the periodontal membrane. *Oral Health* 25: 105–109.

Eaglen, R. H. (1984) Incisor size and diet revised: the view from a platyrrhine perspective. *American Journal of Physical Anthropology* 69: 262–275.

Edgar, W. M. & O'Mullane, D. M. (1996) *Saliva and Dental Health*, 2nd edn. London: British Dental Association.

Edwards, C., Read, J. & Sanson, G. (2000) Characterising sclerophylly: some mechanical properties of leaves from heath and forest. *Oecologia* 123: 158–167.

Ehrlich, P. & Raven, P. H. (1965) Butterflies and plants: a study in coevolution. *Evolution* 18: 586–608.

  (1967) Butterflies and plants. *Scientific American* 216: 104–113.

Emerson, S. B. & Radinsky, L. (1980) Functional analysis of sabertooth cranial morphology. *Paleobiology* 6: 295–312.

Emmons, L. H. (1991) Frugivory in treeshrews (*Tupaia*). *American Naturalist* 138: 642–649.

  (2000) *Tupai: A Field Study of Bornean Treeshrews*. Berkeley, CA: University of California Press.

Emmons, L. H., Nais, J. & Biun, A. (1991) The fruit and dispersers of *Rafflesia keithii* (Rafflesiaceae). *Biotropica* 23: 197–199.

Epstein, B. (1947) The mathematical description of certain breakage functions leading to the logarithmico-normal distribution. *Journal of the Franklin Institute* 244: 471–477.

Eriksson, O., Friis, E. M. & Löfgren, P. (2000) Seed size, fruit size, and dispersal systems in angiosperms from the early Cretaceous to the late Tertiary. *American Naturalist* 156: 47–58.

Evans, A. R. & Sanson, G. D. (1998) The effect of tooth shape on the breakdown of insects. *Journal of Zoology* 246: 391–400.

  (2003) The tooth of perfection: functional and spatial constraints on mammalian tooth shape. *Biological Journal of the Linnean Society* 78: 173–191.

Every, R. F. (1970) Sharpness of teeth in man and other primates. *Postilla* **143**: 1–20.

Farrell, J. (1956) The effect of mastication on the digestion of food. *British Dental Journal* **100**: 149–155.

Fengel, D. & Wegener, G. (1989) *Wood: Chemistry, Ultrastructure, Reactions.* Berlin: Walter de Gruyter.

Fincham, A. G., Luo, W., Morodian-Oldak, J. *et al.* (2000) Enamel biomineralization: the assembly and disassembly of the extracellular organic matrix. In: *Development, Function and Evolution of Teeth* (eds. M. F. Teaford, M. M. Smith & M. W. J. Ferguson), pp. 37–61. Cambridge: Cambridge University Press.

Finney, D. (1971) *Probit Analysis*, 3rd edn. Cambridge: Cambridge University Press.

Fish, D. R. & Mendel, F. C. (1982) Mandibular movement patterns relative to food types in common tree shrews (*Tupaia glis*). *American Journal of Physical Anthropology* **58**: 255–269.

Fleagle, J. G. (1999) *Primate Adaptation and Evolution*, 2nd edn. San Diego, CA: Academic Press.

Fooden, J. & Albrecht, G. H. (1993) Latitudinal and insular variation of skull size in crab-eating macaques (Primates, Cercopithecidae: *Macaca fascicularis*). *American Journal of Physical Anthropology* **92**: 521–538.

Ford, S. M. (1980) Callithricids as phyletic dwarfs, and the place of Callithricidae in Platyrrhini. *Primates* **21**: 31–43.

Fortelius, M. (1985) Ungulate cheek teeth: developmental, functional and evolutionary interrelationships. *Acta Zoologica Fennica* **180**: 1–76.

(1990) Problems with using fossil teeth to estimate body sizes of extinct mammals. In: *Body Size in Mammalian Paleobiology* (eds. J. Damuth & B. J. MacFadden), pp. 207–228. Cambridge: Cambridge University Press.

Fox, C. L., Juan, J. & Albert, R. M. (1996) Phytolith analysis on dental calculus, enamel surface, and burial soil: information about diet and paleoenvironment. *American Journal of Physical Anthropology* **101**: 101–113.

Fox, P. G. (1980) The toughness of tooth enamel, a natural fibrous composite. *Journal of Materials Science* **15**: 3113–3121.

Frank, F. C. & Lawn, B. R. (1967) On the theory of Hertzian fracture. *Proceedings of the Royal Society London series A* **299**: 291–316.

Freeman, P. W. (1979) Specialized insectivory: beetle-eating and moth-eating molossid bats. *Journal of Mammalogy* **60**: 467–479.

(1981a) A multivariate study of the family Molossidae (Mammalia, Chiroptera): morphology, ecology and evolution. *Fieldiana Zoology* (new series) **7**: 1–173.

(1981b) Correspondence of food habits and morphology in insectivorous bats. *Journal of Mammalogy* **62**: 166–173.

(1988) Frugivorous and animalivorous bats (Microchiroptera): dental and cranial adaptations. *Biological Journal of the Linnean Society* **33**: 249–272.

(1992) Canine teeth of bats (Microchiroptera): size, shape and role in crack propagation. *Biological Journal of the Linnean Society* **45**: 97–115.

(2000) Macroevolution in Microchiroptera: recoupling morphology and ecology with phylogeny. *Evolutionary Ecology Research* **2**: 317–335.

Freeman, P. W. & Weins, W. N. (1997) Puncturing ability of bat canine teeth: the tip. In: *Life among the Muses: Papers in Honor of James S. Findley* (eds. T. L. Yates, W. L. Gannon & D. E. Wilson), pp. 225–232. Albuquerque, NM: Museum of Southwestern Biology, University of New Mexico.

Frisch, J. E. (1963) Sexual dimorphism in canines of *Hylobates lar*. *Primates* **4**: 1–10.

Frolich, L. M., LaBarbera, M. & Stevens, W. P. (1994) Poisson's ratio of a crossed fiber sheath: the skin of aquatic salamanders. *Journal of Zoology* **232**: 231–252.

Gabbott, S. E., Aldridge, R. J. & Theron, J. N. (1995) A giant conodont with preserved muscle tissue from the Upper Ordovician of South Africa. *Nature* **374**: 800–803.

Galileo, G. (1638) *Dialogues concerning the Two Sciences*. Translated (1914) by H. Crew & A. de Salvico. Evanston, IL: Northwestern University Press.

Gantt, D. G. & Rafter, J. A. (1998) Evolutionary and functional significance of hominoid tooth enamel. *Connective Tissue Research* **39**: 195–206.

Gardner, R. P. & Austin, L. G. (1962) A chemical engineering treatment of batch grinding. In: *Zerkleinern Symposion* (ed. H. Rumpf), pp. 217–248. Düsseldorf: Verlag Chemie.

Garn, S. M. & Lewis, A. B. (1962) The relationship between third molar agenesis and a reduction in tooth number. *Angle Orthodontist* **33**: 14–18.

Gaudin, A. M. & Meloy, T. P. (1962) Model and a comminution distribution equation for single fracture. *American Institute of Mining Engineers Transactions* **233**: 41–43.

Gautier-Hion, A. (1980) Seasonal variations of diet related to species and sex in a community of *Cercopithecus* monkeys. *Journal of Animal Ecology* **49**: 237–269.

Gautier-Hion, A., Duplantier, J. M., Quris, R. *et al.* (1985) Fruit characters as a basis of fruit choice and seed dispersal in a tropical forest vertebrate community. *Oecologia* **65**: 324–337.

Gibbs, C. H., Mahan, P. E., Lundeen, H. C. *et al.* (1981) Occlusal forces during chewing and swallowing as measured by sound transmission. *Journal of Prosthetic Dentistry* **46**: 443–449.

Gibbs, S., Collard, M. & Wood, B. (2000) Soft tissue characters in higher primate phylogenetics. *Proceedings of the National Academy of Sciences, USA* **97**: 11130–11132.

Gibson, L. J. (1985) The mechanical behaviour of cancellous bone. *Journal of Biomechanics* **18**: 317–328.

Gibson, L. J. & Ashby, M. F. (1999) *Cellular Solids Structure and Properties*, 2nd edn. Cambridge: Cambridge University Press.

Gibson, L. J., Easterling, K. E. & Ashby, M. F. (1981) The structure and mechanics of cork. *Proceedings of the Royal Society London series A* **377**: 99–117.

Gibson, L. J., Ashby, M. F. & Easterling, K. E. (1988) The structure and mechanics of the iris leaf. *Journal of Materials Science* **23**: 3041–3048.

Gilbertson, T. A. (1998) Gustatory mechanisms for the detection of fat. *Current Opinion in Neurobiology* **8**: 447–452.

Gilbertson, T. A., Liu, L., York, D. A. *et al.* (1998) Dietary fat preferences are inversely correlated with peripheral gustatory fatty acid sensitivity. *Annals of the New York Academy of Sciences* **855**: 165–168.

Gipps, J. M. & Sanson, G. D. (1984) Mastication and digestion in *Pseudocheirus*. In: *Possums and Gliders* (eds. A. P. Smith & I. D. Hume), pp. 237–246. Sydney: Australian Mammal Society.

Glantz, P.-O. (1970) The surface tension of saliva. *Odontologisk Revy* **21**: 119–127.

Gordon, I. J. & Illius, A. W. (1994) The nutritional ecology of African ruminants: a reinterpretation. *Journal of Animal Ecology* **65**: 18–28.

Gordon, J. E. (1978) *Structures*. London: Penguin.

(1980) Biomechanics: the last stronghold of vitalism. In: *The Mechanical Properties of Biomaterials* (eds. J. F. V. Vincent & J. D. Currey), pp. 1–11. Cambridge: Cambridge University Press.

(1991) *The New Science of Strong Materials*, 2nd edn. London: Penguin.

Gordon, J. E. & Jeronimidis, G. (1980) Composites with high work of fracture. *Philosophical Transactions of the Royal Society London series A* **294**: 545–550.

Gordon, K. D. (1984) A study of microwear on chimpanzee molars: implications for dental microwear analysis. *Journal of Dental Research* **63**: 195–215.

Gosline, J. M., Lillie, M., Carrington, E. *et al.* (2002) Elastic proteins: biological roles and mechanical properties. *Philosophical Transactions of the Royal Society London Series B* **357**: 121–132.

Gould, S. J. (1966) Allometry and size in ontogeny and phylogeny. *Biological Reviews* **41**: 587–640.

(1975) On the scaling of tooth size in mammals. *American Zoologist* **15**: 351–362.

Grajal, A., Strahl, S. D., Parra, R. *et al.* (1989) Foregut fermentation in the hoatzin, a neotropical leaf-eating bird. *Science* **245**: 1236–1238.

Greaves, W. S. (1973) The inference of jaw motion from tooth wear facets. *Journal of Paleontology* **47**: 1000–1001.

(1978) The jaw-lever system in ungulates: a new model. *Journal of Zoology* **184**: 271–285.

Green, B. G. (1993) Oral astringency: a tactile component of flavor. *Acta Psychologica* **84**: 119–125.

Greenberg, A. R., Mehling, A., Lee, M. *et al.* (1989) Tensile behaviour of grass. *Journal of Materials Science* **24**: 2549–2554.

Griffith, A. A. (1920) Phenomena of rupture and flow in solids. *Philosophical Transactions of the Royal Society London series A* **221**: 163–198.

Grine, F. E. (1981) Trophic differences between 'gracile' and 'robust' australopithecines: a scanning electron microscope analysis of occlusal events. *South African Journal of Science* **77**: 203–230.

Groves, C. P. & Napier, J. R. (1968) Dental dimensions and diet in australopithecines. In: *Proceedings of the 8th Congress of Anthropological and Ethnological Sciences*, vol. 3, pp. 273–276. Tokyo: Science Council of Japan.

Gurney, C. & Hunt, J. (1967) Quasi-static crack propagation. *Proceedings of the Royal Society London series A* **299**: 508–524.

Habelitz, S., Marshall, S. J., Marshall, G. W. Jr *et al.* (2001) The functional width of the dentino-enamel junction determined by AFM-based nanoscratching. *Journal of Structural Biology* **135**: 294–301.

Habelitz, S., Marshall, G. W. Jr, Balooch, M. *et al.* (2002) Nanoindentation and storage of teeth. *Journal of Biomechanics* **35**: 995–998.

Haines, D. J., Berry, D. C. & Poole, D. F. G. (1963) Behavior of tooth enamel under load. *Journal of Dental Research* **42**: 885–888.

Hamilton, G. M. & Goodman, L. E. (1966) The stress field created by a circular sliding contact. *Journal of Applied Mechanics* **33**: 371–376.

Hankins, G. A. (1925) Report on the effects of adhesion between the indenting tool and the material in ball and cone indentation hardness tests. *Proceedings of the Institution of Mechanical Engineers* **1**: 611–645.

Harris, B. (1980) The mechanical behaviour of composite materials. In: *The Mechanical Properties of Biological Materials* (eds. J. F. V. Vincent & J. D. Currey), pp. 37–74. Cambridge: Cambridge University Press.

Harrison, M. J. S. (1986) Feeding ecology of black colobus (*Colobus satanus*) in central Gabon. In: *Primate Ecology and Conservation* (eds. J. G. Else & P. C. Lee), pp. 31–37, Cambridge: Cambridge University Press.

Harvey, P. H., Kavanagh, M. & Clutton-Brock, T. H. (1978) Sexual dimorphism of primate teeth. *Journal of Zoology* **186**: 475–485.

Hatley, T. & Kappelman, J. (1980) Bears, pigs and Plio-Pleistocene hominids: a case for the exploitation of belowground resources. *Human Ecology* **8**: 371–390.

Hayes, V. J., Freedman, L. & Oxnard, C. E. (1995) The differential expression of dental sexual dimorphism in subspecies of *Colobus guereza*. *International Journal of Primatology* **16**: 971–996.

Heath, M. R. (1982) The effect of maximum biting force and bone loss upon masticatory function and dietary selection of the elderly. *International Dental Journal* **32**: 345–356.

(1986) Functional interpretation of patterns of occlusal wear on acrylic teeth. *Restorative Dentistry* **2**: 100–107.

Hector, M. P. & Linden, R. W. A. (1999) Reflexes of salivary secretion. In: *Neural Mechanism of Salivary Gland Secretion* (eds. J. R. Garrett, J. Ekström & L. C. Anderson), pp. 196–217. Basel: Karger.

Helkimo, E., Carlson, G. E. & Helkimo, M. (1978) Chewing efficiency and state of dentition: a methodological study. *Acta Odontologica Scandinavica* **36**: 33–41.

Hellekant, G., Ninomiya, Y., DuBois, G. E. *et al.* (1996) Taste in chimpanzee. I. The summated response to sweeteners and the effect of gymnemic acid. *Physiology and Behavior* **60**: 469–479.

Hendrichs, H. (1965) Vergleichende Untersuchung des Wiederkauverhalten. *Biologisches Zentralblatt* **84**: 651–721.

Herrera, C. M. (2002) Seed dispersal by vertebrates. In: *Plant–Animal Interactions: An Evolutionary Approach* (eds. C. M. Herrera & O. Pellmyr), pp. 185–208. Oxford: Blackwell.

Herring, S. W. (1972) The role of canine morphology in the evolutionary divergence of pigs and peccaries. *Journal of Mammalogy* **53**: 500–512.

(1975) Adaptations for gape in the hippopotamus and its relatives. *Forma e Functio* **8**: 85–100.

Herring, S. W. & Herring, S. E. (1974) The superficial masseter and gape in mammals. *American Naturalist* **108**: 561–575.

Heyes, J. A. & Sealey, D. F. (1996) Textural changes during nectarine (*Prunus persica*) development and ripening. *Scientia Horticulturae* **65**: 49–58.

Hiiemae, K. M. (1978) Mammalian mastication: a review of the activity of the jaw muscles and the movements they produce in chewing. In: *Development, Function and Evolution of Teeth* (eds. P. M. Butler & K. A. Joysey), pp. 359–398. New York: Academic Press.

Hiiemae, K. M. & Ardran, G. M. (1968) A cineradiographic study of feeding in *Rattus norvegicus*. *Journal of Zoology* **154**: 139–154.

Hiiemae, K. M. & Crompton, A. W. (1985) Mastication, food transport, and swallowing. In: *Functional Vertebrate Morphology* (eds. M. Hildebrand, D. M. Bramble, K. F. Liem & D. B. Wake), pp. 262–290. Cambridge, MA: Belknap Press of Harvard University.

Hiiemae, K. M. & Palmer, J. B. (1999) Food transport and bolus formation during complete sequences on foods of different initial consistency. *Dysphagia* **14**: 31–42.

Hill, A. V. (1950) The dimensions of animals and their muscular dynamics. *Scientific Progress* **38**: 209–230.

Hill, D. A. & Lucas, P. W. (1996) Toughness and fiber content of major leaf foods of wild Japanese macaques (*Macaca fuscata yakui*) in Yakushima. *American Journal of Primatology* **38**: 221–231.

Hill, D. A., Lucas, P. W. & Cheng, P. Y. (1995) Bite forces used by Japanese macaques (*Macaca fuscata yakui*) on Yakushima Island, Japan to open aphid-induced galls on *Distylium racemosum* (Hamamelidaceae). *Journal of Zoology* **237**: 57–63.

Hiller, S., Bruce, D. M. & Jeronimidis, G. (1996) A micropenetration technique for mechanical testing of plant cell walls. *Journal of Texture Studies* **27**: 559–587.

Hillerton, J. E., Reynolds, S. E. & Vincent, J. F. V. (1982) On the indentation hardness of insect cuticle. *Journal of Experimental Biology* **96**: 45–52.

Hills, M., Graham, S. H. & Wood, B. A. (1983) The allometry of relative cusp size in hominoid mandibular molars. *American Journal of Physical Anthropology* **62**: 311–316.

Hoaglund, R. G., Rosenfeld, A. R. & Hahn, G. T. (1972) Mechanisms of fast fracture and arrest in steels. *Metallurgical Transactions* **3**: 123–136.

Holland, G. R. (1994). Morphological features of dentine and pulp related to dentine sensitivity. *Archives of Oral Biology* **39** (supplement): 3S–11S.

Holloway, R. L. (1967) Tools and teeth: some speculations regarding canine reduction. *American Anthropologist* **69**: 63–67.

Hopson, J. A. (2001) Origin of mammals. In *Palaeobiology*, vol. 2 (eds. D. E. G. Briggs & P. R. Crowther), pp. 88–94. Oxford: Blackwell.

Howe, H. F. (1989) Scatter- and clump-dispersal and seedling demography: hypothesis and implications. *Oecologia* **79**: 417–426.

Hrycyshyn, A. W. & Basmajian, J. V. (1972) Electromyography of the oral stage of swallowing in man. *American Journal of Anatomy* **133**: 333–340.

Hume, I. D. (1994) Gut morphology, body size and digestive performance in rodents. In: *The Digestive System in Mammals* (eds. D. J. Chivers & P. Langer), pp. 315–323. Cambridge: Cambridge University Press.

Hunter, J. P. & Jernvall, J. (1995) The hypocone as a key innovation in mammalian evolution. *Proceedings of the National Academy of Sciences, USA* **92**: 10718–10722.

Hutchings, J. B. & Lillford, P. J. (1988) The perception of food texture: the philosophy of the breakdown path. *Journal of Texture Studies* **19**: 103–115.

Hylander, W. L. (1975) Incisor size and diet in anthropoids with special reference to Cercopithecidae. *Science* **189**: 1095–1097.

(1977) Morphological changes in human teeth and jaws in a high-attrition environment. In: *Orofacial Growth and Development* (eds. A. A. Dahlberg & T. M. Graber), pp. 301–330. The Hague: Mouton.

(1978) Incisal bite force direction in humans and the functional significance of mammalian mandibular translation. *American Journal of Physical Anthropology* **48**: 1–8.

(1979a) Mandibular function in *Galago crassicaudatus* and *Macaca fascicularis*: an *in vivo* approach to stress analysis of the mandible. *Journal of Morphology* **159**: 253–296.

(1979b) An experimental analysis of temporomandibular joint reaction force in macaques. *American Journal of Physical Anthropology* **51**: 433–456.

(1985) Mandibular function and biomechanical stress and scaling. *American Zoologist* **25**: 315–330.

(1992) Functional anatomy. In: *The Temporomandibular Joint: A Biological Basis for Clinical Practice*. (eds. B. G. Sarnat & W. B. Laskin), pp. 60–92. Philadelphia, PA: W. B. Saunders.

Hylander, W. L., Ravosa, M. J., Ross, C. F. *et al.* (2000) Symphyseal fusion and jaw-adductor muscle force: an EMG study. *American Journal of Physical Anthropology* **112**: 469–492.

Illius, A. W. & Gordon, I. J. (1987) The allometry of food intake in grazing ruminants. *Journal of Animal Ecology* **56**: 989–999.

Inglis, C. E. (1913) Stresses in a plate due to the presence of cracks and sharp corners. *Transactions of the Institution of Naval Architects* **55**: 219–230.

Iwamoto, T. (1979) Feeding ecology. In: *Ecological and Social Studies of Gelada Baboons* (ed. M. Kawai), pp. 279–335. Basel: Karger.

Jablonski, N. G. (1993) *Theropithecus: The Life and Death of a Primate Genus*. Cambridge: Cambridge University Press.

Jablonski, N. G. & Crompton, R. H. (1994) Feeding behavior, mastication and tooth wear in the Western tarsier (Tarsius bancanus). *International Journal of Primatology* **15**: 29–59.

Jackson, A. P., Vincent, J. F. V. & Turner, R. M. (1988) The mechanical design of nacre. *Proceedings of the Royal Society London series B* **234**: 415–440.

Janis, C. M. (1995) Correlations between craniodental morphology and feeding behavior in ungulates: reciprocal illumination between living and fossil taxa. In: *Functional Morphology in Vertebrate Paleontology* (ed. J. J. Thomason), pp. 76–98. Cambridge: Cambridge University Press.

Janis, C. M. & Ehrhardt, D. (1988) Correlation of relative muzzle width and relative incisor width with dietary preference in ungulates. *Zoological Journal of the Linnean Society* **92**: 267–284.

Janis, C. M. & Fortelius, M. (1988) On the means whereby mammals achieve increased functional durability of their dentitions, with special reference to limiting factors. *Biological Reviews* **63**: 197–230.

Janson, C. H. (1983) Adaptation of fruit morphology to dispersal agents in a neotropical forest. *Science* **219**: 187–189.

Janvier, P. (1996). Fishy fragments tip the scales. *Nature* **383**: 757–758.

Janzen, D. H. (1978) The ecology and evolutionary biology of seed chemistry as relates to seed predation. In: *Biochemical Aspects of Plant and Animal Evolution* (ed. J. B. Harborne), pp. 163–206. London: Academic Press.

Janzen, D. H. & Martin, P. S. (1982) Neotropical anachronisms: the fruits the gomphotheres ate. *Science* **215**: 19–27.

Jarman, P. J. (1974) The social organization of antelope in relation to their ecology. *Behaviour* **48**: 215–266.

Jeannorod, M. (1997) *The Cognitive Neuroscience of Action*. Cambridge, MA: Blackwell.

Jenkins, G. N. (1978) *The Physiology and Biochemistry of the Mouth*, 4th edn. Oxford: Blackwell.

Jennings, J. S. & Macmillan, N. H. (1986) A tough nut to crack. *Journal of Materials Science* **21**: 1517–1524.

Jensen, J. L., Lamkin, M. S. & Oppenheim, F. G. (1992) Adsorption of human salivary proteins to hydroxyapatite: a comparison between whole saliva and glandular salivary secretions. *Journal of Dental Research* **71**: 1569–1576.

Jernvall, J. & Selänne, L. (1999) Laser confocal microscopy and geographical information systems in the study of dental morphology. *Palaeontologia Electronica* **2**(1): 18 pp. http://www-odp.tamu.edu/paleo/1999_1/confocal/issue1_99.htm.

Jernvall, J. & Thesleff, I. (2000) Reiterative signaling and patterning during mammalian tooth morphogenesis. *Mechanisms of Development* **92**: 19–29.

Jernvall, J., Hunter, J. P. & Fortelius, M. (2000) Trends in the evolution of molar crown types in ungulate mammals: evidence from the northern hemisphere. In: *Development, Function and Evolution of Teeth* (eds. M. F. Teaford, M. M. Smith & M. Ferguson), pp. 269–281. Cambridge: Cambridge University Press.

Jeronimidis, G. (1980) The fracture behaviour of wood and the relations between toughness and morphology. *Proceedings of the Royal Society London series B* **208**: 447–460.

Jeuniaux, C. (1961) Chitinase: an addition to the list of hydrolases in the digestive tract of vertebrates. *Nature* **192**: 135–136.

Ji, Q., Luo, Z.-X., Wible, J. R. *et al.* (2002) The earliest known eutherian mammal. *Nature* **416**: 816–822.

Jolly, C. J. (1970) The seed-eaters: a new model of hominid differentiation. *Man* (new series) **5**: 1–26.

Jones, E. G., Schwark, H. D. & Callahan, P. A. (1986) Extent of the ipsilateral representation in the ventral posterior medial nucleus of the monkey thalamus. *Experimental Brain Research* **63**: 310–320.

Jones, S. J. (1981). Cement. In: *A Companion to Dental Studies* (eds. A. H. R. Rowe & R. B. Johns), vol. 2, *Dental Anatomy and Embryology* (ed. J. W. Osborn), pp. 193–205. Oxford: Blackwell.

Jones, S. J. & Boyde, A. (1974) Coronal cementogenesis in the horse. *Archives of Oral Biology* **19**: 605–614.

Jordano, P. (1995) Angiosperm fleshy fruits and seed dispersers: a comparative analysis of adaptation and constraints in plant-animal interaction. *American Naturalist* **145**: 163–191.

Jungers, W. E. (1978) On canine reduction in early hominids. *Current Anthropology* **19**: 155–156.

Junqueira, L. C., Toledo, A. M. & Doine, A. L. (1973) Digestive enzymes in the parotid and submandibular glands of mammals. *Anais da Academia Brasileira de Ciencias* **45**: 629–643.

Kasapi, M. D. & Gosline, J. M. (1997) Complexity and fracture control in the equine hoof wall. *Journal of Experimental Biology* **200**: 1639–1659.

Kastelic, J. & Baer, E. (1980) Deformation in tendon collagen. In: *The Mechanical Properties of Biological Materials* (eds. J. F. V. Vincent & J. D. Currey), pp. 397–435. Cambridge: Cambridge University Press.

Kawamura, Y. & Yamamoto, T. (1978) Studies on neural mechanisms of the gustatory–salivary reflex in rabbits. *Journal of Physiology* **285**: 35–47.

Kawasaki, K. & Weiss, K. M. (2003) Mineralized tissue and vertebrate evolution: the secretory calcium-binding phosphoprotein gene cluster. *Proceedings of the National Academy of Sciences, USA* **100**: 4060–4065.

Kay, R. F. (1975a) Allometry and early hominids. *Science* **189**: 63.

(1975b) The functional adaptations of primate molar teeth. *American Journal of Physical Anthropology* **42**: 195–215.

(1978) Molar structure and diet in extant Cercopithecidae. In: *Development, Function and Evolution of Teeth* (eds. P. M. Butler & K. A. Joysey), pp. 309–340. New York: Academic Press.

(1981) The nut crackers: a new theory of the adaptations of the Ramapthecinae. *American Journal of Physical Anthropology* **55**: 141–151.

Kay, R. F. & Covert, H. H. (1984) Anatomy and behavior of extinct primates. In: *Food Acquisition and Processing in Primates* (eds. D. J. Chivers, B. A. Wood & A. Bilsborough), pp. 467–508. New York: Plenum.

Kay, R. F. & Hylander, W. L. (1978) The dental structure of mammalian folivores with special reference to Primates and Phalangeroidea (Marsupialia). In: *The Biology of Arboreal Folivores* (ed. G. G. Montgomery), pp. 173–192. Washington, DC: Smithsonian Institution Press.

Kay, R. N. B. (1987) Weights of salivary glands in ruminant animals. *Journal of Zoology* **211**: 431–436.

Kay, R. N. B., Engelhardt, W. V. & White, R. G. (1980) Digestive physiology of wild ruminants. In: *Digestive Physiology and Metabolism in Ruminants: Proceedings of the 5th International Symposium on Ruminant Physiology* (eds. Y. Ruckebusch & P. Thivend), pp. 743–761. Lancaster, PA: MTP Press.

Kayser, A. (1980) Shortened dental arches and oral function. *Journal of Oral Rehabilitation* **8**: 457–462.

Kemp, A. (1995) *The Hornbills*. Oxford: Oxford University Press.

Kendall, K. (1978a) Complexities of compression failure. *Proceedings of the Royal Society London series A* **361**: 245–263.

(1978b) The impossibility of comminuting small particles by compression. *Nature* **272**: 710–711.

(2001) *Molecular Adhesion*. New York: Kluwer.

Kermack, K. A. & Haldane, J. B. S. (1950) Organic correlation and allometry. *Biometrika* **37**: 30–41.

Keyser, A. W. (2000) The Drimolen skull: the most complete australopithecine cranium and mandible to date. *South African Journal of Science* **96**: 189–193.

Kier, W. M. & Smith, K. K. (1985) Tongues, tentacles and trunks: the biomechanics of movement in muscular-hydrostats. *Zoological Journal of the Linnaean Society* **83**: 307–324.

Kiltie, R. A. (1982) Bite force as a basis for niche differentiation between rain forest peccaries. *Biotropica* **14**: 188–195.

Kindahl, M. (1959) Some aspects of the tooth development in the Soricidae. *Acta Odontologica Scandinavica* **17**: 203–237.

Kinney, J. H., Balooch, M., Marshall, S. J. *et al.* (1996) Hardness and Young's modulus of human peritubular and intertubular dentine. *Archives of Oral Biology* **41**: 9–13.

Kinzey, W. G. (1971) Evolution of the human canine tooth. *American Anthropologist* **73**: 680–694.

(1972) Canine teeth and the monkey, *Callicebus moloch*: lack of sexual dimorphism. *American Journal of Physical Anthropology* **35**: 91–100.

Kinzey, W. G. & Norconk, M. A. (1990) Hardness as a basis of fruit choice in two sympatric primates. *American Journal of Physical Anthropology* **81**: 5–15.

(1993) Physical and chemical properties of fruit and seeds eaten by *Pithecia* and *Chiropotes* in Surinam and Venezuela. *International Journal of Primatology* **14**: 207–227.

Kirk, E. C. & Simons, E. L. (2001) Diets of fossil primates from the Fayum Depression of Egypt: a quantitative analysis of molar shearing. *Journal of Human Evolution* **40**: 203–229.

Kirkham, J. & Robinson, C. (1995) The biochemistry of the fibres of the periodontal ligament. In: *The Periodontal Ligament in Health and Disease*, 2nd edn. (eds. B. K. B. Berkovitz, B. J. Moxham & H. N. Newman), pp. 55–81. London: Mosby-Wolfe.

Kleiber, M. (1932) Body size and metabolism. *Hilgardia* **6**: 315–353.

(1961) *The Fire of Life: An Introduction to Animal Energetics*. New York: Wiley.

Klineberg, I. (1980) Influences of temporomandibular articular receptors on functional jaw movements. *Journal of Oral Rehabilitation* **7**: 307–317.

Klineberg, I. & Wyke, B. O. (1983) Articular reflex control of mastication. In: *Oral Surgery*, vol. 4 (ed. L. W. Kay), pp. 253–258. Copenhagen: Munksgaard.

Korhonen, R. K., Laasanen, M. S., Toyras, J. (2002) Comparison of the equilibrium response of articular cartilage in unconfined compression, confined compression and indentation. *Journal of Biomechanics* **35**: 903–909.

Krishnamani, R. & Mahaney, W. C. (2000) Geophagy among primates: adaptive significance and ecological consequences. *Animal Behaviour* **59**: 899–915.

LaBarbera, M. (1989) Analyzing body size as a factor in ecology and evolution. *Annual Reviews of Ecology and Systematics* **20**: 97–117.

Lacey, R. W. (1994) *Hard to Swallow*. Cambridge: Cambridge University Press.

Laine, P. & Siirilä, H. S. (1971) Oral and manual stereognosis and two-point tactile discrimination of the tongue. *Acta Odontologica Scandinavica* **29**: 197–204.

Lake, G. J. & Yeoh, O. H. (1978) Measurement of rubber cutting resistance in the absence of friction. *International Journal of Fracture* **14**: 509–526.

Lakes, R. S. (1987) Foam structures with a negative Poisson's ratio. *Science* **235**: 1038–1040.

Lambert, J. E. (1999) Seed handling in chimpanzees (*Pan troglodytes*) and redtail monkeys (*Cercopithecus ascanius*): implications for understanding hominoid and cercopithecine fruit-processing strategies and seed dispersal. *American Journal of Physical Anthropology* **109**: 365–386.

Lambrecht, J. R. (1965) The influence of occlusal contact area on chewing performance. *Journal of Prosthetic Dentistry* **15**: 444–450.

Landry, S. O. Jr (1970) The Rodentia as omnivores. *Quarterly Review of Biology* **45**: 351–372.

Lanyon, J. M. & Sanson, G. D. (1986) Koala (*Phascolarctos cinereus*) dentition and nutrition. II. Implications of tooth wear in nutrition. *Journal of Zoology* **209**: 169–181.

Lanyon, L. E. & Rubin, C. T. (1985) Functional adaptation in skeletal structures. In: *Functional Vertebrate Morphology* (eds. M. Hildebrand, D. M. Bramble, K. F. Liem & D. B. Wake), pp. 1–25. Cambridge, MA: Belknap Press of Harvard University.

Laska, M., Kohlmann, S., Hernandez-Salazar, L. T. *et al.* (2001) Gustatory responses to polycose in four species of non-human primates. *Folia Primatologica* **72**: 171–172.

Lauer, C. (1975) A comparison of sexual dimorphism and range of variation in *Papio cynocephalus* and *Gorilla gorilla* dentition. *Primates* **16**: 1–7.

Lavigne, G. J., Kato, T., Kolta, A. *et al.* (2003) Neurobiological mechanisms involved in sleep bruxism. *Critical Reviews in Oral Biology and Medicine* **14**: 30–46.

Lawn, B. R. (1993) *Fracture of Brittle Solids*, 2nd edn. Cambridge: Cambridge University Press.

Lee, C.-S., Lawn, B. R. & Kim, D. K. (2001) Effect of tangential loading on critical conditions for radial cracking in brittle coatings. *Journal of the American Ceramic Society* **84**: 2719–2721.

Lee, V. M. & Linden, R. W. A. (1992) An olfactory-submandibular salivary reflex in humans. *Experimental Physiology* **77**: 221–224.

Lees, C., Vincent, J. F. V. & Hillerton, J. E. (1991) Poisson's ratio in skin. *Bio-Medical Materials and Engineering* **1**: 19–23.

Leigh, E. G. Jr (1999) *Tropical Forest Ecology: A View from Barro Colorado Island.* New York: Oxford University Press.

Leighton, M. (1993) Modeling dietary selectivity by Bornean orangutans: evidence for integration of multiple criteria in fruit selection. *International Journal of Primatology* **14**: 257–313.

Leighton, M. & Leighton, D. R. (1983) Vertebrate responses to fruiting seasonality within a Bornean rain forest. In: *Tropical Rain Forest: Ecology and Management* (eds. S. L. Sutton, T. C. Whitmore & A. C. Chadwick), pp. 181–196. Oxford: Blackwell.

Lermer, C. M. & Mattes, R. D. (1999) Perception of dietary fat: ingestive and metabolic implications. *Progress in Lipid Research* **38**: 117–128.

Levine, A. S. & Silva, S. E. (1980) Absorption of whole peanuts, peanut oil and peanut butter. *New England Journal of Medicine* **303**: 917–918.

Lieberman, D. E. (1993) Life history variables preserved in dental cementum microstructure. *Science* **261**: 1162–1164.

(1996) How and why recent humans grow thin skulls: experimental data on systemic cortical robusticity. *American Journal of Physical Anthropology* **101**: 217–236.

Lieberman, D. E. & Crompton, A. W. (2000) Why fuse the mandibular symphysis? A comparative analysis. *American Journal of Physical Anthropology* **112**: 517–540.

Lightoller, G. H. S. (1925) Facial muscles. The modiolus and muscles surrounding the rima oris with some remarks about the panniculus adiposus. *Journal of Anatomy* **60**: 1–85.

Lillford, P. J. (1991) Texture and acceptability of human foods. In: *Feeding and the Texture of Food* (eds. J. F. V. Vincent & P. J. Lillford), pp. 93–121. Cambridge: Cambridge University Press.

(2000) The materials science of eating and food breakdown. *MRS (Materials Research Society) Bulletin* (December): 38–43.

Linden, R. W. A. (1990) Periodontal mechanoreceptors and their functions. In: *Neurophysiology of the Jaws and Teeth* (ed. A. Taylor), pp. 52–95. London: Macmillan.

Logan, M. & Sanson, G. D. (2002) The association of tooth wear with sociality of free-ranging male koalas (*Phascolarctos cinereus* Goldfuss). *Australian Journal of Zoology* **50**: 621–626.

Lovegrove, B. G. (2000) The zoogeography of mammalian basal metabolic rate. *American Naturalist* **156**: 201–219.

Lowrison, G. C. (1974) *Crushing and Grinding*. London: Butterworth.

Lucas, P. W. (1980) Adaptation and form of the mammalian dentition with special reference to primates and the evolution of Man. Ph.D. thesis. London: University of London.

(1981) An analysis of canine size and jaw shape in some Old and New World non-human primates. *Journal of Zoology* **195**: 437–448.

(1982a) Basic principles of tooth design. In: *Teeth: Form, Function and Evolution* (ed. B. Kurtén), pp. 154–162. New York: Columbia University Press.

(1982b) An analysis of the canine tooth size of Old World higher primates in relation to facial length and body weight. *Archives of Oral Biology* **27**: 493–496.

(1989) A new theory relating seed processing by primates to their relative tooth sizes. In: *The Growing Scope of Human Biology* (eds. L. H. Schmitt, L. Freedman & N. W. Bruce), pp. 37–49. Perth: Centre for Human Biology, University of Western Australia.

(1991) Fundamental physical properties of fruits and seeds in the diet of Southeast Asian primates. In: *Primatology Today* (eds. A. Ehara, T. Kimura, O. Takenaka & M. Iwamoto), pp. 125–128. Amsterdam: Elsevier.

Lucas, P. W. & Corlett, R. T. (1991) Quantitative aspects of the relationship between dentitions and diets. In: *Feeding and the Texture of Food* (eds. J. F. V. Vincent & P. J. Lillford), pp. 93–121. Cambridge: Cambridge University Press.

(1998) Seed dispersal by long-tailed macaques. *American Journal of Primatology* **45**: 29–44.

Lucas, P. W. & Luke, D. A. (1983) Methods for analysing the breakdown of food during human mastication. *Archives of Oral Biology* **28**: 813–819.

(1984a) Optimal mouthful for food comminution in human mastication. *Archives of Oral Biology* **29**: 205–210.

(1984b) Chewing it over: basic principles of food breakdown. In: *Food Acquisition and Processing in Primates* (eds. D. J. Chivers, B. A. Wood & A. Bilsborough), pp. 283–302. New York: Plenum.

(1986) Is food particle size a criterion for the initiation of swallowing? *Journal of Oral Rehabilitation* **13**: 127–136.

Lucas, P. W. & Peters, C. R. (2000) Function of postcanine tooth shape in mammals. In: *Development, Function and Evolution of Teeth* (eds. M. F. Teaford, M. M. Smith & M. Ferguson), pp. 282–289. Cambridge: Cambridge University Press.

Lucas, P. W. & Teaford, M. F. (1994) Functional morphology of colobine teeth. In: *Colobine Monkeys: Their Ecology, Behaviour and Evolution* (eds. A. G. Davies & J. F. Oates), pp. 171–203. Cambridge: Cambridge University Press.

(1995) Significance of silica in leaves eaten by long-tailed macaques (*Macaca fascicularis*). *Folia Primatologica* **64**: 30–36.

Lucas, P. W., Corlett, R. T. & Luke, D. A. (1985) Plio-Pleistocene hominids: an approach combining masticatory and ecological analysis. *Journal of Human Evolution* **14**: 187–202.

(1986a) Postcanine tooth size and diet in anthropoids. *Zeitschrift für Morphologie und Anthropologie* **76**: 253–276.

(1986b) Sexual dimorphism of teeth in anthropoid primates. *Human Evolution* **1**: 23–39.

(1986c) New approach to postcanine tooth size applied to Plio-Pleistocene hominids. In: *Primate Evolution* (eds. J. G. Else & P. C. Lee), pp. 191–201. Cambridge: Cambridge University Press.

Lucas, P. W., Luke, D. A., Voon, F. C. T. *et al.* (1986d) Breakdown patterns produced by human subjects possessing artificial and natural teeth. *Journal of Oral Rehabilitation* **13**: 205–214.

Lucas, P. W., Ow, R. K. K., Ritchie, G. M. *et al.* (1986e) Relationship between jaw movement and food comminution in human mastication. *Journal of Dental Research* **65**: 400–404.

Lucas, P. W., Choong, M. F., Tan, H. T. W. *et al.* (1991a) Fracture toughness of the leaf of the dicotyledonous angiosperm, *Calophyllum inophyllum* L. *Philosophical Transactions of the Royal Society London series B* **334**: 95–106.

Lucas, P. W., Lowrey, T. K., Pereira, B. *et al.* (1991b) The ecology of *Mezzettia leptopoda* Hk. f. et Thoms. (Annonaceae) seeds as viewed from a mechanical perspective. *Functional Ecology* **5**: 345–353.

Lucas, P. W., Oates, C. G. & Lee, W. P. (1993) Fracture toughness of mung bean gels. *Journal of Materials Science* **28**: 1137–1142.

Lucas, P. W., Peters, C. R. & Arrandale, S. (1994) Seed-breaking forces exerted by orangutans with their teeth in captivity and a new technique for estimating forces produced in the wild. *American Journal of Physical Anthropology* **94**: 365–378.

Lucas, P. W., Darvell, B. W., Lee, P. K. D. *et al.* (1995) The toughness of plant cell walls. *Philosophical Transactions of the Royal Society London series B* **348**: 363–372.

Lucas, P. W., Tan, H. T. W. & Cheng, P. Y. (1997) The toughness of secondary cell wall and woody tissue. *Philosophical Transactions of the Royal Society London series B* **352**: 341–352.

Lucas, P. W., Turner, I. M., Dominy, N. J. *et al.* (2000) Mechanical defences to herbivory. *Annals of Botany* **86**: 913–920.

Lucas, P. W., Beta, T., Darvell, B. W. *et al.* (2001) Field kit to characterize physical, chemical and spatial aspects of potential foods of primates. *Folia Primatologica* **72**: 11–15.

Lucas, P. W., Prinz, J. F., Agrawal, K. R. *et al.* (2002) Food physics and oral physiology. *Food Quality and Preference* **13**: 203–213.

Lucas, P. W., Osorio, D., Yamashita, N. *et al.* (2003) Dietary analysis. I. Physics. In: *Field and Laboratory Methods in Primatology* (eds. J. Setchell & D. Curtis), pp. 184–198. Cambridge: Cambridge University Press.

Luke, D. A. & Lucas, P. W. (1983) The significance of cusps. *Journal of Oral Rehabilitation* **10**: 197–210.

(1985) Chewing efficiency in relation to occlusal and other variations in the dentition. *British Dental Journal* **159**: 401–403.

Lumsden, A. G. S. & Osborn, J. W. (1977) The evolution of chewing: a dentist's view of palaeontology. *Journal of Dentistry* **5**: 269–287.

Lund, J. P. (1991) Mastication and its control by the brainstem. *Critical Reviews in Oral Biology and Medicine* **2**: 33–64.

Lundberg, M., Nord, P. G. & Åstrand, P. (1974) Changes in masticatory function after surgical treatment of mandibular prognathism. (Cineradiographic study of bolus position). *Acta Odontologica Scandinavica* **32**: 39–49.

Luo, Z.-X., Cifelli, R. L. & Kielan-Jaworowska, Z. (2001) Dual origin of tribosphenic mammals. *Nature* **409**: 53–57.

Luschei, E. S. & Goldberg, L. J. (1981) Neural mechanisms of mandibular control: mastication and voluntary biting. In: *Handbook of Physiology*, section I, *The Nervous System*, vol. 2, *Motor Control*, part 2 (eds. J. M. Brookhart, V. B. Mountcastle, V. B. Brooks & S. R. Geiger), pp. 1237–1273. Bethesda: American Physiological Society.

Macfadden, B. J., Solounias, N. & Cerling, T. E. (1999) Ancient diets, ecology and extinction of 5-million-year-old horses from Florida. *Science* **283**: 824–827.

Maglio, V. J. (1972) Evolution of mastication in the Elephantidae. *Evolution* **26**: 638–658.

(1973) Origin and evolution of the Elephantidae. *Transactions of the American Philosophical Society of Philadelphia* (new series) **63**: 1–149.

Magnusson, W. E. & Saniotti, T. M. (1987) Dispersal of *Miconia* seeds by the rat *Bolomys lasiurus*. *Journal of Tropical Ecology* **3**: 277–278.

Mai, Y.-W. & Atkins, A. G. (1975) Scale effects and crack propagation in non-linear elastic structures. *International Journal of Mechanical Sciences* **17**: 673–675.

Mai, Y.-W. & Cotterell, B. (1989) On mixed-mode plane stress ductile fracture. In: *Advances in Fracture Research* (eds. K. Salama, K. Ravi-Chandar, D. M. R. Taplin & P. Ramo Rao), pp. 2269–2278. Oxford: Pergamon Press.

Manger, P. R., Woods, T. M. & Jones, E. G. (1995) Representation of the face and intra-oral structures in area 3b of the squirrel monkey (*Saimiri sciureus*) somatosensory cortex, with special reference to the ipsilateral representation. *Journal of Comparative Neurology* **362**: 597–607.

(1996) Representation of face and intra-oral structures in area 3b of macaque monkey somatosensory cortex. *Journal of Comparative Neurology* **371**: 513–521.

Manly, R. S. & Braley, L. C. (1950) Masticatory performance and efficiency. *Journal of Dental Research* **29**: 448–462.

Mao, J. J. (2002) Mechanobiology of craniofacial sutures. *Journal of Dental Research* **81**: 810–816.

Mao, J., Stein, R. B. & Osborn, J. W. (1992) The size and distribution of fiber types in jaw muscles: a review. *Journal of Craniomandibular Disorders: Facial and Oral Pain* **6**: 192–201.

Marshall, D. B. & Lawn, B. R. (1986) Indentation of brittle materials. In: *Microindentation Techniques in Science and Engineering ASTMMSTP 889* (eds. P. J. Blau & B. R. Lawn), pp. 26–46. Philadelphia, PA: American Society for Testing and Materials.

Marshall, G. W. Jr, Balooch, M., Gallagher, R. R. *et al.* (2001) Mechanical properties of the dentinoenamel junction: AFM studies of nanohardness, elastic modulus, and fracture. *Journal of Biomedical Materials Research* **54**: 87–95.

Martin, L. B. (1985) Significance of enamel thickness in hominoid evolution. *Nature* **314**: 260–263.

Martin, R. D. (1979) Phylogenetic aspects of prosimian behaviour. In: *The Study of Prosimian Behaviour* (eds. G. A. Doyle & R. D. Martin), pp. 45–77. London: Academic Press.

(1990) *Primate Origins and Evolution*. Princeton, NJ: Princeton University Press.

Martin, R. D., Chivers, D. J., MacLarnon, A. M. *et al.* (1985) Gastrointestinal allometry in primates and other mammals. In: *Size and Scaling in Primate Biology* (ed. W. L. Jungers), pp. 61–89. New York: Plenum.

Martinez del Rio, C. (1990) Dietary and phylogenetic correlates of intestinal sucrase and maltase activity in birds. *Physiological Zoology* **63**: 987–1011.

Mattes, R. D. (2001) Oral exposure to butter, but not fat replacers elevates post-prandial triaglycerol concentration in humans. *Journal of Nutrition* **131**: 1491–1496.

Matthews, P. B. C. (1972) *Mammalian Muscle Receptors and their Central Actions*. London: Edward Arnold.

Maynard Smith, J. (1979) Game theory and the evolution of behaviour. *Proceedings of the Royal Society London series B* **205**: 475–488.

Maynard Smith, J. & Price, G. R. (1973) The logic of animal conflict. *Nature* **146**: 15–18.

McCann, M. C. & Roberts, K. (1991) Architecture of the primary cell wall. In: *The Cytoskeletal Basis of Plant Growth* (ed. C. W. Lloyd), pp. 109–129. London: Academic Press.

McKey, D. B., Gartlan, J. S., Waterman, P. G. *et al.* (1981) Food selection by black colobus monkeys (*Colobus satanus*) in relation to plant chemistry. *Biological Journal of the Linnean Society* **16**: 114–146.

McMahon, T. A. (1973) Size and scaling in biology. *Science* **179**: 1201–1204.

(1975) Using body size to understand the structural design of animals: quadrapedal locomotion. *Journal of Applied Physiology* **39**: 619–627.

McMahon, T. A. & Greene, P. R. (1979) The influence of track compliance on running. *Journal of Biomechanics* **12**: 893–904.

McMillan, A. S. & Hannam, A. G. (1992) Task-related behaviour of motor units in different regions of the human masseter muscle. *Archives of Oral Biology* **37**: 849–857.

Mendoza, M., Janis, C. M. & Palmqvist, P. (2002) Characterizing complex cranio-dental patterns related to feeding behaviour in ungulates: a multivariate approach. *Journal of Zoology* **258**: 223–246.

Mercader, J., Panger, M. & Boesch, C. (2002) Excavation of a chimpanzee stone tool site in the African Rainforest. *Science* **296**: 1452–1455.

Metcalfe, C. R. & Chalk, L. (1950) *Anatomy of Dicotyledons*, vols. 1 and 2. Oxford: Clarendon Press.

Mills, J. R. E. (1955) Ideal dental occlusion in the Primates. *Dental Practitioner* **6**: 47–61.

Milton, K. (1979) Factors influencing leaf choice by howler monkeys: a test of some hypotheses of food selection by generalist herbivores. *American Naturalist* **114**: 362–378.

Miyazato, M., Chen, J. Y., Ishiguro, E. *et al.* (1994) Studies on the non-destructive measurement of Poisson's ration of watermelon. I. Model development. *Memoirs of the Faculty of Agriculture, Kagoshima University* **30**: 97–104.

Mohsenin, N. N. (1977) Characterization and failure in solid foods with particular reference to fruits and vegetables. *Journal of Texture Studies* **8**: 169–193.

Mongini, F., Tempia-Valenti, G. & Benvegnu, G. (1986) Computer-based assessment of habitual mastication. *Journal of Prosthetic Dentistry* **55**: 638–649.

Moore, S. J. & Sanson, G. D. (1995) A comparison of the molar efficiency of two insect-eating mammals. *Journal of Zoology* **235**: 175–192.

Morgan, E. (1972) *The Descent of Woman*. London: Souvenir.

Mowlana, F. & Heath, M. R. (1993) Assessment of masticatory efficiency: new methods appropriate for clinical research in dental practice. *European Journal of Prosthodontics and Restorative Dentistry* **1**: 121–125.

Moyà Solà, S. & Köhler, M. (1997) The phylogenetic relationships of *Oreopithecus bambolii* Gervais 1872. *Comptes Rendus de l'Academie des Sciences Paris* (série II) **324** : 141–148.

Mühlemann, H. R. (1951) Periodontometry: a method for measuring tooth mobility. *Oral Surgery Oral Medicine Oral Pathology* **4**: 1220–1233.

Murphy, T. (1959) Compensatory mechanisms in facial height adjustment to functional tooth attrition. *Australian Dental Journal* **4**: 312–323.

Murray, C. G. & Sanson, G. D. (1998) Thegosis: a critical review. *Australian Dental Journal* **43**: 192–198.

Murray, P. (1975) The role of cheek pouches in cercopithecine monkey adaptive strategy. In: *Primate Functional Morphology and Evolution* (ed. R. H. Tuttle), pp. 151–194. The Hague: Mouton.

Mysterud, A. (1998) The relative roles of body size and feeding type on activity time of temperate ruminants. *Oecologia* **113**: 442–446.

Neinhuis, C. & Barthlott, W. (1997) Characterization and distribution of water-repellent, self-cleaning plant surfaces. *Annals of Botany* **79**: 667–677.

Nelson, G., Hoon, M. A., Chandrashekar, J. *et al.* (2001) Mammalian sweet taste receptors. *Cell* **106**: 381–390.

Nelson, G., Chandrashekar, J., Hoon, M. A. *et al.* (2002) An amino acid taste receptor. *Nature* **416**: 199–202.

Nguyen, Q. & Zarkadas, C. G. (1989) Comparison of the amino acid composition and connective tissue protein contents of selected bovine skeletal muscles. *Journal of Agricultural and Food Chemistry* **32**: 1279–1286.

Norman, D. B. & Weishampel, D. B. (1991) Feeding mechanisms in some small herbivorous dinosaurs: processes and patterns. In: *Biomechanics in Evolution* (eds. J. M. V. Rayner & R. J. Wootton), pp. 161–181. Cambridge: Cambridge University Press.

Norman, T. L., Vashishth, D. & Burr, D. B. (1992) Effect of groove on bone fracture toughness. *Journal of Biomechanics* **25**: 1489–1492.

Nose, K. (1961) Study on the hardness of human and animal teeth. *Journal of the Kyoto Prefectural University of Medicine* **69**: 1925–1945.

Oates, J. F. (1978) Water-plant and soil consumption by guereza monkeys (*Colobus guereza*): a relationship with minerals and toxins in the diet? *Biotropica* **10**: 241–253.

Olthoff, L. W., van der Bilt, A., Bosman, F. *et al.* (1984) Distribution of particle sizes in food comminuted by human mastication. *Archives of Oral Biology* **29**: 899–903.

Orchardson, R. & Cadden, S. W. (1998) Mastication. In: *The Scientific Basis of Eating* (ed. R. W. A. Linden), pp. 76–121. Basel: Karger.

Oron, U. & Crompton, A. W. (1985) A cineradiographic and electromyographic study of mastication in *Tenrec ecaudatus*. *Journal of Morphology* **185**: 155–182.

Osborn, H. F. (1888) The nomenclature of the mammalian molar cusps. *American Naturalist* **22**: 926–928.

Osborn, J. W. (1961) An investigation into the interdental forces occurring between the teeth of the same arch during clenching of the jaws. *Archives of Oral Biology* **5**: 202–211.

(1969) Dentine hardness and incisor wear in the beaver (*Castor fiber*). *Acta Anatomica* **72**: 123–132.

(1971a) A relationship between the striae of Retzius and prism directions in the transverse plane of the tooth. *Archives of Oral Biology* **16**: 461–470.

(1971b) The ontogeny of tooth succession in *Lacerta vivipara* Jacquin (1787). *Proceedings of the Royal Society London series B* **179**: 261–289.

(1974) Variation in structure and development of enamel. In: *Oral Sciences Reviews*, vol. 3, *Dental Enamel*. Copenhagen: Munksgaard.

(1978) Morphogenetic gradients: fields versus clones. In: *Development, Function and Evolution of Teeth* (eds. P. M. Butler & K. A. Joysey), pp. 171–213. New York: Academic Press.

(1981) *Companion to Dental Studies* (eds. A. H. R. Rowe & R. B. Johns), vol. 2A, *Dental Anatomy and Embryology*. Oxford: Blackwell.

(1982) Helicoidal plane of occlusion. *American Journal of Physical Anthropology* **57**: 273–281.

(1985) The design of the human TMJ: design, function, and failure. *Journal of Oral Rehabilitation* **12**: 279–293.

(1987) Relationship between the mandibular condyle and the occlusal plane during hominid evolution: some of its effects on jaw mechanics. *American Journal of Physical Anthropology* **73**: 193–207.

(1993) A model to describe how ligaments may control symmetrical jaw opening movements in man. *Journal of Oral Rehabilitation* **20**: 585–604.

(1995a) Internal derangement and the accessory ligaments around the temporo-mandibular joint. *Journal of Oral Rehabilitation* **22**: 731–740.

(1995b) Biomechanical implications of lateral pterygoid contribution to biting and jaw opening in humans. *Archives of Oral Biology* **40**: 1099–1108.

Osborn, J. W. & Baragar, F. A. (1985) Predicted pattern of human muscle activity during clenching derived from a computer-assisted model. *Journal of Biomechanics* **18**: 599–612.

Osborn, J. W. & Lumsden, A. G. S. (1978) An alternative to "thegosis" and a re-examination of the ways in which mammalian molars work. *Neues Jahrbuch für Geologie und Paläontologie Abhandlungen* **156**: 371–396.

Osborn, J. W., Baragar, F. A. & Grey, P. (1987) The functional advantage of proclined incisors in man. In: *Teeth Revisited: Proceedings of 7th International Symposium on Dental Morphology* (eds. D. E. Russell, J. P. Santoro & D. Sigognean-Russell). *Memoirs du Musée National d'Histoire Naturelle, Paris* (Série C) **53**: 445–458.

Ostry, D. J., Gribble, P. L., Levin, M. F. *et al.* (1997) Phasic and tonic stretch reflexes in muscles with few muscle spindles: human jaw-opener muscles. *Experimental Brain Research* **116**: 299–308.

Otani, T. & Shibata, E. (2000) Seed dispersal and predation by Yakushima macaques, *Macaca fuscata yakui*, in a warm temperate forest of Yakushima Island, southern Japan. *Ecological Research* **15**: 133–144.

Ottenhoff, F. A., van der Bilt, A., van der Glas, H. W. *et al.* (1996) The relationship between jaw elevator muscle surface electromyogram and simulated food resistance during dynamic condition in humans. *Journal of Oral Rehabilitation* **23**: 270–279.

Overdorff, D. J. & Strait, S. G. (1998) Seed handling by three prosimian primates in Southeastern Madagascar: implications for seed dispersal. *American Journal of Primatology* **45**: 69–82.

Öwall, B. (1978) Interocclusal perception with anaesthetized and unanaesthetized temporomandibular joints. *Swedish Dental Journal* **2**: 199–208.

Öwall, B. & Vorwerk, P. (1974) Analysis of a method for testing oral tactility during chewing. *Odontologisk Revy* **25**: 1–10.

Owen-Smith, N. (1988) *Megaherbivores: The Influence of Very Large Body Size on Ecology*. Cambridge: Cambridge University Press.

Owen-Smith, N., Robbins, C. T. & Hagerman, A. E. (1993) Browse and browsers: interactions between woody plants and mammalian herbivores. *Trends in Ecology and Evolution* **8**: 158–160.

Page, D. H., El-Hosseiny, F. & Winkler, K. (1971) Behaviour of single wood fibres under axial tensile strain. *Nature* **229**: 252–253.

Panger, M. A, Brooks, A. G., Richmond, B. G. *et al.* (2002) Older than the Olduwan? Rethinking the emergence of hominin tool use. *Evolutionary Anthropology* **11**: 235–245.

Paphangkorakit, J. & Osborn, J. W. (1997) Effect of jaw opening on the direction and magnitude of human incisal bite forces. *Journal of Dental Research* **76**: 561–567.

(2000) The effect of normal occlusal forces on fluid movement through human dentine *in vitro*. *Archives of Oral Biology* **45**: 1033–1041.

Parfitt, G. J. (1960) Measurement of the physiological mobility of individual teeth in an axial direction. *Journal of Dental Research* **39**: 608–618.

Peleg, M., Gómez-Brito, L. & Malevski, Y. (1976) Compressive failure patterns of some juicy fruits. *Journal of Food Science* **41**: 1320–1324.

Peleg, M. & Gómez-Brito, L. (1977) Textural changes in ripening plantains. *Journal of Texture Studies* **7**: 457–463.

Pereira, B. P., Lucas, P. W. & Teoh, S. H. (1997) Ranking the fracture toughness of mammalian soft tissues using the scissors cutting test. *Journal of Biomechanics* **30**: 91–94.

Perez-Barberia, F. J. & Gordon, I. J. (1998a) Factors affecting food comminution during chewing in ruminants: a review. *Biological Journal of the Linnean Society* **63**: 233–256.

(1998b) The influence of molar occlusal surface area on the assimilation efficiency, chewing behaviour and diet selection of red deer. *Journal of Zoology* **245**: 307–316.

(2001) Relationship between oral morphology and feeding style in the Ungulata: a phylogenetically controlled evaluation. *Proceedings of the Royal Society London series B* **268**: 1021–1030.

Peters, C. R. (1979) Towards an ecological model of African Plio-Pleistocene hominid adaptations. *American Anthropologist* **81**: 261–278.

(1981) *Australopithecus* vs. *Homo* dietary capabilities: the natural competitive advantage of the megadonts. In: *Perceptions of Human Evolution*, vol. 7 (eds. L. L. Mai, E. Shanklin & R. W. Sussman), pp. 161–181. Los Angeles: UCLA Anthropology.

(1982) Electron-optical microscopic study of incipient dental microdamage from experimental seed and bone crushing. *American Journal of Physical Anthropology* **57**: 283–301.

(1987) Nut-like oil seeds: food for monkeys, chimpanzees, humans and probably ape-men. *American Journal of Physical Anthropology* **73**: 333–363.

Peters, C. R. & Maguire, B. (1981) Wild plant foods of the Makapansgat area: a modern ecosystems analogue for *Australopithecus africanus* adaptations. *Journal of Human Evolution* **10**: 565–583.

Peters, C. R., O'Brien, E. M. & Drummond, R. B. (1992) *Edible Wild Plants of Sub-Saharan Africa*. Kew: Royal Botanical Gardens.

Peters, R. H. (1983) *The Ecological Implications of Body Size*. Cambridge: Cambridge University Press.

Pettifor, E. (2000) From the teeth of the dragon: *Gigantopithecus blacki*. In: *Selected Readings in Physical Anthropology* (ed. P. Scully), pp. 143–149. Dubuque, IA: Kendall/Hunt.

Phua, P. B. & Corlett, R. T. (1989) Seed dispersal by the lesser short-nosed fruit bat (*Cynopterus brachyotis*, Pteropodidae, Megachiroptera). *Malayan Nature Journal* **42**: 251–256.

Picton, D. C. A. (1965) On the part played by the socket in tooth support. *Archives of Oral Biology* **10**: 945–955.

Pilbeam, D. R. & Gould, S. J. (1974) Size and scaling in human evolution. *Science* **186**: 892–901.

(1975) Allometry and early hominids. *Science* 189: 64.

Plavcan, J. M., van Schaik, C. P. & Kappeler, P. M. (1995) Competition, coalitions and canine size in primates. *Journal of Human Evolution* **28**: 245–276.

Poon, T. F. (1974) Physiological studies on fruits of *Nephelium lappaceum* L. B.Sc. dissertation. Singapore: Department of Botany, National University of Singapore.

Popovics, T. E. & Fortelius, M. (1997) On the cutting edge: tooth blade sharpness in herbivorous and faunivorous mammals. *Annales Zoologici Fennici* **34**: 73–88.

Popovics, T. E., Remsberger, J. M. & Herring, S. W. (2002) The fracture behaviour of human and pig molar cusps. *Archives of Oral Biology* **46**: 1–12.

Preston, C. M. & Sayer, B. G. (1992) What's in a nutshell: an investigation of structure by carbon-13 cross-polarization magic-angle spinning nuclear magnetic resonance spectroscopy. *Journal of Agricultural Food Chemistry* **40**: 206–220.

Preston, R. D. (1974) *The Physical Biology of Plant Cell Walls*. London: Chapman & Hall.

Prinz, J. F. (in press) Abrasives in foods and their effect on intra-oral processing: a two-colour chewing gum study. *Journal of Oral Rehabilitation*.

Prinz, J. F. & Lucas, P. W. (1995) Swallow thresholds in humans. *Archives of Oral Biology* **40**: 401–403.

(1997). An optimization model for mastication and swallowing in mammals. *Proceedings of the Royal Society London series B* **264**: 1715–1721.

(2000) Saliva tannin interactions. *Journal of Oral Rehabilitation* **27**: 991–994.

(2001) '*The first bite of the cherry*': intra-oral manipulation prior to the first bite in humans. *Journal of Oral Rehabilitation* **28**: 614–617.

Prinz, J. F., Silwood, C. J. L., Claxson, A. W. D. *et al.* (2003) Simulated digestion status of intact and exoskeletally punctured insects and insect larvae: a spectroscopic investigation. *Folia Primatologica* **74**: 12–26.

Prothero, D. R. & Sereno, P. C. (1982) Allometry and paleoecology of medial Miocene dwarf rhinoceroses from the Texas Gulf Coastal Plain. *Paleobiology* **8**: 16–30.

Purnell, M. A. (1995) Microwear on conodont elements and macrophagy in the first vertebrates. *Nature* **374**: 798–800.

Purslow, P. P. (1983) Measurement of the fracture toughness of extensible connective tissues. *Journal of Materials Science* **18**: 3591–3598.

(1985) The physical basis of meat texture: observations on the fracture behaviour of cooked bovine *M. semitendinosus* during heating. *Meat Science* **12**: 39–60.

(1991a) Notch-sensitivity of nonlinear materials. *Journal of Materials Science* **26**: 4468–4476.

(1991b) Measuring meat texture and understanding its structural basis. In: *Feeding and the Texture of Food* (eds. J. F. V. Vincent & P. J. Lillford), pp. 35–56. Cambridge: Cambridge University Press.

Purslow, P. P., Bigi, A., Ripamonti, A. *et al.* (1984) Collagen fibre orientation around a crack in biaxially stretched aortic media. *International Journal of Biological Macromolecules* **6**: 21–25.

Rajaram, A. (1986) Tensile properties and fracture of ivory. *Journal of Materials Science Letters* **5**: 1077–1080.

Ralls, K. (1977) Mammals in which females are larger than males. *Quarterly Review of Biology* **51**: 245–275.

Rasmussen, S. T., Patchin, R. E., Scott, D. B. *et al.* (1976) Fracture properties of human enamel and dentine. *Journal of Dental Research* **55**: 154–164.

Raven, J. A. (1983) The transport and function of silicon in plants. *Biological Reviews* **58**: 179–207.

Ravosa, M. J. (1991) Structural allometry of the prosimian mandibular corpus and symphysis. *Journal of Human Evolution* **20**: 3–20.

Rees, J. S. & Jacobsen, P. H. (1997) Elastic modulus of the periodontal ligament. *Biomaterials* **18**: 995–999.

Rees, L. A. (1954) The structure and function of the mandibular joint. *British Dental Journal* **96**: 125–133.

Rensberger, J. M. (1973) An occlusion model for mastication and dental wear in herbivorous mammals. *Journal of Palaeontology* **47**: 515–528.

(2000) Pathways to functional differentiation in mammalian enamel. In: *Development, Function and Evolution of Teeth* (eds. M. F. Teaford, M. M. Smith & M. W. J. Ferguson), pp. 252–268. Cambridge: Cambridge University Press.

Rensberger, J. M. & von Koenigswald, W. (1980) Functional phylogenetic interpretation of enamel microstructure in rhinoceroses. *Paleobiology* **6**: 447–495.

Renson, C. E. & Braden, M. (1971) The experimental deformation of human dentine by indenters. *Archives of Oral Biology* **16**: 563–572.

(1975) Experimental determination of the rigidity modulus, Poisson's ratio and elastic limit of shear of human dentine. *Archives of Oral Biology* **20**: 43–47.

Rich, T. H., Flanner, T. F., Trusler, P. *et al.* (2002) Evidence that monotremes and ausktribosphenids are not sistergroups. *Journal of Vertebrate Paleontology* **22**: 466–469.

Ridley, H. N. (1930) *The Dispersal of Plants around the World*. Ashford: Reeve.

Ringel, R. L. & Ewanowski, S. J. (1965) Oral perception. I. Two-point discrimination. *Journal of Speech and Hearing Research* **8**: 389–397.

Robbins, C. T., Spalinger, D. E. & Van Hoven, W. (1995) Adaptation of ruminants to browse and grass diets: are anatomical-based browser–grazer interpretations valid? *Oecologia* **103**: 208–213.

Robbins, M. W. (1977) Biting loads generated in the laboratory rat. *Archives of Oral Biology* **22**: 43–47.

Robinson, C., Brookes, S. J., Bonass, W. A. *et al.* (1997) Enamel maturation. In: *Dental Enamel* (Ciba Foundation Symposium 205), pp. 156–174. Chichester: Wiley.

Robinson, J. T. (1956) The dentition of the Australopithecinae. *Memoirs of the Transvaal Museum* **9**: 1–179.

Rogers, M. E., Maisels, F., Williamson, E. A. *et al.* (1990) Gorilla diet in the Lopé Reserve, Gabon: a nutritional analysis. *Oecologia* **84**: 326–339.

Rolls, E. T., Critchley, H. D., Browning, A. S. *et al.* (1999) Responses to the sensory properties of fat of neurons in the primate orbitofrontal cortex. *Journal of Neuroscience* **19**: 1532–1540.

Romer, A. S. (1966) *Vertebrate Paleontology*, 3rd edn. Chicago, IL: University of Chicago Press.

Rose, H. E. & Sullivan, R. M. E. (1961) *Vibration Mills and Vibration Milling*. London: Constable.

Rose, K. D., Walker, A. & Jacobs, L. (1981) Function of the mandibular tooth comb in living and extinct mammals. *Nature* **289**: 583–585.

Rosenberger, A. L. & Kinzey, W. G. (1976) Functional patterns of molar occlusion in platyrrhine primates. *American Journal of Physical Anthropology* **45**: 281–298.

Roth, V. L. (1990) Insular dwarf elephants: a case study in body mass estimation and ecological inference. In: *Body Size in Mammalian Paleobiology* (ed. J. Damuth & B. J. MacFadden), pp. 151–179. Cambridge: Cambridge University Press.

Rubin, C., Turner, S., Bain, S. *et al.* (2001) Extremely low level mechanical signals are anabolic to trabecular bone. *Nature* **412**: 603–604.

Runham, N. W., Thornton, P. R., Shaw, D. A. *et al.* (1969) Mineralization and hardness of the radular teeth of the limpet *Patella vulgata. Zeitschrift für Zellforschung und Mikroscopische Anatomie Abteilung Histochemie* **99**: 608–626.

Ryan, J. M. (1986) Comparative morphology and evolution of cheek pouches in rodents. *Journal of Morphology* **190**: 27–41.

Rybczynski, N. & Reisz, R. R. (2001) Earliest evidence for efficient oral processing in a terrestrial herbivore. *Nature* **411**: 684–687.

St Hoyme, L. E. & Horitzer, R. T. (1971) Significance of canine wear in pongid evolution. *American Journal of Physical Anthropology* **35**: 145–147.

Sato, K., Yoshinaka, R., Sato, M. *et al.* (1986) Collagen content in the muscles of fish in association with their swimming movement and meat texture. *Bulletin of the Japanese Society of Scientific Fisheries* **52**: 1595–1600.

Savage, R. J. G. (1977) Evolution in carnivorous mammals. *Palaeontology* **20**: 237–271.

Scapino, R. P. (1965) The third joint of the canine jaw. *Journal of Morphology* **116**: 23–50.

Schofield, R. M. S., Nesson, M. H. & Richardson, K. A. (2002) Tooth-hardness increases with zinc-content in mandibles of young adult leaf-cutter ants. *Naturwissenschaften* **89**: 579–583.

Schmidt-Neilsen, K. (1972) *How Animals Work.* Cambridge: Cambridge University Press.

Schwartz, J. H. (1974) Premolar loss in the primates: a theoretical reinvestigation. In: *Prosimian Biology* (eds. R. D. Martin, G. A. Doyle & A. C. Walker), pp. 621–640. Pittsburgh, PA: University of Pittsburgh Press.

Sclafani, A. (1991) Starch and sugar tastes in rodents: an update. *Brain Research Bulletin* **27**: 383–386.

Semaw, S., Renne, P., Harris, J. W. K. *et al.* (1997) 2.5 million-year-old stone tools from Gona, Ethiopia. *Nature* **385**: 333–336.

Sereno, P. C. (1997) Origin and evolution of dinosaurs. *Annual Review of Earth and Planetary Sciences* **25**: 435–489.

(1999) The evolution of dinosaurs. *Science* **284**: 2137–2147.

Shama, F. & Sherman, P. (1973) Evaluation of some textural properties of foods with the Instron universal testing machine. *Journal of Texture Studies* **4**: 344–353.

Sharp, S. J., Ashby, M. F. & Fleck, N. A. (1993) Material response under static and sliding indentation loads. *Acta Metallurgica et Materialia* **41**: 685–692.

Sharpe, P. T. (2000) Homeobox genes in initiation and shape of teeth during development in mammalian embryos. In: *Development, Function and Evolution of Teeth* (eds. M. F. Teaford, M. M. Smith & M. W. J. Ferguson), pp. 3–12. Cambridge: Cambridge University Press.

Shaw, D. M. (1917) Form and function of teeth: a theory of "maximum shear". *Journal of Anatomy* **52**: 97–106.

Sheikh-Ahmad, J. Y. & McKenzie, W. M. (1997) Measurement of tool wear and dulling in the machining of particleboard. In: *Proceedings of the 13th International Wood Machining Seminar*, Vancouver, Canada, pp. 659–670.

Sheine, W. S. (1979) The effect of variations in molar morphology on masticatory effectiveness and digestion of cellulose in prosimian primates. PhD thesis. Duke University, Durham, NC.

Sheine, W. S. & Kay, R. F. (1977) An analysis of chewed food particle size and its relationship to molar structure in the primates *Cheirogaleus medius* and *Galago senegalensis* and the insectivoran *Tupaia glis*. *American Journal of Physical Anthropology* **47**: 15–29.

Shellis, R. P. (1981) Comparative histology of dental tissues. In: *Companion to Dental Studies* (eds. A. H. R. Rowe & R. B. Johns), *Dental Anatomy and Embryology* (ed. J. W. Osborn), vol. 2A, pp. 158–165. Oxford: Blackwell.

Shellis, R. P. & Dibden, G. H. (2000) Enamel microporosity and its functional implications. In: *Development, Function and Evolution of Teeth* (eds. M. F. Teaford, M. M. Smith & M. W. J. Ferguson), pp. 242–268. Cambridge: Cambridge University Press.

Shipley, L. A. & Spalinger, D. E. (1992) Mechanics of browsing in dense food patches: effects of plant and animal morphology on intake rate. *Canadian Journal of Zoology* **70**: 1743–1753.

Shipley, L. A., Gross, J. E., Spalinger, D. E. *et al.* (1994) The scaling of intake rate of mammalian herbivores. *American Naturalist* **143**: 1055–1082.

Sim, B. J., Lucas, P. W., Pereira, B. P. *et al.* (1993) Mechanical and sensory assessment of the texture of refrigerator-stored spring roll pastry. *Journal of Texture Studies* **24**: 27–44.

Simpson, G. G. (1936) Studies of the earliest mammalian dentitions. *Dental Cosmos* **78**: 940–953.

(1941) *The Function of Saber-Like Canines in Carnivorous Mammals. American Museum Novitates* no. 1130.

(1953) *The Major Features of Evolution.* New York: Columbia University Press.

Smith, J. M. & Savage, R. J. G. (1959) The mechanics of mammalian jaws. *School Science Review* **141**: 289–301.

Smith, K. K. (1992) The evolution of the mammalian pharynx. *Zoological Journal of the Linnean Society* **104**: 313–349.

Smith, M. M. and Coates, M. I. (2000) Evolutionary origins of teeth and jaws: developmental models and phylogenetic patterns. In: *Development, Function and Evolution of Teeth* (eds. M. F. Teaford, M. M. Smith & M. W. J. Ferguson), pp. 133–151. Cambridge: Cambridge University Press.

Smith, M. M. & Johanson, Z. (2003) Separate evolutionary origins of teeth from evidence in fossil jawed vertebrates. *Science* **29**: 1235–1236.

Smith, M. M. and Sansom, I. J. (2000) Evolutionary origins of dentine in the fossil record of early vertebrates: diversity, development and function. In: *Development, Function and Evolution of Teeth* (eds. M. F. Teaford, M. M. Smith & M. W. J. Ferguson), pp. 65–81. Cambridge: Cambridge University Press.

Smith, R. J. (1978) Mandibular biomechanics and temporomandibular joint function in primates. *American Journal of Physical Anthropology* **49**: 341–350.

(1983) The mandibular corpus of female primates: taxonomic, dietary and allometric correlates of interspecific variations in size and shape. *American Journal of Physical Anthropology* **61**: 315–330.

(1984) Comparative functional morphology of maximum mandibular opening (gape) in primates. In: *Food Acquisition and Processing in Primates* (eds. D. J. Chivers, B. A. Wood & A. Bilsborough), pp. 231–255. New York: Plenum.

Sofaer, J. A. (1973) A model relating developmental interaction and differential evolutionary reduction of tooth size. *Evolution* **27**: 427–434.

(1977) Co-ordinated growth of successively initiated tooth germs in the mouse. *Archives of Oral Biology* **22**: 71–72.

Sofaer, J. A., Chung, C. S., Niswander, J. D. *et al.* (1971a) Developmental interaction, size and agenesis among maxillary incisors. *Human Biology* **43**: 36–45.

Sofaer, J. A., Bailit, H. L. & MacLean, J. (1971b) A developmental basis for differential tooth reduction during hominid evolution. *Evolution* **25**: 509–517.

Solounias, N. & Moelleken, S. M. C. (1993) Dietary adaptation of some extinct ruminants determined by premaxillary shape. *Journal of Mammalogy* **74**: 1059–1071.

Solounias, N., Teaford, M. F. & Walker, A. (1988) Interpreting the diet of extinct ruminants: the case of a grazing giraffe. *Paleobiology* **14**: 287–300.

Spears, I. R., van Noort, R., Crompton, R. H. (1993) The effects of enamel anisotropy on the distribution of stress in a tooth. *Journal of Dental Research* **72**: 1526–1531.

Spencer, M. (1998) Force production in the primate masticatory system: electromyographic tests of biomechanical hypotheses. *Journal of Human Evolution* **34**: 25–54.

Spielman, A. I. (1990) Interaction of saliva and taste. *Journal of Dental Research* **69**: 838–843.

Stirton, R. A. (1947) Observations on evolutionary rates in hypsodonty. *Evolution* **1**: 32–41.

Strait, S. G. (1993a) Molar morphology and food texture among small-bodied insectivorous mammals. *Journal of Mammalogy* **74**: 391–402.

(1993b) Differences in occlusal morphology and molar size in frugivores and faunivores. *Journal of Human Evolution* **25**: 471–484.

(1993c) Molar microwear in extant small-bodied faunivorous mammals: an analysis of feature density and pit frequency. *American Journal of Physical Anthropology* **92**: 163–79.

Strait, S. G. & Overdorff, D. J. (1996) Physical properties of fruits eaten by Malagasy primates. *American Journal of Physical Anthropology*, Supplement **22**: 224.

Strait, S. G. & Vincent, J. F. V. (1998) Primate faunivores: physical properties of prey items. *International Journal of Primatology* **19**: 867–878.

Su, H. H. & Lee, L. L. (2001) Food habits of Formosan rock macaques (*Macaca cyclopis*) in Jentse, northeastern Taiwan, assessed by fecal analysis and behavioral observation. *International Journal of Primatology* **22**: 359–377.

Swanson, S. A. V. (1980) Articular cartilage. In: *The Mechanical Properties of Biological Materials* (eds. J. F. V. Vincent & J. D. Currey), pp. 377–395. Cambridge: Cambridge University Press.

Synge, J. L. (1933) The tightness of the teeth, considered as a problem concerning the equilibrium of a thin incompressible membrane. *Philosophical Transactions of the Royal Society London series A* **231**: 435–477.

Szalay, F. S. (1975) Hunting–scavenging protohominids. *Man* (new series) **10**: 420–429.

Szczesniak, A. S. (1963) Classification of textural characteristics. *Journal of Food Science* **28**: 385–389.

Tabor, D. (1951) *The Hardness of Metals*. Oxford: Clarendon Press.

Tanne, K., Tanaka, E. & Sakuda, M. (1991) The elastic modulus of the temporomandibular joint disc from adult dogs. *Journal of Dental Research* **70**: 1545–1548.

Tavaré, S., Marshall, C. R., Will, O. *et al.* (2002) Using the fossil record to estimate the age of the last common ancestor of extant primates. *Nature* **416**: 726–729.

Teaford, M. F. (1988) A review of dental microwear and diet in modern mammals. *Scanning Microscopy* **2**: 1149–1166.

(1994) Dental microwear and dental function. *Evolutionary Anthropology* **3**: 17–30.

Teaford, M. F. & Walker, A. (1983) Dental microwear in adult and still-born guinea pigs (*Cavia porcellus*). *Archives of Oral Biology* **28**: 1077–1081.

Teng, S., Xu, Y., Cheng, M. *et al.* (1991) Biomechanical properties and collagen fiber orientation of temporomandibular joint discs in dogs. II. Tensile mechanical properties of discs. *Journal of Craniomandibular Disorders: Facial and Oral Pain* **5**: 107–114.

Terborgh, J. (1983) *New World Primates: A Study in Comparative Ecology.* Princeton, NJ: Princeton University Press.

(1986) Keystone plant resources in the tropical forest. In: *Conservation Biology: Science of Scarcity and Diversity*, (ed. M. Soulé), pp. 330–344. Sunderland, MA: Sinauer.

Theimer, T. C. (2003) Intraspecific variation in seed size affects scatterhoarding behaviour of an Australian tropical rain-forest rodent. *Journal of Tropical Ecology* **19**: 95–98.

Thexton, A. J. & Crompton, A. W. (1998) The control of swallowing. In: *The Scientific Basis of Eating* (ed. R. W. A. Linden), pp. 168–222. Basel: Karger.

Tittelbach, T. J. & Mattes, R. D. (2001) Oral stimulation influences postprandial triaglycerol concentrations in humans: nutrient specificity. *Journal of the American College of Nutrition* **20**: 485–493.

Tomos, A. D. & Leigh, R. A. (1999) The pressure probe: a versatile tool in plant cell physiology. *Annual Review of Plant Physiology and Plant Molecular Biology* **50**: 447–472.

Trulsson, M. & Essick, G. K. (1997) Low-threshold mechanoreceptive afferents in human lingual nerve. *Journal of Neurophysiology* **77**: 737–748.

Turner, I. M. (2001) *The Ecology of Trees in the Tropical Rain Forest.* Cambridge: Cambridge University Press.

Uden, P. & Van Soest, P. J. (1982) The determination of digesta particles in some herbivores. *Animal Feed Science and Technology* **7**: 35–44.

Ungar, P. S. (1992) Incisal microwear and feeding behavior of four Sumatran anthropoids. Ph.D. thesis. State University of New York at Stony Brook.

(1994) Patterns of ingestive behavior and anterior tooth use differences in sympatric anthropoid primates. *American Journal of Physical Anthropology* **95**: 197–219.

(1996) Relationship of incisor size to diet and anterior tooth use in sympatric Sumatran anthropoids. *American Journal of Primatology* **38**: 145–146.

Ungar, P. S. & Grine, F. E. (1991) Incisor size and wear in *Australopithecus africanus* and *Paranthropus robustus. Journal of Human Evolution* **20**: 313–340.

Ungar, P. S. & Teaford, M. F. (1996) A preliminary examination of non-occlusal dental microwear in anthropoids: implications for the study of fossil primates. *American Journal of Physical Anthropology* **100**: 101–113.

(2002) *Human Diet: Its Origin and Evolution.* Westport, CT: Bergin & Garvey.

Ungar, P. S. & Williamson, M. D. (2000) Exploring the effects of tooth wear on functional morphology: a preliminary study using dental topographic analysis. *Palaeontologia Electronica* **3**: 18 pp. http://www-odp.tamu.edu/paleo/2000_1/gorilla/main.htm.

Ungar, P. S., Teaford, M. F., Glander, K. E. *et al.* (1995) Dust accumulation in the canopy: implications for the study of dental microwear in primates. *American Journal of Physical Anthropology* **97**: 93–99.

Ungar, P. S., Fennell, K. J., Gordon, K. *et al.* (1997) Neanderthal incisor beveling. *Journal of Human Evolution* **32**: 407–421.

Utz, K. H. (1986) Untersuchungen über die interokklusale tactile Feinsensi-bilitat naturischer Zahne mit Hilfe von Aluminium-oxid-teilchen. *Deutsch Zahnärztliche Zeitschrift* **41**: 313–315.

van den Braber, W., van der Glas, H. W., van der Bilt, A. *et al.* (2001) Chewing efficiency of pre-orthognathic surgery patients: selection and breakage of food particles. *European Journal of Oral Science* **109**: 306–311.

van der Bilt, A., Olthoff, L. W., van der Glas, H. W. *et al.* (1987) A mathematical description of the comminution of food in human mastication. *Archives of Oral Biology* **32**: 579–588.

van der Glas, H. W., van der Bilt, A., Olthoff, L. W. *et al.* (1987) Measurement of selection chances and breakage functions during chewing in man. *Journal of Dental Research* **66**: 1547–1550.

(1992) A selection model to estimate the interaction between food particles and the postcanine teeth in human mastication. *Journal of Theoretical Biology* **155**: 103–120.

van Reenen, J. F. (1982) The effects of attrition on tooth dimensions of San (Bush-men). In: *Teeth: Form, Function and Evolution* (ed. B. Kurtén), pp. 182–203. New York: Columbia University Press.

van Roosmalen, M. G. M. (1980) Habitat preferences, diet, feeding strategy and social organization of the black spider monkey (*Ateles paniscus paniscus* Lin-naeus 1958) in Surinam. Ph.D. thesis. Rijksuniversiteit voor Natuurbeheer, Leersum.

van Schaik, C. P., Fox, E. A. & Sitompul, A. F. (1996) Manufacture and use of tools in wild Sumatran orangutans. *Naturwissenschaften* **83**: 186–188.

Van Soest, P. J. (1994) *Nutritional Ecology of the Ruminant*, 2nd edn. Ithaca, NY: Cornell University Press.

(1996) Allometry and ecology of feeding behavior and digestive capacity in herbivores: a review. *Zoo Biology* **15**: 455–479.

Van Valen, L. (1960) A functional analysis of hypsodonty. *Evolution* **14**: 531–532.

Van Valkenburgh, B. (1988) Incidence of tooth breakage among large, predatory mammals. *American Naturalist* **131**: 291–300.

(1990) Skeletal and dental predictors of body mass in carnivores. In: *Body Size in Mammalian Paleobiology* (eds. J. Damuth & B. J. MacFadden), pp. 181–205. Cambridge: Cambridge University Press.

(1996) Feeding behavior in free-ranging, large African carnivores. *Journal of Mammalogy* **77**: 240–254.

Van Valkenburgh, B. & Hertel, F. (1993) Tough times at La Brea: tooth breakage in large carnivores of the late Pleistocene. *Science* **261**: 456–459.

Van Valkenburgh, B. & Ruff, C. B. (1987) Canine tooth strength and killing behaviour in large carnivores. *Journal of Zoology, London* **212**: 379–397.

Van Valkenburgh, B., Teaford, M. F. & Walker, A. (1990) Molar microwear and diet in large carnivores. *Journal of Zoology* **22**: 319–340.

van Vliet, T. (2002) On the relation between texture perception and fundamental mechanical properties of liquids and time dependent solids. *Food Quality and Preference* **13**: 227–236.

Vickers, Z. (1981) Relationships of chewing sounds to judgements of crispness, crunchiness and hardness. *Journal of Food Science* **47**: 121–124.

Vincent, J. F. V. (1980) Insect cuticle: a paradigm for natural composites. In: *The Mechanical Properties of Biomaterials* (eds. J. F. V. Vincent & J. D. Currey), pp. 183–210. Cambridge: Cambridge University Press.

(1981) Morphology and design of the extensible intersegmental membrane of the female migratory locust. *Tissue and Cell* **13**: 831–853.

(1982) The mechanical design of grass. *Journal of Materials Science* **17**: 856–860.

(1990) Fracture properties of plants. *Advances in Botanical Research* **17**: 235–287.

(1991) Strength and fracture of grasses. *Journal of Materials Science* **26**: 1947–1950.

(1992) *Biomaterials*. Oxford: IRL Press.

Vincent, J. F. V. & Hillerton, J. E. (1979) The tanning of insect cuticle: a critical review and a revised mechanism. *Journal of Insect Physiology* **25**: 653–658.

Vincent, J. F. V. & Khan, A. A. (1993) Anisotropy in the fracture properties of apple flesh as investigated by crack-opening tests. *Journal of Materials Science* **28**: 45–51.

Vincent, J. F. V. & Sibbing, F. A. (1992) How the grass carp (*Ctenopharnygodon idella*) chooses and chews its food: some clues. *Journal of Zoology* **226**: 435–444.

Vincent, J. F. V., Jeronimidis, G., Khan, A. A. & Luyten, H. (1991) The wedge fracture test: a new method for measurement of food texture. *Journal of Texture Studies* **22**: 45–57.

Vincent, J. F. V., Saunders, D. E. J. & Beyts, P. (2002) The use of stress intensity factor to quantify "hardness" and "crunchiness" objectively. *Journal of Texture Studies* **33**: 149–159.

Vinyard, C. J., Wall, C. E., Williams, S. H. *et al.* (2003) Comparative functional analysis of skull morphology of tree-gouging primates. *American Journal of Physical Anthropology* **120**: 158–170.

Visser, M. (1991) *The Rituals of Dinner*. London: Penguin.

von Koenigswald, W. (1982) Enamel structure in the molars of Arvicolidae (Rodentia, Mammalia), a key to functional morphology and phylogeny. In: *Teeth: Form, Function and Evolution* (ed. B. Kurtén), pp. 109–122. New York: Columbia University Press.

Voon, F. C. T., Lucas, P. W., Luke, D. A. *et al.* (1986) A simulation approach to understanding the masticatory process. *Journal of Theoretical Biology* **119**: 251–262.

Wagner, T., Neinhuis, C. & Barthlott, W. (1996) Wettability and contaminability of insect wings as a function of their surface sculptures. *Acta Zoologica* **77**: 213–225.

Walker, A. C. (1981) Diet and teeth: dietary hypotheses and human evolution. *Philosophical Transactions of the Royal Society London series B* **292**: 57–76.

(1984) Mechanisms of honing in the male baboon canine. *American Journal of Physical Anthropology* **65**: 47–60.

Walker, A. C., Hoech, H. N. & Perez, L. (1978) Microwear of mammalian teeth as an indicator of diet. *Science* **201**: 908–910.

Walker, P. L. (1979) The adaptive significance of pongid lip mobility. *Ossa* **6**: 277–284.

Walker, P. & Murray, P. (1975) An assessment of masticatory efficiency in a series of anthropoid primates with special reference to the Colobinae and Cercopithecinae. In: *Primate Functional Morphology and Evolution* (ed. R. H. Tuttle), pp. 135–150. The Hague: Mouton.

Waterman, P. G. (1984) Food acquisition and processing as a function of plant chemistry. In: *Food Acquisition and Processing in Primates* (eds. D. J. Chivers, B. A. Wood & A. Bilsborough), pp. 177–211. New York: Plenum.

Waterman, P. G. & Kool, K. M. (1994) Colobine food selection and plant chemistry. In: *Colobine Monkeys: Their Ecology, Behaviour and Evolution* (eds. A. G. Davies & J. F. Oates), pp. 251–284. Cambridge: Cambridge University Press.

Waterman, P. G. & Mole, S. (1994) *Analysis of Phenolic Plant Metabolites*. Oxford: Blackwell.

Waters, N. E. (1975) Aspects of dental biomechanics. In: *Scientific Aspects of Dental Materials* (ed. J. A. von Fraunhofer), pp. 1–47. London: Butterworth.

   (1980) Some mechanical and physical properties of teeth. In: *The Mechanical Properties of Biological Materials* (eds. J. F. V. Vincent & J. D. Currey), pp. 99–135. Cambridge: Cambridge University Press.

Weijs, W. (1975) Mandibular movements of the albino rat during feeding. *Journal of Morphology* **154**: 107–124.

Weijs, W. & Dantuma, R. (1981) Functional anatomy of the masticatory apparatus in the rabbit. *Netherlands Journal of Zoology* **31**: 99–147.

Wendeln, M. C., Runkle, J. R. & Kalko, E. K. V. (2000) Nutritional values of 14 fig species and bat feeding preferences in Panama. *Biotropica* **32**: 489–501.

White, S. N., Luo, W., Paine, M. L. *et al.* (2001) Biological organization of hydroxyapatite crystallites into a fibrous continuum toughens and controls anisotropy in human enamel. *Journal of Dental Research* **80**: 321–326.

Whitten, A. J. (1982) Diet and feeding behaviour of kloss gibbons on Siberut Island, Indonesia. *Folia Primatologica* **37**: 177–208.

Wictorin, L., Hedegård, B. & Lundberg, M. (1971) Cineradiographic studies of bolus position during chewing. *Journal of Prosthetic Dentistry* **26**: 236–246.

Wilding, R. J. (1993) The association between chewing efficiency and occlusal contact area in man. *Archives of Oral Biology* **38**: 589–596.

Williams, G. (1956) The relationship between the length of the jaw and the length of the molar series in some eutherian mammals. *Journal of Zoology* **126**: 51–56.

Williamson, L. & Lucas, P. W. (1995) The effect of moisture content on the mechanical properties of a seed shell. *Journal of Materials Science* **30**: 162–166.

Wilsea, M., Johnson, K. L. & Ashby, M. F. (1975) Indentation of foamed plastics. *International Journal of Mechanical Sciences* **17**: 457–460.

Wilson, E. O. (1975) *Sociobiology: The New Synthesis*. Cambridge, MA: Belknap Press of Harvard University.

Wing, S. L., Hickey, L. J. & Swisher, C. C. (1993) Implications of an exceptional fossil flora for Late Cretaceous vegetation. *Nature* **363**: 342–344.

Wong, K. (2003) An ancestor to call our own. *Scientific American* (January): 42–51.

Wood, B. A. (1984) Tooth size and shape and their relevances to studies of human evolution. *Philosophical Transactions of the Royal Society London series B* **292**: 57–64.

Wood, B. A. & Abbott, S. A. (1983) Analysis of the dental morphology of Plio-Pleistocene hominids. I. Mandibular molars: crown area measurements and morphological traits. *Journal of Anatomy* **136**: 197–219.

Wood, B. A. & Collard, M. (1999a) The human genus. *Science* **284**: 65–71.

(1999b) Is *Homo* defined by culture? *Proceedings of the British Academy* **99**: 11–23.

Wood, B. A., Abbott, S. A. & Graham, S. H. (1983) Analysis of the dental morphology of Plio-Pleistocene hominids. II. Mandibular molars: study of cusp areas, fissure pattern and cross-sectional shape of the crown. *Journal of Anatomy* **137**: 287–314.

Wood, B. A., Abbott, S. A. & Uytterschaut, H. (1988) Analysis of the dental morphology of Plio-Pleistocene hominids. IV. Mandibular postcanine root morphology. *Journal of Anatomy* **156**: 107–139.

Wrangham, R. W., Conklin, N. L., Chapman, C. A. *et al.* (1992) The significance of fibrous foods for Kibale Forest chimpanzees. In: *Foraging Strategies and Natural Diets of Monkeys, Apes, and Humans* (eds. A. Whiten & E. M. Widdowson), pp. 11–18. Oxford: Clarendon Press.

Wrangham, R. W., Jones, J. H., Laden, G. *et al.* (1999) The raw and the stolen: cooking and the ecology of human origins. *Current Anthropology* **40**: 567–594.

Wright, W. & Illius, A. (1995) A comparative study of the fracture properties of five grasses. *Functional Ecology* **9**: 269–278.

Wright, W. & Vincent, J. F. V. (1996) Herbivory and the mechanics of fracture in plants. *Biological Reviews* **71**: 401–413.

Yamada, H. (1970) *The Strength of Biological Materials* (ed. F. G. Evans). Baltimore, MD: Williams & Wilkins.

Yamashita, N. (1998) Functional dental correlates of food properties in five Malagasy lemur species. *American Journal of Physical Anthropology* **106**: 169–188.

(2000) Mechanical thresholds as a criterion for food selection in two prosimian primate species. In: *Plant Biomechanics 2000* (eds. H.-C. Spatz & T. Speck), pp. 590–595. Stuttgart: Georg Thieme.

(2003) Food procurement and tooth use in two sympatric lemur species. *American Journal of Physical Anthropology* **221**: 125–133.

Young, G. C., Karatajute, V. N. & Smith, M. M. (1996) A possible Late Cambrian vertebrate from Australia. *Nature* **383**: 810–812.

Yucker, A. S., Matthews, K. L. & Sharpe, P. T. (1998) Transformation of tooth type induced by inhibition of BMP signaling. *Science* **282**: 1136–1138.

Yurkstas, A. A. (1965) The masticatory act: a review. *Journal of Prosthetic Dentistry* **15**: 248–260.

Yurkstas, A. A. & Manly, R. S. (1949) Measurement of occlusal contact area effective in mastication. *American Journal of Orthodontics* **35**: 185–195.

Zahradnik, R. T. & Moreno, E. C. (1975) Structural features of human dental enamel as revealed by isothermal water vapour sorption. *Archives of Oral Biology* **20**: 317–325.

Zarkadas, C. G., Karatzas, C. N., Khalili, A. D. *et al.* (1988) Quantitative determination of the myofibrillar proteins and connective tissue content in selected porcine skeletal muscles. *Journal of Agricultural and Food Chemistry* **36**: 1131–1146.

Zhang, X., Gubbels, G. H. M., Terpstra, R. A. *et al.* (1997) Toughening of calcium hydroxyapatite with silver particles. *Journal of Materials Science* **32**: 235–243.

Zingeser, M. R. (1969) Cercopithecoid honing mechanisms. *American Journal of Physical Anthropology* **31**: 205–214.

(1973) Dentition of *Brachyteles arachnoides* with reference to alouattine and ateline affinities. *Folia Primatologica* **20**: 351–390.

Zuccotti, L. F., Williamson, M. D., Limp, W. F. *et al.* (1998) Modeling primate occlusal topography using geographical information systems technology. *American Journal of Physical Anthropology* **107**: 137–142.

Zwell, M. & Pilbeam, D. R. (1972) The single-species hypothesis, sexual dimorphism and variability in early hominids. *Yearbook of Physical Anthropology* **16**: 69–79.

Index